Rotational Brownian Motion and Dielectric Theory

Rotational Brownian Motion and Dielectric Theory

JAMES McCONNELL

Senior Professor of Theoretical Physics
Dublin Institute for Advanced Studies

1980

ACADEMIC PRESS

A Subsidiary of Harcourt Brace Jovanovich, Publishers

London New York Toronto Sydney San Francisco

ACADEMIC PRESS INC. (LONDON) LTD
24–28 Oval Road
London NW1

US edition published by
ACADEMIC PRESS INC.
111 Fifth Avenue
New York, New York 10003

British Library Cataloguing in Publication Data

McConnell, J R
Rotational Brownian motion and dielectric theory.
1. Dielectrics, Liquid 2. Brownian movements
I. Title
537.2'43 QC585.8.L56 80–40736
ISBN 0–12–481850–1

Printed in Great Britain at the Alden Press
Oxford London and Northampton

Foreword

The Debye theory of the response of a dielectric material to an external electric field with circular frequency ω is characterised by the expression $(1 + i\omega\tau)^{-1}$. Debye calculated the relaxation time τ for the case of a dilute solution of dipolar molecules in a non polar liquid. This formula – though not the particular value of τ – was, however, successfully applied to many other materials. The above expression is the Fourier transform of an exponential decay in time, $\exp(-t/\tau)$, which, at least approximately, holds in very many cases and is responsible for the great success of the Debye formula.

For very high frequencies such that $\omega\tau \gg 1$, the Debye formula leads to a coefficient of absorption per unit length which is independent of frequency which, clearly, does not correspond to experimental facts. This implies that for very short times, $t \ll \tau$, the spontaneous decay of a polarisation cannot be exponential in time. Qualitatively this is easily understood as an exponential decay can arise only on an average over many collisions of the polar molecule with the disordered thermal motion of the molecules of the solvent. It becomes thus necessary to treat in some detail the rotational Brownian motion of a polar molecule.

Work on this problem has been going on for quite a number of years, but it is only in recent years that systematic and general solutions have been obtained. The relevant mathematical task is very formidable. The present monograph is the first comprehensive presentation of these achievements. The results are given in a form in which they can directly be compared with experiments. It will be noticed that considerable gaps exist in the experimental data, and it may be hoped that a stimulus to close these gaps will arise from this book. The availability of sources for frequencies in the mm wave region does now close the gap between the infrared and the microwave regions and thus permits systematic investigations over the whole frequency spectrum.

August 1980
University of Liverpool

Professor H. Fröhlich, F.R.S.

Preface

The theory of Brownian motion as proposed by Einstein (1905) inspired Langevin (1908) to formulate a mathematical theory of translational Brownian motion. Debye (1913) applied the method of Einstein to study the rotational Brownian motion of molecules that have permanent electric dipoles. In his investigations Debye neglected effects of the inertia of the Brownian particles, which Langevin had attempted to include in his study of translational motion. When the Debye theory was applied to the absorption of electromagnetic rays in liquid dielectrics, serious disagreement with experimental results was encountered.

A theory of rotational Brownian motion which takes account of inertial effects was slow in evolving. This was due partly to the absence of a satisfactory general theory of Brownian motion and partly to calculational difficulties. The investigations of Fokker (1914), Planck (1917), Wiener (1930), Uhlenbeck and Ornstein (1930), Kolmogorov (1931) and Doob (1942) contributed greatly towards setting up a rigorous mathematical formulation of Brownian motion. Sack (1957a, b) provided a method, based on partial differential equations, of calculating in linear approximation polarization and relaxation effects arising from the rotational Brownian motion of polar molecules which are linear or spherical. During the last few years investigations based on Langevin-type equations succeeded in including inertial effects in the study of rotational motion of a Brownian particle of any shape (Ford *et al.*, 1979), and so extended the results of Perrin (1934). The development of submillimetre spectroscopy during the past twenty years has done much to provide accurate comparisons between the theory and experiments on dielectric absorption and dispersion.

In spite of some remaining conceptual and mathematical difficulties, it seems that the time has now been reached when a self-contained account of the theory can be written. Moreover a number of research workers have expressed their desire to see such an account published. The present volume is a development of a course of lectures given at the Dublin Institute for Advanced Studies to university staff members and senior students, most of whom were actively engaged in the study of dielectrics. The experience of giving this course has helped me to form some idea of the extent to which it is necessary to explain concepts that are familiar only to mathematicians and theoretical physicists, of the need to present mathematical theorems on stochastic processes with a reasonable amount of rigour while quoting references for a more thorough treatment, and of the desirability of concentrating on aspects of the theory that are closely related to experiment.

It is hoped that this book will be of service not only to mathematicians but also to chemists, physicists and electrical engineers who have a working knowledge of undergraduate mathematics and statistical mechanics, and that it will provide them with an intelligible account of the theory of rotational Brownian motion and of its applications not only to dielectric relaxation phenomena but also to certain nuclear magnetic relaxation processes. It is also hoped that the book may serve readers, who have no special interest in its central theme, as a source of introductory information and reference for incidental topics such as the representation of linear operators, rotations in three-dimensional space, random variables, random processes, correlation functions, spectral densities, stochastic differential equations, solution of nonlinear differential equations, diffusion processes. Experimentalists engaged on dielectric dispersion and absorption research may wish to consider suggestions made at the end of the book for future experiments.

In the preparation of this book I have been constantly advised by Professor B. K. P. Scaife, who both placed his long experience in dielectric research at my disposal and read a considerable part of the manuscript. I am indebted to Professor R. A. Sack and Dr. G. W. Chantry for permission to quote from their papers. Among others, from whose advice I have benefited, I may mention especially M. S. Beevers, R. G. Bennett, J. R. Birch, J. H. Calderwood, M. W. Evans, G. W. Ford, J. T. Lewis, V. J. McBrierty and G. R. Wilkinson.

Some of the preparatory calculations were performed while I was visiting professor at the University of Salford, where I was helped by discussions with staff members and visiting scientists. Finally I wish to thank Mrs. M. Farrelly, who typed the manuscript with great care.

Dublin
August, 1980

James McConnell

Contents

1 The Debye Theory of Rotational Brownian Motion

1.1 Two-dimensional Debye Theory

A convenient starting point for the study of rotational Brownian motion is a paper published by Debye (1913) that deals with the theory of dielectrics. In it he developed his new idea (Debye, 1912) that dielectrics contain not only electrically bound electrons but also molecules with dipoles of constant electric moment. These are called *polar molecules.* Neglecting the dipole–dipole interactions he assumed that all these molecules behave on the average in the same way, and so it was sufficient for him to consider the behaviour of a single molecule. Debye took the molecule to be a sphere of radius a with a dipole at its centre, and he examined the case of the molecule being subject to a time dependent electric field F in a fixed direction.

In order to simplify the calculations Debye at first supposed that the dipole axis rotates in a fixed plane that contains the direction of F. The direction of the axis is specified by the angle θ which it makes with that of F. Let us consider a circular cylinder as in Fig. 1.1, whose axis passes through the origin of the coordinate system and is perpendicular to the plane of rotation of the dipole. The dimensions of the cylinder are such that, while it is macroscopically small, it contains many polar molecules moving independently. Since the polar molecules have been assumed to behave on the average in the same way, they may be regarded as members of an ensemble. We can therefore make an ensemble average over the polar molecules in the cylinder. We denote by $w(\theta, t)d\theta$ the number of these molecules with polar axes lying in the angular interval $(\theta, \theta + d\theta)$ at time t. The first objective of Debye was to derive a partial differential equation for w.

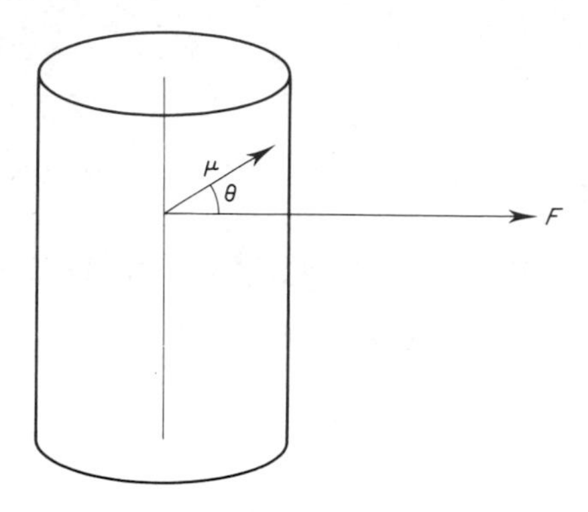

Fig. 1.1 The orientation of a molecule with dipole moment μ in an electric field F.

Let us suppose that the dielectric is either a gas or liquid in dilute solution, the solvent consisting of nonpolar molecules whose linear dimensions are small compared with the radius a of the polar molecule. The thermal motion of the environment produces torques which cause rotational Brownian motion of the polar molecules. According to the law of Fick (1855) this gives rise to a current of polar molecules $-\lambda(\partial w/\partial\theta)$ in the direction of θ increasing, λ being a constant to be determined later.

There is another current due to the field F and the frictional drag on a polar molecule. If μ is the moment of the dipole, the field produces a couple of moment $-\mu F \sin\theta$ on the molecule. Indeed, we see from Fig. 1.2 that if we regard the dipole as a combination of charges $\pm e$ at distance μ/e apart, the field produces a couple of

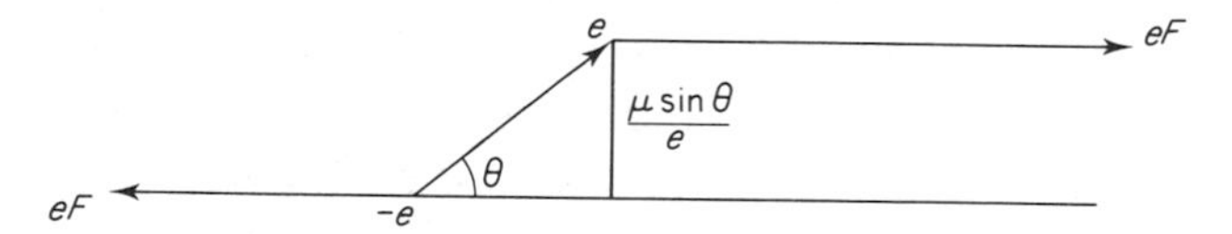

Fig. 1.2 The torque produced by a field F on a dipole μ.

moment $-eF(\mu/e)\sin\theta$. Since the linear dimensions of particles in the heat bath environment are small compared with a, the frictional problem is a hydrodynamical one. Thus the drag consists of a couple $\zeta\dot{\theta}$ resisting the increase of θ, ζ being a *frictional constant*. In fact, if it is supposed that the macroscopic Rayleigh–Stokes theory holds for the microscopic system, then

$$\zeta = 8\pi a^3 \eta \tag{1,1.1}$$

where η is the viscosity of the surrounding fluid (Lamb, 1932, p. 589). If I is the moment of inertia of a polar molecule, its equation of motion resulting from F and the frictional couple is

$$I\ddot{\theta} = -\mu F \sin\theta - \zeta\dot{\theta}. \tag{1,1.2}$$

Debye neglects the term on the left-hand side, so that (1,1.2) becomes

$$\dot{\theta} = -\frac{\mu F \sin\theta}{\zeta}. \tag{1,1.3}$$

This leads to a current $w\dot{\theta}$ equal to $-w\mu F \sin\theta/\zeta$. Adding this to $-\lambda(\partial w/\partial\theta)$ we have at the angle θ the total current

$$-\frac{w\mu F \sin\theta}{\zeta} - \lambda\frac{\partial w}{\partial\theta} \tag{1,1.4}$$

in the direction of increasing θ.

We can deduce a partial differential equation for w. During a short interval of time δt the increase in the number of molecules whose polar axes lie in $(\theta, \theta + d\theta)$ is $(\partial w/\partial t)\delta t d\theta$. According to (1,1.4) the number of these particles has increased at θ by

$$-\left(\frac{w\mu F \sin\theta}{\zeta} + \lambda\frac{\partial w}{\partial\theta}\right)\delta t. \tag{1,1.5}$$

However, particles with polar axes lying in the interval $(\theta, \theta + d\theta)$ have also been leaving and the number that have left during δt is (1, 1.5) evaluated at $\theta + d\theta$. Since in general

$$f(\theta, t) - f(\theta + d\theta, t) = -\frac{\partial f(\theta, t)}{\partial\theta} d\theta,$$

it follows that the net increase in the number of particles whose polar axes lie in $(\theta, \theta + d\theta)$ is

$$\frac{\partial}{\partial\theta}\left(\frac{w\mu F \sin\theta}{\zeta} + \lambda\frac{\partial w}{\partial\theta}\right) d\theta\,\delta t.$$

Equating this to $(\partial w/\partial t)\delta t d\theta$ we have the equation

$$\frac{\partial w}{\partial t} = \frac{\partial}{\partial \theta}\left(\frac{\mu F \sin \theta w}{\zeta} + \lambda \frac{\partial w}{\partial \theta}\right). \qquad (1,1.6)$$

To determine the constant λ we consider the special case of a time independent F. Then the w for the ensemble of polar molecules is a Maxwell–Boltzmann distribution. Since the potential energy of the polar molecule is $-\mu F \cos \theta$ (Jeans, 1933, p. 377) and the kinetic energy is $\frac{1}{2} I \dot{\theta}^2$,

$$w = \text{const. } \exp\{[-\tfrac{1}{2} I \dot{\theta}^2 + \mu F \cos \theta]/(kT)\}, \qquad (1,1.7)$$

where k is the Boltzmann constant 1.38×10^{-23} J K^{-1} (1.38×10^{-16} erg deg^{-1}) and T the absolute temperature. We then have a stationary state, so that $(\partial w/\partial t)$ vanishes and from (1,1.6)

$$\frac{\mu F \sin \theta w}{\zeta} + \lambda \frac{\partial w}{\partial \theta} = \text{constant.}$$

On substituting into this equation from (1,1.7) we find that $\lambda = kT/\zeta$, so that (1,1.6) is expressible as

$$\zeta \frac{\partial w}{\partial t} = \frac{\partial}{\partial \theta}\left(\mu F \sin \theta w + kT \frac{\partial w}{\partial \theta}\right). \qquad (1,1.8)$$

We say that this is the *diffusion equation* for $w(\theta, t)$ and that kT/ζ is the *diffusion coefficient*. When solving (1,1.8) we shall impose the normalizing condition

$$\int_0^{2\pi} w(\theta, t) d\theta = 1. \qquad (1,1.9)$$

We then call w the probability density function. The probability of finding a molecule with dipole axis pointing in the interval $(\theta, \theta + d\theta)$ is $w d\theta$.

Debye applies (1,1.8) first of all to the case where F is the periodic function $F_0 e^{i\omega t}$ and $\mu F_0 \ll kT$. Then the electric moment of the dielectric is proportional to the field, to a close approximation. To solve (1,1.8) for the present problem we may expand w as a series in powers of F_0, which we restrict to two terms:

$$w = w_0 + F_0 w_1 .$$

Debye deduces differential equations for w_0 and w_1. A short calculation gives

$$w = A\left(1 + \frac{\mu F_0 e^{i\omega t} \cos\theta}{kT + i\omega\zeta}\right) + BF_0, \tag{1,1.10}$$

where A and B are constants. For vanishingly small values of ω the probability density must have the form of (1,1.7) with $F = F_0$ and the $\dot{\theta}^2$ term ignored because w is not a function of $\dot{\theta}$. On expanding the exponential to terms of order $\mu F_0/(kT)$ we have from (1,1.7)

$$w = \frac{1}{2\pi}\left(1 + \frac{\mu F_0 \cos\theta}{kT}\right), \tag{1,1.11}$$

where the multiplying constant factor has been determined by (1,1.9). On comparing (1,1.11) with (1,1.10) we see that $A = 1/(2\pi)$, $B = 0$, so that (1,1.10) becomes

$$w = \frac{1}{2\pi}\left(1 + \frac{\mu F_0 e^{i\omega t} \cos\theta}{kT + i\omega\zeta}\right). \tag{1,1.12}$$

When as in (1,1.11) we neglect powers of higher order than the first in $\mu F_0/(kT)$, we say that we are working in the *linear approximation.* We also say that the neglected terms correspond to *saturation effects.* Usually our calculations for dielectric phenomena will be made in the linear approximation. Equation (1,1.12) is the consequence of such a calculation.

Debye also applied (1,1.8) to a process known as dielectric relaxation which we shall now describe. As a result of the presence for a long time of a constant electric field F_0 the dipole moment has attained a steady value. At time $t = 0$ the field is removed, and the system under the influence of the thermal motion of the environment tends to revert to a random arrangement. This process is called *dielectric relaxation* and the term *after effect* is often employed to qualify physical quantities related to the relaxation process. When the field is operative, w is the steady state solution (1,1.11). When the field is removed, (1,1.8) becomes

$$\zeta\frac{\partial w}{\partial t} = kT\frac{\partial^2 w}{\partial \theta^2}. \tag{1,1.13}$$

We have to solve this equation with the initial condition (1,1.11).

On putting

$$w = \frac{1}{2\pi}\left(1 + \frac{\mu F_0 f(t) \cos\theta}{kT}\right)$$

we find that the required solution of (1,1.13) is

$$w = \frac{1}{2\pi}\left(1 + \frac{\mu F_0 \exp(-kTt/\zeta)\cos\theta}{kT}\right) \qquad (t > 0). \tag{1,1.14}$$

The first term in w gives the asymptotic random distribution; the second term gives the distribution induced by the field F_0. The amplitude $\mu F_0 \exp(-kTt/\zeta)/(kT)$ of the induced distribution decays to $(1/e)$ its initial value in a time $\zeta/(kT)$, which is called the *relaxation time* for the problem. Returning to (1,1.12) we see that the frequency of the exciting field makes a significant change in the distribution function when ω becomes of order of magnitude of the reciprocal of the relaxation time.

We may employ (1,1.14) to calculate the ensemble average over the polar molecules of $\cos\theta$, denoted by $\langle\cos\theta\rangle$, for the relaxation process. This is obtained by multiplying $\cos\theta$ by the probability density $w(\theta, t)$ and integrating with respect to θ from 0 to 2π. We find immediately that

$$\langle\cos\theta\rangle = \frac{\mu F_0 \exp(-kTt/\zeta)}{2kT}. \tag{1,1.15}$$

1.2 Three-dimensional Debye Theory

Debye (1929, Chap. 5) extended his investigations to the case where the dipole axis of the polar molecule is allowed to rotate freely in three-dimensional space. The unnormalized w is defined by taking in the dielectric a sphere with centre at the origin of coordinates and saying that $w d\Omega$ is the number of polar molecules whose axes point in the element of solid angle $d\Omega$. In three dimensions the orientation of the dipole axis is specified by two polar angles, namely, θ which now ranges from 0 to π and ϕ which goes from 0 to 2π. In terms of these variables

$$d\Omega = \sin\theta d\theta d\phi. \tag{1,2.1}$$

We take a line through the origin and in the direction of the electric field F as the spherical polar axis of the coordinate system, so that θ is the angle between the dipole axis and F.

Since the configuration is axially symmetric with respect to the direction of the field, w is still just a function of θ and t, and we may now take the element of solid angle to be $2\pi \sin \theta d\theta$. The number of molecules whose axes enter the element of solid angle during the time interval δt is

$$2\pi \sin \theta d\theta \frac{\partial w}{\partial t} \delta t = \Delta_1 + \Delta_2, \tag{1,2.2}$$

where Δ_1 is the contribution due to F and Δ_2 that due to Brownian motion. In performing the calculations Debye assumes that the directions of the dipole axes move only by infinitesimal amounts during δt. From (1,1.3) he now deduces

$$\Delta_1 = \frac{2\pi\mu F}{\zeta} \frac{\partial}{\partial \theta} (\sin^2 \theta w) d\theta \delta t.$$

He also obtains

$$\Delta_2 = \frac{2\pi kT \sin \theta}{\zeta} \left(\cot \theta \frac{\partial w}{\partial \theta} + \frac{\partial^2 w}{\partial \theta^2} \right) d\theta \delta t,$$

so that (1,2.2) yields

$$\zeta \frac{\partial w}{\partial t} = \frac{1}{\sin \theta} \frac{\partial}{\partial \theta} \left[\sin \theta \left(kT \frac{\partial w}{\partial \theta} + \mu F \sin \theta w \right) \right]. \tag{1,2.3}$$

This is the three-dimensional diffusion equation, kT/ζ being the diffusion coefficient as before. The normalization condition is now

$$1 = \int w(\theta, t) d\Omega = \int_0^{2\pi} d\phi \int_0^{\pi} w(\theta, t) \sin \theta d\theta,$$

by (1,2.1); that is,

$$\int_0^{\pi} w(\theta, t) \sin \theta d\theta = \frac{1}{2\pi}. \tag{1,2.4}$$

The probability density function is the solution of (1,2.3) that obeys (1,2.4).

Debye obtained for the probability density function in linear approximation when $F = F_0 e^{i\omega t}$

$$w = \frac{1}{4\pi}\left(1 + \frac{2\mu F_0 e^{i\omega t}\cos\theta}{2kT + i\omega\zeta}\right). \qquad (1,2.5)$$

Then one obtains for the relaxation problem, when the constant field F_0 has been switched off at $t = 0$, that

$$w = \frac{1}{4\pi}\left(1 + \frac{\mu F_0 \exp(-2kTt/\zeta)\cos\theta}{kT}\right) \quad (t > 0). \qquad (1,2.6)$$

The relaxation time is now τ_D defined by

$$\tau_D = \frac{\zeta}{2kT}, \qquad (1,2.7)$$

which is known as the *Debye relaxation time*. It is one half the relaxation time found from (1,1.14) in the two-dimensional investigation. The ensemble average of $\cos\theta$ for the relaxation process obtained from (1,2.6) and (1,2.1),

$$\int_0^{2\pi} d\phi \int_0^{\pi} w\cos\theta\sin\theta d\theta = \frac{\mu F_0 e^{-t/\tau_D}}{2kT}\int_0^{\pi}\cos^2\theta\sin\theta d\theta,$$

so that

$$\langle\cos\theta\rangle = \frac{\mu F_0 e^{-t/\tau_D}}{3kT}. \qquad (1,2.8)$$

Apart from the Debye time there are two other times that we shall frequently encounter and which we shall now define. If in (1,1.2) we put F equal to zero so that only the frictional couple acts on the sphere,

$$I\ddot{\theta} = -\zeta\dot{\theta}.$$

This has the solution

$$\dot{\theta} = \dot{\theta}_0 \exp(-\zeta t/I),$$

where $\dot{\theta}_0$ is the angular velocity for $t = 0$. The angular velocity drops to $(1/e)$ its initial value in time I/ζ, which we call the *friction time* and denote by τ_F:

$$\tau_F = \frac{I}{\zeta}. \qquad (1,2.9)$$

We also define the *free rotation time* τ_R by

$$\tau_R = \left(\frac{I}{kT}\right)^{1/2}. \tag{1,2.10}$$

A molecule rotating with angular velocity τ_R^{-1} has rotational energy $\frac{1}{2}kT$, which is the mean kinetic energy when the system is in statistical equilibrium. We see from (1,2.7) that

$$\tau_D = \frac{\tau_R^2}{2\tau_F}. \tag{1,2.11}$$

1.3 Polarization of Dielectrics

In our studies of the polarization of dielectrics by electric fields we shall assume that the intensities of the fields are sufficiently weak that we may superimpose effects; in other words, we assume that we have a *linear theory* in which the polarization caused by the field $\mathbf{E}_1 + \mathbf{E}_2$ is the vector sum of the polarizations caused by $\mathbf{E}_1$ and by $\mathbf{E}_2$. This will be the case when we are working in the linear approximation, explained in Section 1.1. In order to simplify the exposition we furthermore assume throughout that we are dealing with a spherical portion of dielectric, that the dielectric material is isotropic and that it is uniformly polarized. These assumptions will allow us to take the polarization to be in the direction of the electric intensity. We choose the radius of the sphere to be such that the sphere contains many molecules while at the same time its radius is small compared with the wave lengths of periodic fields in which we are interested, so that the field is approximately constant in magnitude throughout the sphere. Since the molecular radius is of order 10^{-8} cm and since the smallest wave length that will interest us is about 10^{-3} cm, it is possible to find a suitable radius for the sphere. The sphere is a macroscopic object, just as polarization is a macroscopic concept.

When the dielectric is subject to an external field, polarization effects may be induced in different ways (Fröhlich, 1968, Section 14). The positive and negative charges in each molecule are displaced relative to one another, and this gives rise to *displacement polarization*, which is a translational effect. Secondly, if the molecules possess permanent dipoles, the dipole axes tend to orient somewhat in the direction of the field. This is a rotational effect which causes *orientational polarization*, also called *dipolar polarization.* Our future investigations will be directed to a great extent towards a study of orientational polarization.

The polarization $P(t)$ is defined as the induced dipole moment per unit volume. Let us now obtain an equation for $P(t)$ when the polarization is due to rotational effects only. Suppose that a field in a fixed direction is acting on the dielectric, and that the dielectric consists of polar molecules each with a dipole of moment μ. If θ is the angle between the direction of the field and a dipole axis, the dipole moment of each molecule will have components $\mu \cos \theta$, $\mu \sin \theta$ in the direction of the field and in a perpendicular direction, respectively. On summing over the molecules, the components in the perpendicular directions will cancel, by symmetry. Thus $P(t)$ is got by summing the moments $\mu \cos \theta$. Now, by definition, the probability density function w is normalized so as to satisfy $\int w d\tau = 1$, where $d\tau$ is the volume element in the space of the variables, other than the time, of which w is a function. Hence

$$P(t) = N\mu \int w \cos \theta d\tau, \tag{1,3.1}$$

where N is the number of polar molecules per unit volume. Moreover the polarization is in the direction of the field.

Let us employ (1,3.1) and the results of the last section to derive expressions for the polarization in the Debye theory. If the external field is $F_0 e^{i\omega t}$, we write $P(\omega)e^{i\omega t}$ for $P(t)$; if the external field is F_0, we write $P(0)$ for $P(t)$; if we are dealing with the relaxation effect, we write $P_a(t)$ for $P(t)$, the suffix a denoting after effect. In the present context $d\tau$ is $\sin \theta d\theta d\phi$, which when integrated over ϕ yields $2\pi \sin \theta d\theta$. We then deduce from (1,2.5) that

$$\begin{aligned} P(\omega)e^{i\omega t} &= \frac{N\mu}{2} \int_0^\pi \left[1 + \frac{2\mu F_0 e^{i\omega t} \cos \theta}{2kT + i\omega\zeta}\right] \cos \theta \sin \theta d\theta \\ &= \frac{N\mu}{3} \frac{2\mu F_0 e^{i\omega t}}{2kT + i\omega\zeta}, \end{aligned}$$

so that, from (1,2.7)

$$P(\omega) = \frac{N\mu^2 F_0}{3kT} \frac{1}{1 + i\omega\tau_D}. \tag{1,3.2}$$

We may obtain $P(0)$ from this just by putting ω equal to zero, and therefore

$$P(0) = \frac{N\mu^2 F_0}{3kT}, \tag{1,3.3}$$

$$\frac{P(\omega)}{P(0)} = \frac{1}{1 + i\omega\tau_D}. \tag{1,3.4}$$

We remark for future reference that, if in place of (1,2.5) w were given by

$$w = A[1 + L(\omega)\mu F_0 e^{i\omega t} \cos\theta/(kT)], \tag{1,3.5}$$

we would clearly obtain

$$P(\omega) = \frac{N\mu^2 F_0}{3kT} L(\omega), \tag{1,3.6}$$

$$\frac{P(\omega)}{P(0)} = \frac{L(\omega)}{L(0)}. \tag{1,3.7}$$

Finally eqn (1,2.6) for the relaxation process gives

$$P_a(t) = \frac{N\mu^2 F_0}{3kT} e^{-t/\tau_D}. \tag{1,3.8}$$

For the case of the spherical molecules with their polar axes rotating in one plane we obtain from (1,3.1), (1,1.12) and (1,1.14)

$$P(\omega) = \frac{N\mu^2 F_0}{2kT} \frac{1}{1 + 2i\omega\tau_D}, \tag{1,3.9}$$

$$P(0) = \frac{N\mu^2 F_0}{2kT}, \tag{1,3.10}$$

$$P_a(t) = \frac{N\mu^2 F_0}{2kT} e^{-t/2\tau_D}, \tag{1,3.11}$$

$$\frac{P(\omega)}{P(0)} = \frac{1}{1 + 2i\omega\tau_D}. \tag{1,3.12}$$

2

Dielectric Polarizability and Permittivity

2.1 Complex Polarizability

In the previous section we took a spherical portion of an isotropic dielectric and discussed the macroscopic concept of polarization in the context of a linear theory. In the present chapter we shall introduce for the same situation other macroscopic physical quantities like response function, after-effect function, complex polarizability, complex permittivity, absorption coefficient and refractive index. We shall introduce the microscopic concept of orientational correlation function, and shall relate this to polarizability and permittivity through an equation of Kubo.

Let us suppose that at time $t = 0$ a field of unit intensity is switched on (Scaife, 1971, Chap. 1). Then there is induced in the dielectric a dipole moment in the direction of the field $a(t)$, say, which is called the *unit step response function.* The graph of $a(t)$ as a function of t is depicted roughly in Fig. 2.1. If the field switched on at time 0 is $F(t)$ at time t, the induced dipole moment $M(t)$ at time t ($> t'$) is obtained by summing the contribution to the

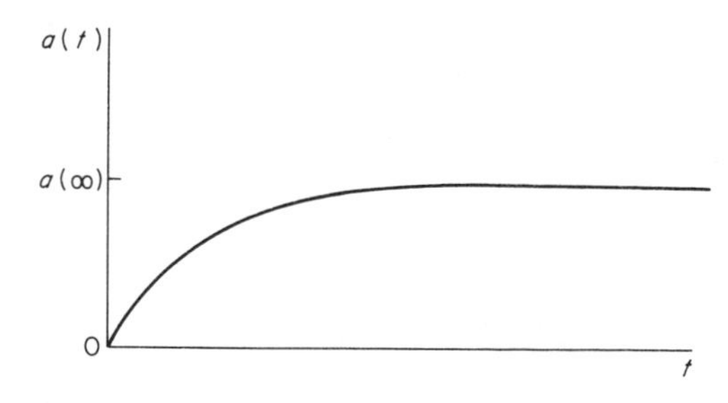

Fig. 2.1 The unit step response function $a(t)$ as a function of t.

moment from the increments $(dF(t')/dt')\delta t'$ to the field at all previous times. Hence $M(t)$ is given by

$$M(t) = \int_0^t \frac{dF(t')}{dt'} a(t-t')dt'. \tag{2,1.1}$$

On integrating by parts and putting $t - t' = x$ we obtain

$$M(t) = \int_0^t F(t-x)\frac{da(x)}{dx}dx. \tag{2,1.2}$$

Let us next consider a different situation in which a field of unit intensity which had been operating for a very long time is switched off at $t = 0$. The *after-effect function* $b(t)$ defined by

$$b(t) = \begin{cases} a(\infty) - a(t) & \text{for } t > 0 \\ 0 & \text{for } t < 0 \end{cases} \tag{2,1.3}$$

gives the dipole moment in the direction of the field at time t.

For the field

$$F(t) = \begin{cases} 0 & \text{for } t < 0 \\ F_0 e^{i\omega t} & \text{for } t > 0 \end{cases}$$

eqn (2,1.2) gives

$$M(t) = F_0 e^{i\omega t} \int_0^t e^{-i\omega x} \frac{da(x)}{dx} dx. \tag{2,1.4}$$

We see from Fig. 2.1 that, when x is very large, $da(x)/dx$ is negligibly small. Hence for a *steady state process* or *stationary process*, that is, one for which t is very great, the integrals $\int_t^\infty \cos \omega x\,(da(x)/dx)dx$, $\int_t^\infty \sin \omega x\,(da(x)/dx)dx$ are negligible and we may replace (2,1.4) by

$$M(t) = F_0 e^{i\omega t} \int_0^\infty e^{-i\omega x} \frac{da(x)}{dx} dx. \tag{2,1.5}$$

Let us write

$$\alpha(\omega) = \int_0^\infty e^{-i\omega x} \frac{da(x)}{dx} dx, \tag{2,1.6}$$

$$\alpha'(\omega) = \int_0^\infty \cos \omega x \frac{da(x)}{dx} dx, \alpha''(\omega) = \int_0^\infty \sin \omega x \frac{da(x)}{dx} dx, \quad (2,1.7)$$

so that

$$\alpha(\omega) = \alpha'(\omega) - i\alpha''(\omega). \quad (2,1.8)$$

We call $\alpha(\omega)$ the *complex polarizability*. Since we had taken the field to be $F_0 e^{+i\omega t}$, we must now write $-i$ before $\alpha''(\omega)$ in order that certain physical quantities derived below from $\alpha(\omega)$ will be positive, as they must. We see that $\alpha'(\omega)$ is an even function and that $\alpha''(\omega)$ is an odd function of ω. From (2,1.5) and (2,1.6)

$$M(t) = F_0 e^{i\omega t} \alpha(\omega). \quad (2,1.9)$$

We see from (2,1.6) that $\alpha(\omega)$ is the Laplace transform of $da(x)/dx$ for the transformation variable $i\omega$ (Appendix C). Then from (2,1.3) and (2,1.6)

$$\begin{aligned} \alpha(\omega) &= -\int_0^\infty \frac{db(t)}{dt} e^{-i\omega t} dt \\ &= b(0) - i\omega \int_0^\infty b(t) e^{-i\omega t} dt \\ &= \int_0^\infty \frac{da(x)}{dx} dx - i\omega \int_0^\infty b(t) e^{-i\omega t} dt, \end{aligned}$$

so that

$$\alpha(\omega) = \alpha(0) - i\omega \int_0^\infty b(t) e^{-i\omega t} dt. \quad (2,1.10)$$

This relates the complex polarizability with the after-effect function for a stationary process. Since $\omega = 0$ gives a static field, we write α_s for $\alpha(0)$. We see from (2,1.6) that α_s is real. Equation (2,1.10) may be expressed as

$$\alpha(\omega) = \alpha_s - i\omega \int_0^\infty b(t) e^{-i\omega t} dt. \quad (2,1.11)$$

We now relate complex polarizability and after-effect function with the quantities $P(t), P(\omega), P(0), P_a(t)$ defined at the end of Chapter 1. If

V is the volume of the dielectric sphere under consideration, the polarization $P(t)$ being the induced dipole moment per unit volume is given by

$$P(t) = \frac{M(t)}{V}. \tag{2,1.12}$$

It follows from (2,1.12), (2,1.9) and (2,1.3) that

$$VP(\omega) = F_0\alpha(\omega), \tag{2,1.13}$$

$$VP(0) = F_0\alpha_s, \tag{2,1.14}$$

$$VP_a(t) = F_0 b(t), \tag{2,1.15}$$

since this is the dipole moment for the after-effect resulting from the field F_0. We deduce from (2,1.11) that

$$P(\omega) = P(0) - i\omega\int_0^\infty P_a(t)e^{-i\omega t}\,dt. \tag{2,1.16}$$

Let us apply these results to the three-dimensional Debye theory. On comparing (1,3.2), (1,3.3), (1,3.8) with (2,1.13)–(2,1.15) we see that

$$\alpha(\omega) = \frac{NV\mu^2}{3kT}\frac{1}{1+i\omega\tau_D},$$

$$\alpha_s = \frac{NV\mu^2}{3kT},$$

$$b(t) = \frac{NV\mu^2}{3kT}e^{-t/\tau_D},$$

so that

$$\frac{\alpha(\omega)}{\alpha_s} = \frac{1}{1+i\omega\tau_D}, \tag{2,1.17}$$

$$b(t) = \alpha_s e^{-t/\tau_D}. \tag{2,1.18}$$

2.2 The Kubo Relation

In the last section we considered a dipole moment $M(t)$ which had been induced by an electric field in a fixed direction. Let us now suppose that a three-dimensional dipole moment $\mathbf{M}(t)$ has been caused by the processes described in Section 1.3. We shall introduce certain quantities related to $\mathbf{M}(t)$ and derive some useful results involving them.

Choosing a long interval of time τ we first define the transform $\widetilde{\mathbf{M}}(\omega, \tau)$ of $\mathbf{M}(t)$ by

$$\widetilde{\mathbf{M}}(\omega, \tau) = \int_{-\tau/2}^{\tau/2} \mathbf{M}(t) e^{-i\omega t}\, dt. \tag{2,2.1}$$

This is in fact the Fourier transform of

$$[u(t + \tfrac{1}{2}\tau) - u(t - \tfrac{1}{2}\tau)]\,\mathbf{M}(t),$$

where $u(s)$ is the step function:

$$u(s) = \begin{cases} 1 & \text{for } s \geqslant 0 \\ 0 & \text{for } s < 0. \end{cases}$$

Hence on inverting the Fourier transform we obtain

$$[u(t + \tfrac{1}{2}\tau) - u(t - \tfrac{1}{2}\tau)]\,\mathbf{M}(t) = \frac{1}{2\pi} \int_{-\infty}^{\infty} \widetilde{\mathbf{M}}(\omega, \tau) e^{i\omega t}\, d\omega. \tag{2,2.2}$$

Since in the present case $\mathbf{M}(t)$ is real, it follows from (2,2.1) that

$$\widetilde{\mathbf{M}}(-\omega, \tau) = (\widetilde{\mathbf{M}}(\omega, \tau))^*, \tag{2,2.3}$$

where the asterisk will be used to designate complex conjugate.

We next introduce the time average $\rho(t)$ on the scalar product $(\mathbf{M}(t') \cdot \mathbf{M}(t' + t))$ over t' by the equation

$$\rho(t) = \overline{(\mathbf{M}(t') \cdot \mathbf{M}(t' + t))}, \tag{2,2.4}$$

the overhead bar denoting time average. We perform the averaging process by recording many values of the scalar product during the interval τ, calculating their average and allowing τ to become indefinitely long; thus

$$\rho(t) = \lim_{\tau \to \infty} \frac{1}{\tau} \int_{-\tau/2}^{\tau/2} (\mathbf{M}(t') \cdot \mathbf{M}(t' + t))\, dt'.$$

Now we see from (2,2.2) that

$$\mathbf{M}(t) = \lim_{\tau\to\infty} \frac{1}{2\pi}\int_{-\infty}^{\infty} \widetilde{\mathbf{M}}(\omega,\tau)e^{i\omega t}d\omega,$$

and therefore

$$\rho(t) = \lim_{\tau\to\infty} \frac{1}{4\pi^2\tau}\int_{-\tau/2}^{\tau/2} dt' \int_{-\infty}^{\infty} e^{i\omega t'}d\omega \int_{-\infty}^{\infty} e^{i\omega'(t'+t)} \times$$

$$(\mathbf{M}(\omega,\tau)\cdot\mathbf{M}(\omega',\tau))d\omega'$$

$$= \lim_{\tau\to\infty} \frac{1}{4\pi^2\tau}\int_{-\infty}^{\infty} e^{i\omega' t}d\omega' \times$$

$$\int_{-\infty}^{\infty} (\widetilde{\mathbf{M}}(\omega,\tau)\cdot\widetilde{\mathbf{M}}(\omega',\tau))d\omega \int_{-\tau/2}^{\tau/2} e^{i(\omega+\omega')t'}dt'.$$

In general

$$\lim_{\tau\to\infty}\int_{-\tau/2}^{\tau/2} e^{i\lambda t'}dt' = \lim_{\tau\to\infty}\int_{-\tau/2}^{\tau/2} (\cos\lambda t' + i\sin\lambda t')dt'$$

$$= 2\lim_{\tau\to\infty}\int_{0}^{\tau/2} \cos\lambda t' dt'$$

$$= 2\lim_{\tau\to\infty}\frac{\sin\frac{1}{2}\lambda\tau}{\lambda}.$$

This quantity has the properties of $2\pi\delta(\lambda)$, where $\delta(\lambda)$ is the *Dirac delta function* which satisfies the relations

$$\delta(\lambda) = \begin{cases} 0 & \text{for } \lambda \neq 0 \\ \infty & \text{for } \lambda = 0, \end{cases}$$

$$\int_{-\infty}^{\infty} \delta(\lambda)d\lambda = 1, \quad \delta(-\lambda) = \delta(\lambda)$$

and

$$\int_{-\infty}^{\infty} f(\lambda)\delta(\lambda - a)d\lambda = f(a) \tag{2,2.5}$$

for any function $f(\lambda)$ that is well-behaved in the neighbourhood of $\lambda = a$. Hence for τ sufficiently large we may replace $\int_{-\tau/2}^{\tau/2} e^{i(\omega+\omega')t'}dt'$

by $2\pi\delta(\omega + \omega')$, so that

$$\rho(t) = \lim_{\tau\to\infty} \int_{-\infty}^{\infty} \frac{(\widetilde{\mathbf{M}}(\omega, \tau) \cdot \widetilde{\mathbf{M}}(-\omega, \tau))}{2\pi\tau} e^{-i\omega t} d\omega$$

$$= \lim_{\tau\to\infty} \int_{-\infty}^{\infty} \frac{(\widetilde{\mathbf{M}}(\omega, \tau) \cdot \widetilde{\mathbf{M}}(\omega, \tau)^*)}{2\pi\tau} e^{-i\omega t} d\omega,$$

by (2,2.5) and (2,2.3).

We define the spectral density $\mathscr{M}(\omega)$ of $\mathbf{M}(t)$ by

$$\mathscr{M}(\omega) = \lim_{\tau\to\infty} \frac{(\widetilde{\mathbf{M}}(\omega, \tau) \cdot \widetilde{\mathbf{M}}(\omega, \tau)^*)}{\tau}.$$

It is clearly a real and even function of ω, and so we may express $\rho(t)$ by the equations

$$\rho(t) = \frac{1}{2\pi} \int_{-\infty}^{\infty} e^{i\omega t} \mathscr{M}(\omega) d\omega \tag{2,2.6}$$

$$\rho(t) = \frac{1}{\pi} \int_{0}^{\infty} \cos \omega t \, \mathscr{M}(\omega) d\omega. \tag{2,2.7}$$

The second equation shows that $\rho(t)$ is an even function of t. If we put t equal to zero in (2,2.4), $\rho(t)$ becomes the time average of M^2 and (2,2.6) gives

$$\overline{M^2} = \frac{1}{2\pi} \int_{-\infty}^{\infty} \mathscr{M}(\omega) d\omega. \tag{2,2.8}$$

This completes the discussion for a general dipole moment $\mathbf{M}(t)$.

At this stage we interpret $\mathbf{M}(t)$ as the dipole moment due to the rotational Brownian motion of the polar molecules. For this interpretation of $\mathbf{M}(t)$ we may invoke the *fluctuation-dissipation theorem.* This theorem has been expounded by many authors including Callen and Welton (1951), Kubo (1957), Landau and Lifschitz (1958, Chap. 12), Bernard and Callen (1959), and Scaife (1959, 1963). A critical study of papers on the subject was made by Case (1972). The proofs of the theorem involve physical assumptions and are often based on quantum-mechanical methods. For our present purposes the theorem is expressible (Scaife, 1963, eqn (73)) by

$$\mathscr{M}(\omega) = \frac{6kT\alpha''(\omega)}{\omega}, \tag{2,2.9}$$

where as before k is the Boltzmann constant and T the absolute temperature. This equation relates the spectral density of the dipole moment caused by the rotational Brownian motion of the polar molecules to the imaginary part of the complex polarizability, introduced in the previous section by the application of an external field and working in a linear theory.

We are interested in relating $\rho(t)$ for $t > 0$ to the after-effect function $b(t)$. Equations (2,2.6) and (2,2.9) yield

$$\rho(t) = \frac{3kT}{\pi} \int_{-\infty}^{\infty} \frac{\alpha''(\omega)}{\omega} e^{i\omega t} d\omega.$$

On inverting this we obtain

$$\frac{\alpha''(\omega)}{\omega} = \frac{1}{6kT} \int_{-\infty}^{\infty} \rho(t) e^{-i\omega t} dt,$$

so that

$$\alpha''(\omega) = \frac{\omega}{3kT} \int_0^{\infty} \rho(t) \cos \omega t dt.$$

On the other hand equating imaginary parts of (2,1.11) gives

$$\alpha''(\omega) = \omega \int_0^{\infty} b(t) \cos \omega t dt,$$

and hence

$$\int_0^{\infty} \left[b(t) - \frac{\rho(t)}{3kT} \right] \cos \omega t dt = 0. \tag{2,2.10}$$

We want to show that this equation leads to the vanishing of $b(t) - \rho(t)/(3kT)$. For an arbitrary function $f(t)$ we have, on introducing Dirac delta functions as above,

$$\int_0^{\infty} d\omega \int_0^{\infty} dt' f(t') \cos \omega t' \cos \omega t$$

$$= \frac{1}{2}\int_0^\infty dt' \int_0^\infty d\omega[\cos\omega(t'+t) + \cos\omega(t'-t)]f(t')$$

$$= \frac{\pi}{2}\int_0^\infty f(t')[\delta(t'+t) + \delta(t'-t)]\,dt'$$

$$= \begin{cases} \frac{\pi}{2}f(t) & \text{for } t > 0 \\ \frac{\pi}{2}f(-t) & \text{for } t < 0. \end{cases} \qquad (2,2.11)$$

Now putting

$$b(t) - \frac{\rho(t)}{3kT} = s(t), \qquad \int_0^\infty s(t')\cos\omega t' dt' = v(\omega)$$

we deduce from (2,2.11) that for $t > 0$

$$\frac{\pi}{2}s(t) = \int_0^\infty v(\omega)\cos\omega t\, d\omega$$

and from (2,2.10) that $v(\omega)$ vanishes. Hence $s(t)$ vanishes and

$$b(t) = \frac{\rho(t)}{3kT}. \qquad (2,2.12)$$

We have from (2,1.6) that

$$\alpha(0) = \int_0^\infty \frac{da(x)}{dx}dx = a(\infty) - a(0) = b(0),$$

by (2,1.3). Combining this result with (2,2.4) and (2,2.12) we deduce that

$$\alpha(0) = \frac{\overline{M^2}}{3kT}. \qquad (2,2.13)$$

On substitution of (2,2.13) and (2,2.4) we express (2,1.10) as

$$\alpha(\omega) = \frac{1}{3kT}\left[\overline{M^2} - i\omega\int_0^\infty \overline{(\mathbf{M}(t')\cdot\mathbf{M}(t'+t))}e^{-i\omega t}dt\right]. \qquad (2,2.14)$$

This is one form of the *Kubo relation* (Kubo, 1957). Since we are dealing with a steady state the time average $\overline{(\mathbf{M}(t') \cdot \mathbf{M}(t'+t))}$ is independent of t' and may be taken as $\overline{(\mathbf{M}(0) \cdot \mathbf{M}(t))}$.

In Section 1.1 we explained that we may take ensemble averages over polar molecules in a dielectric, and expressions for such ensemble averages of $\cos\theta$ were given in (1,1.15) and (1,2.8). We shall frequently have to deal with another type of ensemble average, namely, the ensemble average of a physical system. The ensemble consists of the system in question together with a very large number of copies of it, with which it is in thermal equilibrium. By *ensemble average* we shall in future understand an average over this ensemble at a specified instant of time, and we shall denote the ensemble average by angular brackets. If the system consists of a single polar molecule in a nonpolar environment, the ensemble average becomes identical with the ensemble average of Section 1.1.

To obtain alternative forms of the Kubo relation we apply the *ergodic theorem* (Landau and Lifschitz, 1958, Section1) that the time average is equal to the ensemble average in the absence of an external field. Then we have

$$\overline{(\mathbf{M}(0) \cdot \mathbf{M}(t))} = \langle(\mathbf{M}(0) \cdot \mathbf{M}(t))\rangle.$$

This is called the correlation function of $\mathbf{M}(t)$. Correlation functions will be discussed in Section 4.4. We see that

$$\begin{aligned} \overline{M^2} &= \overline{(M(t))^2} = \overline{(\mathbf{M}(t) \cdot \mathbf{M}(t))} \\ &= \langle(\mathbf{M}(t) \cdot \mathbf{M}(t))\rangle = \langle(\mathbf{M}(0) \cdot \mathbf{M}(0))\rangle. \end{aligned}$$

Hence (2,2.13) may be written

$$\alpha_s = \frac{\langle(M(t))^2\rangle}{3kT}, \qquad (2,2.15)$$

which is the *Kirkwood–Fröhlich theorem* (Kirkwood, 1939; Fröhlich, 1948). Equation (2,2.14) is expressible in the following forms:

$$\alpha(\omega) = \frac{1}{3kT}\left[\langle M^2\rangle - i\omega\int_0^\infty \langle(\mathbf{M}(0) \cdot \mathbf{M}(t))\rangle e^{-i\omega t}dt\right], \qquad (2,2.16)$$

$$\alpha(\omega) = \alpha_s - \frac{i\omega}{3kT}\int_0^\infty \langle(\mathbf{M}(0) \cdot \mathbf{M}(t))\rangle e^{-i\omega t}dt, \qquad (2,2.17)$$

$$\frac{\alpha(\omega)}{\alpha_s} = 1 - i\omega\int_0^\infty \frac{\langle(\mathbf{M}(0)\cdot\mathbf{M}(t))\rangle}{\langle(\mathbf{M}(0)\cdot\mathbf{M}(0))\rangle} e^{-i\omega t} dt, \qquad (2,2.18)$$

where (2,2.15) has been used in deriving (2,2.17) and (2,2.18). The ensemble averages are taken for the rotational Brownian motion of the polar molecules, which is caused by the thermal motion of the particles in the environment and the frictional drag.

It is found experimentally that, when a very high frequency field acts on polar molecules, the dipole axes do not tend to orient towards the field direction (Smyth, 1955, p. 15). To render this result theoretically plausible let us take the axis of rotation of a polar molecule to be in a fixed direction and suppose that the molecule is subject to a couple $M_0 e^{i\omega t}$. The equation of motion

$$I\ddot{\theta} = M_0 e^{i\omega t}$$

will have the solution

$$\theta = -\frac{M_0 e^{i\omega t}}{I\omega^2}.$$

For very high frequencies θ will vanish and so there will be no orientational polarization. When one is considering polarization as coming not only from rotational effects, it is customary to denote by α_∞ the polarizability for a frequency so high that the permanent dipole can no longer orient, so that the polarization effect is due entirely to mechanisms other than dipolar polarization. Then the complex polarizability for orientational polarization is $\alpha(\omega) - \alpha_\infty$ on the assumption that we can separate the rotational from the other polarization effects.

The value of $\alpha(\omega) - \alpha_\infty$ is the right-hand side of (2,2.16), where $\mathbf{M}(t)$ refers to orientational polarization only. Then

$$\mathbf{M}(t) = \sum_{i=1}^{NV} \boldsymbol{\mu}_i(t) \qquad (2,2.19)$$

summed over the individual moments $\boldsymbol{\mu}_i(t)$ of the dipoles of the sphere of volume V that we have under consideration. Hence on substituting (2,2.19) into the right-hand side of (2,2.16) we obtain $\alpha(\omega) - \alpha_\infty$. If we are dealing with a gas dielectric at normal pressures, the random orientations of the dipoles are independent, that is to say

$$\langle(\boldsymbol{\mu}_i(t)\cdot\boldsymbol{\mu}_k(t'))\rangle = (\langle\boldsymbol{\mu}_i(t)\rangle\cdot\langle\boldsymbol{\mu}_k(t')\rangle) = 0 \qquad (k \neq i),$$

and therefore

$$\langle(\mathbf{M}(0)\cdot\mathbf{M}(t))\rangle = \sum_{i=1}^{NV} \langle(\boldsymbol{\mu}_i(0)\cdot\boldsymbol{\mu}_i(t))\rangle.$$

When we are dealing with a pure gas every molecule of which has dipole moment μ, the mean values of the terms in the last sum are all equal and so

$$\langle(\mathbf{M}(0)\cdot\mathbf{M}(t))\rangle = NV\langle(\boldsymbol{\mu}(0)\cdot\boldsymbol{\mu}(t))\rangle,$$

where $\boldsymbol{\mu}(t)$ is the dipole moment at time t of any molecule of the gas. If $\mathbf{n}(t)$ is a unit vector in the direction of $\boldsymbol{\mu}(t)$, so that $\boldsymbol{\mu}(t) = \mu\mathbf{n}(t)$, then

$$\langle(\mathbf{M}(0)\cdot\mathbf{M}(t))\rangle = N\mu^2 V\langle(\mathbf{n}(0)\cdot\mathbf{n}(t))\rangle,$$

$$\langle M^2\rangle = \langle(\mathbf{M}(0)\cdot\mathbf{M}(0))\rangle = N\mu^2 V.$$

On substitution into the right-hand side of (2,2.16) and replacing the left-hand side by $\alpha(\omega)-\alpha_\infty$ we obtain

$$\alpha(\omega)-\alpha_\infty = \frac{N\mu^2 V}{3kT}\left[1 - i\omega\int_0^\infty \langle(\mathbf{n}(0)\cdot\mathbf{n}(t))\rangle e^{-i\omega t}dt\right]. \qquad (2,2.20)$$

Putting ω equal to zero gives

$$\alpha_s - \alpha_\infty = \frac{N\mu^2 V}{3kT}, \qquad (2,2.21)$$

$$\frac{\alpha(\omega)-\alpha_\infty}{\alpha_s-\alpha_\infty} = 1 - i\omega\int_0^\infty \langle(\mathbf{n}(0)\cdot\mathbf{n}(t))\rangle e^{-i\omega t}dt. \qquad (2,2.22)$$

If polarization effects other than orientational polarization are neglected, we replace α_∞ by zero and obtain

$$\frac{\alpha(\omega)}{\alpha_s} = 1 - i\omega\int_0^\infty \langle(\mathbf{n}(0)\cdot\mathbf{n}(t))\rangle e^{-i\omega t}dt. \qquad (2,2.23)$$

While the correlation functions in (2,2.18) are macroscopic quantities, that in (2,2.23) refers to a single molecule and is therefore microscopic. Equations (2,2.20), (2,2.22) and (2,2.23) relate the microscopic correlation function to the macroscopic complex polarizability. The conditions that we needed to establish these equations may often by obeyed fairly well by liquid dielectrics in very dilute solution in a nonpolar solvent. Since we find on integrating by parts that

$$\int_0^\infty \frac{d}{dt}\langle\langle \mathbf{n}(0)\cdot\mathbf{n}(t)\rangle\rangle e^{-i\omega t}dt$$

$$= -1 + i\omega\int_0^\infty \langle\langle \mathbf{n}(0)\cdot\mathbf{n}(t)\rangle\rangle e^{-i\omega t}dt,$$

we may express (2,2.22) as

$$\frac{\alpha(\omega)-\alpha_\infty}{\alpha_s-\alpha_\infty} = -L_{i\omega}\left\{\frac{d}{dt}\langle\langle \mathbf{n}(0)\cdot\mathbf{n}(t)\rangle\rangle\right\}$$

in the notation of Appendix C. Similar results in terms of Laplace transforms may be found for (2,2.18) and (2,2.23).

On comparing (2,2.23) with (2,1.11) when only orientational polarization is considered, we have

$$\int_0^\infty [b(t)-\alpha_s\langle\langle \mathbf{n}(0)\cdot\mathbf{n}(t)\rangle\rangle]e^{-i\omega t}dt = 0.$$

On equating to zero the real part of the left-hand side we get

$$\int_0^\infty [b(t)-\alpha_s\langle\langle \mathbf{n}(0)\cdot\mathbf{n}(t)\rangle\rangle]\cos\omega t\,dt = 0.$$

Then using the argument employed for (2,2.10) we deduce that

$$b(t) = \alpha_s\langle\langle \mathbf{n}(0)\cdot\mathbf{n}(t)\rangle\rangle. \tag{2,2.24}$$

It follows from (2,1.14) and (2,1.15) that

$$P_a(t) = P(0)\langle\langle \mathbf{n}(0)\cdot\mathbf{n}(t)\rangle\rangle. \tag{2,2.25}$$

Equations (2,2.24) and (2,2.25) relate the correlation function for the rotational Brownian motion of the unit vector in the direction of the dipole axis of a polar molecule with the after-effect function and the after-effect polarization.

To connect the above with the Debye theory we see that (2,1.18),

$$b(t) = \alpha_s e^{-t/\tau_D},$$

when compared with (2,2.24) gives

$$\langle\langle \mathbf{n}(0) \cdot \mathbf{n}(t) \rangle\rangle = e^{-t/\tau_D} \qquad (t \geqslant 0). \tag{2,2.26}$$

It follows that

$$1 - i\omega \int_0^\infty \langle\langle \mathbf{n}(0) \cdot \mathbf{n}(t) \rangle\rangle e^{-i\omega t} dt$$

$$= 1 - i\omega \int_0^\infty \exp\left[-t(i\omega + \tau_D^{-1})\right] dt$$

$$= 1 - \frac{i\omega}{i\omega + \tau_D^{-1}} = \frac{1}{1 + i\omega\tau_D}. \tag{2,2.27}$$

Since in the Debye theory only orientational polarization is considered, eqn (2,2.23) applies. On substitution of (2,2.27) into it we find that

$$\frac{\alpha(\omega)}{\alpha_s} = \frac{1}{1 + i\omega\tau_D},$$

in agreement with (2,1.17).

2.3 Complex Permittivity, Absorption and Dispersion

We next introduce the concept of permittivity for an isotropic dielectric. Associated with a static field F_0 in a fixed direction there is an *electric displacement* D_0 related to F_0 by

$$D_0 = F_0 + 4\pi P(0).$$

Writing

$$1 + \frac{4\pi P(0)}{F_0} = \epsilon_s$$

we have

$$D_0 = \epsilon_s F_0. \tag{2,3.1}$$

We call ϵ_s the *static relative permittivity* rather than dielectric constant, since it depends on temperature and pressure. It is a real and positive quantity.

The electric displacement due to a periodic field in a fixed direction $F_0 \cos \omega t$, which has been operative for a very long time, will have frequency $\omega/(2\pi)$ but it will not be in phase with the field. Thus it will be expressible as a linear combination of $\cos \omega t$ and $\sin \omega t$. To obtain this combination one may put $F_0 e^{i\omega t} = F$, replace (2,3.1) by

$$D = \epsilon(\omega)F, \tag{2,3.2}$$

where the complex $\epsilon(\omega)$ is called the *complex relative permittivity*, and equate real parts of each side. We write

$$\epsilon(\omega) = \epsilon'(\omega) - i\epsilon''(\omega), \tag{2,3.3}$$

where $\epsilon'(\omega)$ and $\epsilon''(\omega)$ are real. The quantities $\epsilon'(\omega)$, $\epsilon''(\omega)$ are the *relative permittivity* and the *loss factor*, respectively. Since (2,3.2) must reduce to (2,3.1) for zero frequency

$$\epsilon'(0) = \epsilon_s, \quad \epsilon''(0) = 0. \tag{2,3.4}$$

Let us examine the plane wave solution of the equation

$$\nabla^2 F - \frac{1}{c^2}\frac{\partial^2 D}{\partial t^2} = 0, \tag{2,3.5}$$

which comes from Maxwell's equations for an isotropic medium. We put

$$F = G \exp\left[i\omega\left(t - \frac{\nu(\omega)}{c}x\right)\right], \tag{2,3.6}$$

where G is real, so that on taking real parts the electric intensity for fixed x will depend on time through $\cos \omega t$. In the last equation $\nu(\omega)$ is the *complex refractive index*, and we write

$$\nu(\omega) = n(\omega) - i\kappa(\omega), \tag{2,3.7}$$

$$\frac{\omega}{c}\nu(\omega) = k = k' - ik'', \tag{2,3.8}$$

so that

$$k' = \frac{\omega}{c}n(\omega), \quad k'' = \frac{\omega}{c}\kappa(\omega), \tag{2,3.9}$$

$$F = Ge^{-k''x} \exp\left[i\omega\left(t - \frac{n(\omega)}{c}x\right)\right],$$

$$\operatorname{Re} F = Ge^{-k''x} \cos\left[\omega\left(t - \frac{n(\omega)}{c}x\right)\right],$$

where Re denotes "real part of". We see that $n(\omega)$ is the *refractive index* of the plane wave and that the energy density averaged over a cycle is $G^2 e^{-2k''x}/(16\pi)$. The intensity of the radiation decreases by a factor $1/e$ in a distance $1/(2k'')$. We therefore call $2k''$ the *absorption coefficient*, and often write it $a(\omega)$. The absorption coefficient is usually expressed in *neper* cm^{-1}. This is the reciprocal of the number of centimetres in which the intensity will drop by a factor $1/e$.

We now relate permittivity to *dispersion* in the dielectric, that is, the phenomenon that the refractive index depends on the frequency. On substituting (2,3.2) and (2,3.6) into (2,3.5) we have

$$\epsilon(\omega) = (\nu(\omega))^2. \tag{2,3.10}$$

Then from (2,3.7)

$$n(\omega) = \text{Re}\,(\epsilon(\omega))^{1/2}, \tag{2,3.11}$$

where we must choose the square root so that $n(\omega)$ is positive. Let us write

$$\epsilon'(\omega) = |\epsilon(\omega)|\cos\delta, \quad \epsilon''(\omega) = |\epsilon(\omega)|\sin\delta, \tag{2,3.12}$$

so that

$$\epsilon(\omega) = |\epsilon(\omega)|(\cos\delta - i\sin\delta) = |\epsilon(\omega)|e^{-i\delta},$$

$$(\epsilon(\omega))^{1/2} = |\epsilon(\omega)|^{1/2}\exp[-i\delta/2] \qquad (-\pi<\delta<\pi). \tag{2,3.13}$$

From (2,3.12) we deduce that

$$\cos\tfrac{1}{2}\delta = \left(\frac{|\epsilon(\omega)| + \epsilon'(\omega)}{2|\epsilon(\omega)|}\right)^{1/2},$$

and then from (2,3.11) and (2,3.13) that

$$n(\omega) = \left(\frac{|\epsilon(\omega)| + \epsilon'(\omega)}{2}\right)^{1/2}. \tag{2,3.14}$$

This expresses the dispersion in the dielectric in terms of $\epsilon(\omega)$.

From (2,3.10) and (2,3.8)

$$\epsilon(\omega) = \frac{c^2}{\omega^2}(k'^2 - k''^2 - 2ik'k''),$$

and therefore

$$\epsilon'(\omega) = \frac{c^2}{\omega^2}(k'^2 - k''^2), \quad \epsilon''(\omega) = \frac{2c^2 k'k''}{\omega^2}, \tag{2,3.15}$$

$$|\epsilon(\omega)| = \frac{c^2}{\omega^2}(k'^2 + k''^2),$$

$$|\epsilon(\omega)| - \epsilon'(\omega) = \frac{2c^2 k''^2}{\omega^2}.$$

Since $a(\omega) = 2k''$, we may express the absorption coefficient in terms of the permittivity by

$$a(\omega) = \frac{2^{1/2}\omega}{c}(|\epsilon(\omega)| - \epsilon'(\omega))^{1/2}. \qquad (2,3.16)$$

We may note that, since k' from (2,3.9) and k'' on account of its physical significance are both positive or zero, so by (2,3.15) $\epsilon''(\omega)$ is positive or zero.

We may employ (2,3.14) and (2,3.16) to express $\epsilon(\omega)$ in terms of $n(\omega)$ and $a(\omega)$. Indeed they give immediately

$$\epsilon'(\omega) = (n(\omega))^2 - \frac{c^2(a(\omega))^2}{4\omega^2}, \qquad (2,3.17)$$

$$|\epsilon(\omega)| = (n(\omega))^2 + \frac{c^2(a(\omega))^2}{4\omega^2},$$

from which it follows that

$$(\epsilon''(\omega))^2 = |\epsilon(\omega)|^2 - (\epsilon'(\omega))^2 = \frac{c^2(n(\omega)a(\omega))^2}{\omega^2}.$$

Since $\epsilon''(\omega)$ is non-negative,

$$\epsilon''(\omega) = \frac{cn(\omega)a(\omega)}{\omega}. \qquad (2,3.18)$$

From (2,3.17) and (2,3.18) it follows that

$$\epsilon(\omega) = (n(\omega))^2 - \frac{c^2(a(\omega))^2}{4\omega^2} - \frac{icn(\omega)a(\omega)}{\omega}.$$

The graph of $\epsilon''(\omega)$*v.* $\epsilon'(\omega)$ is called the *Cole–Cole plot* (Cole and Cole, 1941). This plot is very valuable for comparing theoretical results with experiment. We have seen that $\epsilon''(\omega) \geqslant 0$, so the plot is in the upper half-plane.

The complex relative permittivity $\epsilon(\omega)$ will be related to the complex polarizability $\alpha(\omega)$ in a way that depends on the shape of the dielectric in question. As before we take a sphere of volume V of uniformly polarized isotropic dielectric material. Since we agreed already in Section 1.3 that the radius a, say, of the sphere is to be taken small compared with the wave length of the external periodic field $F_0 e^{i\omega t}$, we may presume at any instant that the field is constant throughout the sphere and may apply the method of electrostatics to obtain the field inside and the disturbed field outside. Taking the origin at the centre of the sphere and polar coordinate axis in the direction of F_0 we must find the potential V_i inside the sphere and the potential V_0 outside the sphere such that they obey

$$V_0 = V_i, \quad \frac{\partial V_0}{\partial r} = \epsilon(\omega)\frac{\partial V_i}{\partial r} \quad \text{for } r = a, \tag{2,3.19}$$

$$V_0 = -F_0 e^{i\omega t} r \cos\theta \quad \text{for } r \to \infty.$$

On writing

$$V_0 = -F_0 e^{i\omega t} r \cos\theta + \frac{A e^{i\omega t}\cos\theta}{r^2},$$

$$V_i = -B e^{i\omega t} r \cos\theta,$$

in order to have V_i finite everywhere and V_0 satisfying the condition for $r \to \infty$, performing the differentiations and inserting into (2,3.19) we obtain

$$V_0 = -F_0 e^{i\omega t} r \cos\theta + \frac{\epsilon(\omega) - 1}{\epsilon(\omega) + 2}\,\frac{F_0 a^3 \cos\theta e^{i\omega t}}{r^2}.$$

The second term is the potential due to a dipole of moment

$$\frac{\epsilon(\omega) - 1}{\epsilon(\omega) + 2}\,\frac{3V}{4\pi} F_0 e^{i\omega t}$$

placed at the centre of the sphere with its axis in the direction of the external field. The complex polarizability $\alpha(\omega)$ of the sphere is therefore given by

$$\alpha(\omega) = \frac{\epsilon(\omega) - 1}{\epsilon(\omega) + 2}\,\frac{3V}{4\pi}. \tag{2,3.20}$$

Just as in the previous section we denoted by α_∞ the polarizability at frequencies where dipolar polarization is no longer effective, so now we denote by ϵ_∞ the relative permittivity for such frequencies.

Thus ϵ_∞ is the contribution to ϵ_s that comes from non-orientational polarization. Then from (2,3.20)

$$\alpha(\omega) - \alpha_\infty = \frac{3V}{4\pi}\left\{\frac{\epsilon(\omega)-1}{\epsilon(\omega)+2} - \frac{\epsilon_\infty - 1}{\epsilon_\infty + 2}\right\}$$

$$= \frac{9V}{4\pi}\frac{\epsilon(\omega)-\epsilon_\infty}{(\epsilon(\omega)+2)(\epsilon_\infty+2)},$$

and hence

$$\frac{\alpha(\omega)-\alpha_\infty}{\alpha_s - \alpha_\infty} = \frac{\epsilon(\omega)-\epsilon_\infty}{\epsilon_s - \epsilon_\infty}\,\frac{\epsilon_s + 2}{\epsilon(\omega)+2}. \qquad (2,3.21)$$

In order that the assumption of Section 1.1, that the dipole–dipole interactions can be neglected, may be considered reasonable we suppose that we are dealing with a gas or a very dilute solution of a liquid dielectric in a nonpolar solvent. We may then assume that

$$0 \leqslant |\epsilon'(\omega) - 1| \ll 1, \quad |\epsilon''(\omega)| \ll 1, \qquad (2,3.22)$$

so that both $\epsilon(\omega) + 2$ and $\epsilon_s + 2$ are very nearly equal to 3 and

$$\frac{\alpha(\omega)-\alpha_\infty}{\alpha_s - \alpha_\infty} = \frac{\epsilon(\omega)-\epsilon_\infty}{\epsilon_s - \epsilon_\infty}. \qquad (2,3.23)$$

For this situation eqn (2,2.22) is true and it is expressible, by (2,3.23), as

$$\frac{\epsilon(\omega)-\epsilon_\infty}{\epsilon_s - \epsilon_\infty} = 1 - i\omega\int_0^\infty \langle(\mathbf{n}(0)\cdot\mathbf{n}(t))\rangle e^{-i\omega t}dt. \qquad (2,3.24)$$

Thus, if we know the correlation function $\langle(\mathbf{n}(0)\cdot\mathbf{n}(t))\rangle$ for rotational Brownian motion of a polar molecule and the constants ϵ_0, ϵ_∞ we can find $\epsilon'(\omega)$, $\epsilon''(\omega)$ and therefore the Cole–Cole plot from (2,3.24). Then eqn (2,3.14) and (2,3.16) will provide dispersion and absorption curves.

2.4 Difficulties in the Debye Theory of Dielectric Relaxation

We saw in (2,2.27) that for the Debye theory

$$1 - i\omega\int_0^\infty \langle(\mathbf{n}(0)\cdot\mathbf{n}(t))\rangle e^{-i\omega t}dt = \frac{1}{1+i\omega\tau_D}.$$

When this is substituted into (2,3.24) we obtain

$$\frac{\epsilon(\omega) - \epsilon_\infty}{\epsilon_s - \epsilon_\infty} = \frac{1}{1 + i\omega\tau_D}. \tag{2,4.1}$$

On equating real and imaginary parts of each side we deduce that

$$\epsilon'(\omega) = \epsilon_\infty + \frac{\epsilon_s - \epsilon_\infty}{1 + \omega^2 \tau_D^2}, \tag{2,4.2}$$

$$\epsilon''(\omega) = \frac{(\epsilon_s - \epsilon_\infty)\omega\tau_D}{1 + \omega^2 \tau_D^2}. \tag{2,4.3}$$

The variations of $\epsilon'(\omega)$ and $\epsilon''(\omega)$ as functions of ω are depicted in Fig. 2.2 and Fig. 2.3. The real part of $\epsilon(\omega)$ decreases monotonically from ϵ_s to ϵ_∞. Then $\epsilon''(\omega)$ rises from zero at $\omega = 0$, attains a maximum value $\frac{1}{2}(\epsilon_s - \epsilon_\infty)$ when ω is equal to the reciprocal of the Debye time and decreases slowly to zero as ω tends to infinity.

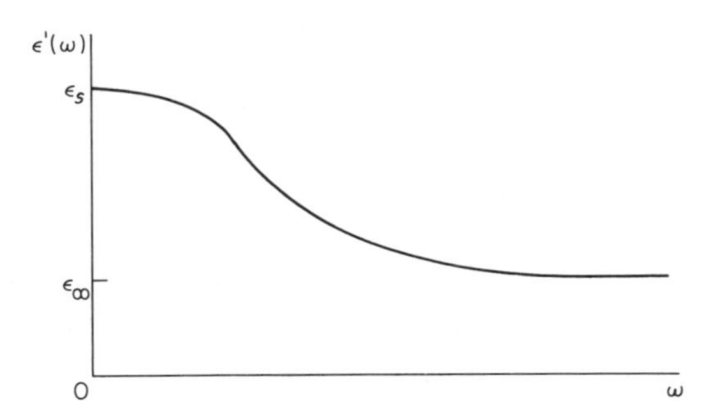

Fig. 2.2 Variation of the relative permittivity $\epsilon'(\omega)$ as a function of ω in the Debye theory.

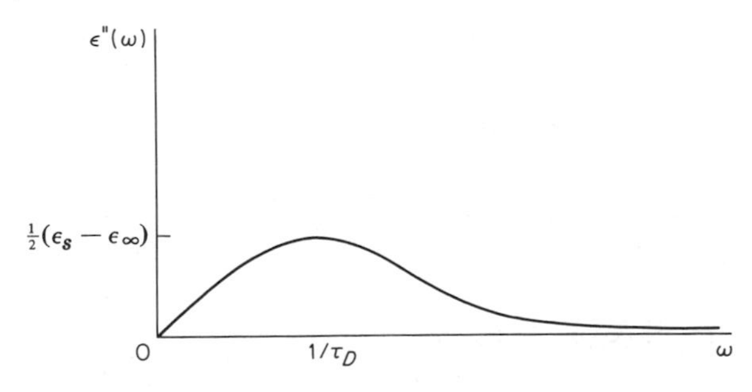

Fig. 2.3 Variation of the loss factor $\epsilon''(\omega)$ as a function of ω in the Debye theory.

We see from (2,4.2) and (2,4.3) that

$$\epsilon'(\omega) - \tfrac{1}{2}(\epsilon_s + \epsilon_\infty) = \frac{(\epsilon_s - \epsilon_\infty)(1 - \omega^2\tau_D^2)}{2(1 + \omega^2\tau_D^2)},$$

$$[\epsilon'(\omega) - \tfrac{1}{2}(\epsilon_s + \epsilon_\infty)]^2 + [\epsilon''(\omega)]^2 = [\tfrac{1}{2}(\epsilon_s - \epsilon_\infty)]^2.$$

The Cole–Cole plot is therefore a semicircle in the first quadrant with centre at $(\tfrac{1}{2}(\epsilon_s + \epsilon_\infty), 0)$ and radius $\tfrac{1}{2}(\epsilon_s - \epsilon_\infty)$. It is drawn in Fig. 2.4. As ω increases from 0 to ∞, the point on the semicircle moves round from $(\epsilon_s, 0)$ to $(\epsilon_\infty, 0)$. The maximum point on the semicircle is reached when $\omega\tau_D = 1$.

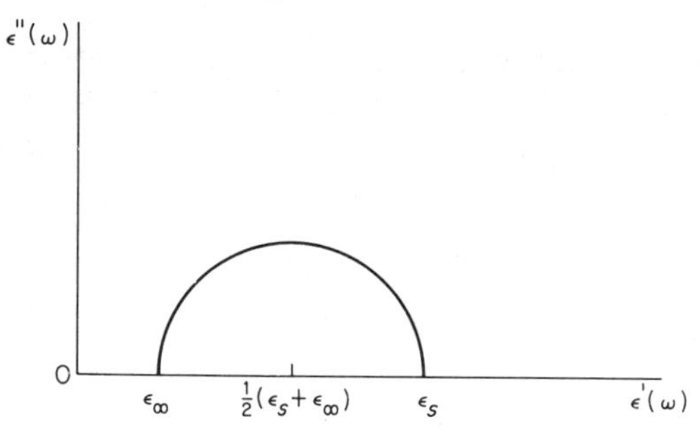

Fig. 2.4 The Cole–Cole plot in the Debye theory.

The refractive index $n(\omega)$ and absorption coefficient $a(\omega)$ may be obtained from (2,3.14) and (2,3.16). It is seen immediately that

$$n(0) = \epsilon_s^{1/2}, \quad a(0) = 0.$$

For values of ω such that $\omega\tau_D \gg 1$ we approximate (2,4.2) and (2,4.3) by

$$\epsilon'(\omega) = \epsilon_\infty + \frac{\epsilon_s - \epsilon_\infty}{\omega^2\tau_D^2}, \quad \epsilon''(\omega) = \frac{\epsilon_s - \epsilon_\infty}{\omega\tau_D}, \tag{2,4.4}$$

and deduce that approximately

$$|\epsilon(\omega)|^2 = \epsilon_\infty^2 + \frac{2\epsilon_\infty(\epsilon_s - \epsilon_\infty)}{\omega^2\tau_D^2} + \frac{(\epsilon_s - \epsilon_\infty)^2}{\omega^2\tau_D^2},$$

so that

$$|\epsilon(\omega)| = \epsilon_\infty + \frac{\epsilon_s - \epsilon_\infty}{\omega^2\tau_D^2} + \frac{(\epsilon_s - \epsilon_\infty)^2}{2\epsilon_\infty\omega^2\tau_D^2}.$$

From this equation, (2,4.4). (2,3.14) and (2,3.16) it is deduced that

$$n(\omega) = \epsilon_\infty^{1/2}\left[1 + \frac{(\epsilon_s - \epsilon_\infty)(\epsilon_s + 3\epsilon_\infty)}{8\epsilon_\infty^2\,\omega^2\tau_D^2}\right] \qquad (\omega\tau_D \gg 1), \quad (2{,}4.5)$$

$$a(\omega) = \frac{\epsilon_s - \epsilon_\infty}{c\tau_D\,\epsilon_\infty^{1/2}} \qquad (\omega\tau_D \gg 1). \qquad (2{,}4.6)$$

The graphs of $n(\omega)$ and $a(\omega)$ as functions of ω are sketched in Fig. 2.5 and Fig. 2.6. The part of the graph of $a(\omega)$ that is nearly horizontal is called the *Debye plateau.*

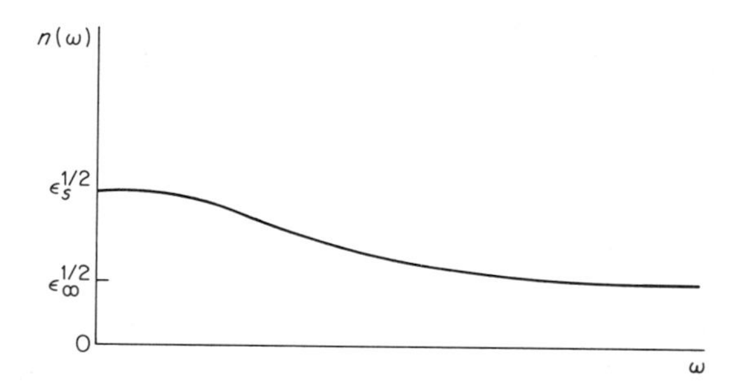

Fig. 2.5 The refractive index $n(\omega)$ as a function of ω in the Debye theory.

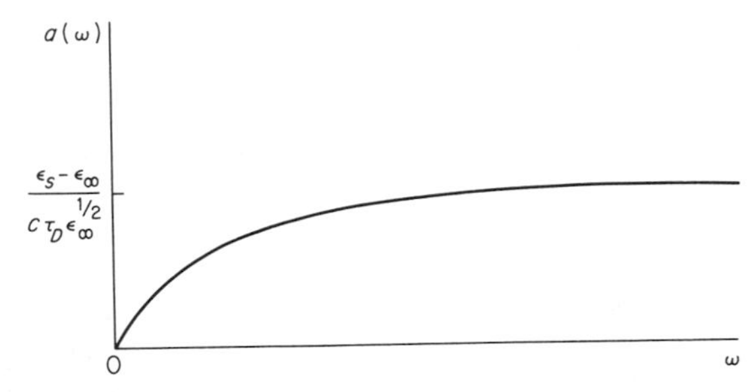

Fig. 2.6 The absorption coefficient $a(\omega)$ as a function of ω in the Debye theory.

While experimentalists are often more inclined to present their results as absorption curves rather than Cole–Cole plots, experimental plots by Goulon *et al.* (1973) for pure chloroform and by Chantry (1977) for pure chlorobenzene are the semicircular Debye ones up to a certain frequency, after which the experimental points go to the left of the semicircle. For liquid chloroform at temperatures $-60°C$ and $25°C$ there is agreement with the Debye plots up to frequencies of order 100 GHz, that is 100×10^9 Hz, one hertz being the frequency corresponding to one complete revolution, or 2π radians, per second. Admittedly in comparing the consequences of the Debye theory with experiment one is frequently faced with the difficulty that the dielectric under examination does not satisfy the condition postulated in the theory that the environment of the polar molecules consists of nonpolar molecules whose linear dimensions are small compared with those of polar molecules. However, even allowing for discrepancies that could be traced to the violation of this condition, there is still serious disagreement with experiment. If the experiments at lower frequencies provide a sufficient number of points to determine an arc of the semicircle, this is often used to determine ϵ_s and ϵ_∞ for substitution into eqn (2,3.24) for the complex permittivity.

There is also a very marked discrepancy between the consequences of the Debye theory and experimental results for absorption in the high-frequency region. In contrast with Fig. 2.6 the experimental curve attains a high maximum and decreases towards zero (Goulon *et al.*, 1973; Gerschel *et al.*, 1976; Chantry, 1977). There are other absorption difficulties, e.g. according to the Debye theory water should be opaque to visible light (Scaife, 1973, p. 295).

Before attempting to do something about these difficulties we shall recall certain simplifying assumptions that were made by Debye in his original theory, namely,

(i) the dipole-dipole interaction between polar molecules may be neglected;

(ii) the $I\ddot{\theta}$-term of (1,1.2) is neglected in (1,1.3) and subsequent equations;

(iii) the polar molecules are spherical.

There exists no adequate treatment of the intermolecular interactions of the first point, and we shall not investigate them. Debye (1929, p. 82) interprets the neglect of the $I\ddot{\theta}$ term as implying that the acceleration effect is negligibly small. It may also be interpreted as ignoring the effects of the inertia of the polar molecules. We shall see in Section 14.2 how the inclusion of inertial effects obviates at least some of the absorption difficulties. Finally the assumption of a spherical shape for polar molecules is often untenable; for example, the well-studied chlorobenzene molecule is approximately a plate. We shall therefore aim at including inertial effects in the study of rotational Brownian motion of a rigid polar molecule of arbitrary shape with a view to applying the results of this study to the theory of dielectric phenomena.

2.5 Early Attempts to Include Inertial Effects

Rocard (1933) tried to take account of inertial effects in orientational polarization by writing (1,1.2) with $F = F_0 \sin \omega t$ and ζ given by (1,1.1):

$$I\ddot{\theta} + \zeta\dot{\theta} + \mu F_0 \sin \omega t \sin \theta = 0 \qquad (\zeta = 8\pi a^3 \eta). \quad (2,5.1)$$

He retained the $I\ddot{\theta}$ term in his subsequent calculations. The resulting differential equation for the probability density function w is nonlinear and Rocard expressed w in the form

$$\frac{1}{4\pi}\{1 + \cos \theta [a_1 \sin (\omega t + \phi) + b_1 \cos (\omega t + \phi)]$$

$$+ P_2(\cos \theta)[c + a_2 \sin (2\omega t + 2\phi) + b_2 \cos (2\omega t + 2\phi)]\},$$

where P_2 is the Legendre polynomial of second order, $\tan \phi = -I\omega/\zeta$ and a_1, b_1, a_2, b_2, c are given explicitly in his paper. He used this probability density to calculate the ensemble average of $\cos \theta$, and thence deduced a result which, if given for complex polarizability, would read

$$\alpha(\omega) = \frac{\dfrac{\mu^2}{3kT}}{\left(1 + \dfrac{i\zeta\omega}{2kT}\right)\left(1 + \dfrac{iI\omega}{\zeta}\right)}. \quad (2,5.2)$$

This shows that

$$\frac{\alpha(\omega)}{\alpha_s} = \frac{1}{\left(1 + \dfrac{i\zeta\omega}{2kT}\right)\left(1 + \dfrac{iI\omega}{\zeta}\right)}. \quad (2,5.3)$$

The real part of the right-hand side of (2,5.2) leads to the expression (8) in Rocard's paper. If we put I equal to zero and refer back to (1,2.7) for the definition of the Debye relaxation time, we obtain

$$\alpha(\omega) = \frac{\dfrac{\mu^2}{3kT}}{1 + i\omega\tau_D}.$$

This gives (2,1.17), which was derived for the Debye theory. Equation (2,5.3) may be expressed as

$$\frac{\alpha(\omega)}{\alpha_s} = \frac{1}{(1 + i\omega\tau_D)(1 + i\omega\tau_F)}, \tag{2,5.4}$$

where τ_F is the friction time defined in (1,2.9).

Essentially the same investigation was carried out by Dmitriev and Gurevich (1946).

A fresh attempt to incorporate inertial effects into the Debye theory was made by Powles (1948). He expressed (1,1.2) as

$$I\ddot{\theta} + \zeta\dot{\theta} = M = -\mu F_0 e^{i\omega t} \sin\theta, \tag{2,5.5}$$

of which a solution is

$$\dot{\theta} = \frac{M}{\zeta + i\omega I}. \tag{2,5.6}$$

He then followed the line of argument of Debye, taking (2,5.6) in place of (1,1.3), and arrived at the equation

$$\frac{\partial w}{\partial t} = \frac{1}{\sin\theta}\frac{\partial}{\partial\theta}\left\{\sin\theta\left[\frac{kT\dfrac{\partial w}{\partial\theta}}{\zeta + \lambda i\omega I} - \frac{Mw}{\zeta + i\omega I}\right]\right\}, \tag{2,5.7}$$

where λ is a real parameter, $0 \leqslant \lambda \leqslant 1$, introduced arbitrarily. This equation is found to be satisfied by

$$w = A[1 + C(\omega)\mu F_0 e^{i\omega t}\cos\theta/(kT)],$$

where in our notation

$$C(\omega) = [1 + i\omega\tau_D + (1-\lambda)i\omega\tau_F - \tau_F\tau_D\omega^2]^{-1}.$$

It follows from (1,3.5), (1,3.7), (2,1.13) and (2,1.14) that

$$\frac{\alpha(\omega)}{\alpha_s} = \frac{1}{1 + i\omega\tau_D + (1-\lambda)i\omega\tau_F - \tau_F\tau_D\omega^2}, \tag{2,5.8}$$

which shows the modification of the Debye result (2,1.17).

If we put $\lambda = 0$ in (2,5.8), we obtain (2,5.4). If we put $\lambda = 1$ in (2,5.8), we obtain

$$\frac{\alpha(\omega)}{\alpha_s} = \frac{1}{1 + i\omega\tau_D - \tau_F \tau_D \omega^2}. \tag{2,5.9}$$

In rotational Brownian motion of a polar molecule the angle θ changes continuously but the impacts produce discontinuous changes in $\dot{\theta}$. While discontinuous changes do not occur in Nature, they have to be accepted, as they were by Newton, in order to provide mathematical solutions of physical problems involving collisions. Since $\dot{\theta}$ is discontinuous, it has no time derivative and the equations (2,5.1) and (2,5.5), as they stand, are meaningless. The investigations of Rocard and of Powles were therefore inadequate. Though it may seem pointless to talk about deductions from non-existent equations, the results for complex polarizability given in (2,5.4) and (2,5.8) will be found useful for comparison with results that we shall derive later.

3
Probability Distributions

3.1 Programme for a Systematic Investigation of Rotational Brownian Motion

The concept of rotational Brownian motion was introduced in Chapter 1 in the context of the Debye theory where a polar molecule under the influence of an external electric field is tossed around by torques arising from the thermal motion of its environment and is slowed by a frictional drag. We shall retain this concept in our future investigations. Debye set up the diffusion equation (1,1.8) for the probability density function as a consequence of the first order differential equation (1,1.3) and of what was essentially an application of Fick's law.

Debye applied his results for rotational Brownian motion to the study of dispersion and absorption in liquid dielectrics. We saw in Section 2.3 how these phenomena could be described in terms of complex permittivity. Moreover, this macroscopic quantity can be related through eqn (2,3.24) to the microscopic correlation function $\langle\langle \mathbf{n}(0) \cdot \mathbf{n}(t) \rangle\rangle$, when the dielectric is a gas, or a liquid in very dilute nonpolar solution. This important result is independent of how the correlation function is calculated.

On account of the conflict between implications of the Debye theory and experiment, it was proposed in Section 2.4 to include inertial effects in the study of rotational Brownian motion and of its consequences for dielectric phenomena. We shall calculate correlation functions not only for $\mathbf{n}(t)$ but also for angular velocity components and for spherical harmonics Y_{jm}, so that the theory may also be applied to nuclear magnetic relaxation phenomena.

The attempts of Rocard and of Powles to take account of inertial effects suffer thc disability that they are based on non-existent differential equations. Besides it will be found, when inertial effects are included, that the equation obeyed by the probability density function is something more complicated than a diffusion equation.

In this and in the three subsequent chapters the mathematical foundations for a study of random processes will be laid. In Chapter 7 general equations for the rotational Brownian motion of a rigid body will be derived and the concept of the stochastic rotation operator will be introduced. This concept is found to be most useful for the calculation in Chapters 8–11 of correlation functions for Brownian particles of different shapes, including entirely asymmetric particles, which could be taken as models of a polar molecule. The correlation functions so obtained are used in Chapters 12 and 13 for the derivation of expressions for complex polarizabilities, spectral densities and correlation times. Finally the consequences of our calculations will be compared in Chapter 14 with the results of experiments on liquid and gas dielectrics.

We embark on this programme with a study of random variables and of their probability distributions.

3.2 Distributions of a Random Variable

A *random variable* or a *stochastic variable*, is one that takes on its values according to some probability law. The variable must be defined on some space, which may be continuous or discrete: for example, if the random variable is the position of a particle that moves in a box, the space is that enclosed by the box; if the random variable is the spin angular momentum of an electron, the space may be defined as the set $\{h/(4\pi), -h/(4\pi)\}$, where h is the Planck constant. To be more precise, we define a random variable as a function on a space with a probability defined on each element of the space.

We now consider how probability itself may be defined. We denote by $P(X; A)$ the probability that an observed value of the random variable X will lie in A. In the case of the particle in the box we could take for X the three-dimensional velocity, and then the space on which X is defined is a velocity space. This is of course different from the space enclosed by the box. We next introduce the *distribution function* $F(X;x)$ by the relation

$$F(X;x) = P[X \leqslant x], \tag{3,2.1}$$

that is, the probability that the observed value of X will be less than or equal to a real number x. In the last equation x could be multi-dimensional; for the example where X is the three-dimensional velocity $\mathbf{v}$ of a particle in a box $P[X \leqslant x]$ is an abbreviation for $P[v_1 \leqslant a_1, v_2 \leqslant a_2, v_3 \leqslant a_3]$, say. We shall, however, continue to use a single variable in the present section. If it is possible to write

$$F(X;x) = \int_{-\infty}^{x} f(X;x')dx', \tag{3,2.2}$$

the probability that an observed value of X lies in the interval $(x, x + dx)$ will be $f(X; x)dx$. We then say that X is a *continuous random variable* and that $f(X, x)$ is its *probability density function.* In contrast to this, a *discrete random variable* X will have a *probability mass function* $p(X;x)$ such that

$$P(X;A) = \sum p(X;x), \tag{3,2.3}$$

the summation being over all x's that belong to A.

In Section 1.1 we had the continuous stochastic variable θ. It is stochastic because it varies in a random manner. It is continuous because θ cannot jump suddenly from one value to another, and therefore the probability of the polar axis lying in the interval $(\theta, \theta + d\theta)$ at time t is expressible as $w(\theta, t)d\theta$. Since θ ranges from 0 to 2π, we would take 0 for the lower limit of integration in (3,2.2) and would restrict the upper limit to 2π. When dealing with rotations in three dimensions we have to interpret the dx' in (3,2.2) as the element of area of a unit sphere $\sin\theta d\theta d\phi$, which when integrated over ϕ gives $2\pi \sin\theta d\theta$, θ now ranging from 0 to π. Thus we have the normalizing condition $2\pi\int_0^\pi w(\theta, t)\sin\theta d\theta = 1$ of (1,2.4).

The distribution of a continuous random variable with which we shall be chiefly concerned is the *Gaussian distribution* or *normal distribution* $N(m, \sigma^2)$ defined by

$$F(X;x) = \int_{-\infty}^{x} \frac{\exp \dfrac{-(s-m)^2}{2\sigma^2}}{(2\pi\sigma^2)^{1/2}} ds \qquad (-\infty < m < \infty, \sigma > 0), \tag{3,2.4}$$

so that the probability density function

$$f(X;x) = \frac{\exp\left\{-\dfrac{(x-m)^2}{2\sigma^2}\right\}}{(2\pi\sigma^2)^{1/2}}. \qquad (3,2.5)$$

If we put $z = cx$, where c is any real constant, eqn (3,2.4) becomes

$$F\left(Z;\frac{z}{c}\right) = \int_{-\infty}^{z/c} \frac{\exp\left\{-\dfrac{(s-m)^2}{2\sigma^2}\right\}}{(2\pi\sigma^2)^{1/2}} ds.$$

On changing the integration variable from s to t, where $t = cs$, we have

$$F(Z;z) = \int_{-\infty}^{z} \frac{\exp\left\{\dfrac{-(t-cm)^2}{2c^2\sigma^2}\right\}}{(2\pi c^2\sigma^2)^{1/2}} dt.$$

Hence the distribution of cX is $N(cm,\ c^2\sigma^2)$. Similarly, if we substitute y for $x-m$, we find that the distribution of $X-m$ is $N(0,\ \sigma^2)$. On putting $c = \sigma^{-1}$ we deduce from these two results that the distribution of $(X-m)/\sigma$ is $N(0,\ 1)$. We can thus reduce any normal distribution to $N(0,\ 1)$.

To provide an example of a probability mass function we consider a procedure in which the outcome is either success or failure, the probability of success being q and that of failure consequently being $1-q$. Thus, if we cast a die, we might define a success as occurring when the number 4 comes up and then $q = 1/6$. The random variable X is the number of successes. Let us carry out n independent trials and calculate the probability of obtaining x successes. This implies that failure occurs $n-x$ times. The probability of obtaining x successes in x independent trials is q^x, and the probability of obtaining $n-x$ failures in $n-x$ independent trials is $(1-q)^{n-x}$. Since we are not interested in the order in which the successes occur, the required probability is $q^x(1-q)^{n-x}$ multiplied by the number of combinations of n taken x at a time, and this is the coefficient $n![x!(n-x)!]^{-1}$ that occurs in the binomial expansion. Thus we find the probability mass function for the *binomial distribution*,

$$p_n(x) = \frac{n!}{x!(n-x)!} q^x (1-q)^{n-x}. \qquad (3,2.6)$$

Let us write $nq = m$ and allow n to become very great while m remains finite, so that q becomes very small. Then from (3,2.6)

$$\lim_{n \to \infty} p_n(x) = \lim_{n \to \infty} \frac{n!}{x!(n-x)!} \left(\frac{m}{n}\right)^x \left(1 - \frac{m}{n}\right)^n \left(1 - \frac{m}{n}\right)^{-x}$$

$$= \lim_{n \to \infty} \frac{e^{-m} m^x}{x!} \frac{n!}{(n-x)! n^x},$$

since $\lim_{n \to \infty} (1 - (m/n))^n = e^{-m}$, $\lim_{n \to \infty} (1 - (m/n))^{-x} = 1$. Now

$$\lim_{n \to \infty} \frac{n!}{(n-x)! n^x} = \lim_{n \to \infty} \frac{n(n-1) \ldots (n-x+1)}{n^x}$$

$$= \lim_{n \to \infty} 1 \left(1 - \frac{1}{n}\right) \ldots \left(1 - \frac{x-1}{n}\right) = 1,$$

so writing $\lim_{n \to \infty} p_n(x)$ as $p(m, x)$ we have

$$p(m, x) = \frac{e^{-m} m^x}{x!}. \qquad (3,2.7)$$

This is the probability mass function of the *Poisson distribution.* It gives the probability of occurrence of x events per unit time, when a large number of events are occurring randomly at the average rate of m per unit time.

We return to the consideration of the general properties of a random variable X. Its mean value, denoted by $\langle X \rangle$, is defined for X continuous and with a probability density function $f(X; x)$ by

$$\langle X \rangle = \int_{-\infty}^{\infty} x f(X; x) dx, \qquad (3,2.8)$$

and for X discrete and with a probability mass function $p(X; x)$ by

$$\langle X \rangle = \sum x p(X; x) \qquad (3,2.9)$$

summed over the x's as in (3,2.3). Since the only mean values with which we shall be concerned are ensemble averages, introduced in Section 2.2, we denote mean values also by angular brackets. To cover both the cases of X continuous and X discrete we write

$$\langle X \rangle = \int_{x=-\infty}^{\infty} x dF(X;x),$$

which is to be understood as meaning (3,2.8) for X continuous and (3,2.9) for X discrete. The mean value of a constant c is c itself, and the mean value of cX is $c\langle X \rangle$. When $\langle X \rangle$ vanishes, X is said to be a *centred random variable.* Similarly the mean value $\langle g(X) \rangle$ of a function $g(X)$ of a random variable X is defined by

$$\langle g(X) \rangle = \int_{x=-\infty}^{\infty} g(x) dF(X;x),$$

which is to be interpreted for X continuous and X discrete as

$$\langle g(X) \rangle = \int_{-\infty}^{\infty} g(x) f(X;x) dx, \qquad (3,2.10)$$

$$\langle g(X) \rangle = \sum g(x) p(X;x), \qquad (3,2.11)$$

respectively. It is obvious that the mean value of the sum of two or more random variables is the sum of their mean values.

The *variance* $V(X)$ of X is defined as the mean value of $(X - \langle X \rangle)^2$; that is,

$$\begin{aligned} V(X) &= \langle X^2 - 2X\langle X \rangle + \langle X \rangle^2 \rangle \\ &= \langle X^2 \rangle - 2\langle X \rangle\langle X \rangle + \langle X \rangle^2 \\ &= \langle X^2 \rangle - \langle X \rangle^2, \end{aligned}$$

where we have used (3,2.10) and (3,2.11) and the knowledge that $\langle X \rangle$ is a constant. When X is centred, $V(X)$ is just $\langle X^2 \rangle$. The *standard deviation* of X is the positive square root of its variance. The *nth moment* of X is $\langle X^n \rangle$. Thus the mean value of a random variable is its first moment and the variance of a centred random variable is its second moment. The *moment-generating function* $\theta(X;k)$ is defined by

$$\theta(X;k) = \langle \exp(kX) \rangle = \int_{x=-\infty}^{\infty} e^{kx} dF(X;x).$$

If all the moments up to the sth exist, we may differentiate under the integral sign up to s times obtaining

$$\frac{d^p \theta(X;k)}{dk^p} = \int_{x=-\infty}^{\infty} x^p e^{kx} dF(X;x), \quad (p = 1, 2, \ldots, s)$$

which for $k = 0$ equals $\langle X^p \rangle$. This enables us to make the Maclaurin expansion

$$\theta(X;k) = 1 + k\langle X \rangle + \frac{k^2}{2!}\langle X^2 \rangle + \ldots$$

$$+ \frac{k^s}{s!}\langle X^s \rangle + k^s \lambda(k), \tag{3,2.12}$$

where $\lambda(k)$ tends to zero with k so that the sth derivative of $\theta(X;k)$ will give $\langle X^s \rangle$ for $k = 0$.

In place of the moment-generating function one commonly uses the *characteristic function* $\phi(X;u)$ defined for a real parameter u by

$$\phi(X;u) = \langle \exp(iuX) \rangle. \tag{3,2.13}$$

If X is a continuous random variable with probability density function $f(X;x)$,

$$\phi(X;u) = \int_{-\infty}^{\infty} e^{iux} f(X;x) dx. \tag{3,2.14}$$

Then, if $\int_{-\infty}^{\infty} |\phi(X;u)| du$ is finite, we may take the Fourier inversion of (3,2.14):

$$f(X;x) = \frac{1}{2\pi} \int_{-\infty}^{\infty} e^{-iux} \phi(X;u) du \tag{3,2.15}$$

(Goldberg, 1962, p. 10). If X is discrete or continuous,

$$\phi(X;u) = \int_{x=-\infty}^{\infty} e^{iux} dF(X;x).$$

Lévy (1925, p. 166) has shown that

$$F(x) - F(0) = \frac{1}{2\pi} \mathscr{P} \int_{-\infty}^{\infty} \phi(X;u) \frac{1 - e^{-ixu}}{iu} du,$$

where $\mathscr{P}$ denotes principal value. An important corollary of Lévy's theorem is that there is a one-to-one correspondence between a probability distribution function and a characteristic function (Jeffreys, 1967, p. 92).

We calculate mean values and the characteristic function for the Gaussian distribution $N(m, \sigma^2)$. From (3,2.5), (3,2.8) and (3.2.10)

$$\langle X \rangle = \frac{1}{(2\pi\sigma^2)^{1/2}} \int_{-\infty}^{\infty} x \exp\{-\tfrac{1}{2}[(x-m)/\sigma]^2\}dx$$

$$= \frac{1}{(2\pi\sigma^2)^{1/2}} \int_{-\infty}^{\infty} (z+m) \exp\{-\tfrac{1}{2}(z/\sigma)^2\}dz = m,$$

$$\langle X^2 \rangle = \frac{1}{(2\pi\sigma^2)^{1/2}} \int_{-\infty}^{\infty} (z^2 + 2zm + m^2) \exp\{-\tfrac{1}{2}(z/\sigma)^2\}dz$$

$$= \frac{1}{(2\pi\sigma^2)^{1/2}} \int_{-\infty}^{\infty} z^2 \exp\{-\tfrac{1}{2}(z/\sigma)^2\}dz + \langle X \rangle^2,$$

$$V(X) = \frac{1}{(2\pi\sigma^2)^{1/2}} \int_{-\infty}^{\infty} z^2 \exp\{-\tfrac{1}{2}(z/\sigma)^2\}dz = \sigma^2,$$

on integration by parts. Thus m is the mean value, σ the standard deviation and $m^2 + \sigma^2$ the second moment of X. We likewise find that

$$\langle X^3 \rangle = 3m\sigma^2 + m^3,$$

$$\langle X^4 \rangle = 3\sigma^4 + 6m^2\sigma^2 + m^4,$$

and it is clear that the nth moment may be expressed as a homogeneous polynomial of order n in m and σ. From (3,2.5) and (3,2.13) we have the characteristic function

$$\phi(X;u) = \frac{1}{(2\pi\sigma^2)^{1/2}} \int_{-\infty}^{\infty} e^{iux} \exp\{-\tfrac{1}{2}(x-m)^2/\sigma^2\}dx$$

$$= \frac{e^{ium}}{(2\pi\sigma^2)^{1/2}} \int_{-\infty}^{\infty} e^{iuz} \exp\{-\tfrac{1}{2}(z/\sigma)^2\}dz$$

$$= \frac{2e^{ium}}{(2\pi\sigma^2)^{1/2}} \int_{0}^{\infty} \cos uz \exp\{-\tfrac{1}{2}(z/\sigma)^2\}dz.$$

On employing the result

$$\int_0^{\infty} \exp(-a^2x^2)\cos bx dx = \frac{\pi^{1/2}\exp\{-b^2/(4a^2)\}}{2a} \qquad (a > 0), \tag{3,2.16}$$

we deduce that

$$\phi(X;u) = \exp\{imu - \tfrac{1}{2}\sigma^2u^2\}\cdot \tag{3,2.17}$$

Since

$$\int_{-\infty}^{\infty} |\phi(X;u)|du = \int_{-\infty}^{\infty} \exp\{-\tfrac{1}{2}\sigma^2 u^2\}du = \frac{(2\pi)^{1/2}}{\sigma},$$

which is finite, we can use (3,2.15) to get (3,2.5), as is easily verified. A random variable which has a Gaussian distribution is called a *Gaussian random variable.* The necessary and sufficient condition that a random variable be Gaussian is that its characteristic function be of the form $\exp\{imu - \frac{1}{2}\sigma^2 u^2\}$. When the Gaussian variable is centred, the characteristic function assumes the form $\exp(-\frac{1}{2}\sigma^2 u^2)$.

Let us now calculate mean values and the characteristic function for the Poisson distribution determined by (3,2.7). We write this equation

$$p(X;x) = \frac{e^{-m} m^x}{x!} \quad (x = 0, 1, 2, \ldots)$$

to conform with the notation of (3,2.3). Then from (3,2.9) and (3,2.11)

$$\langle X \rangle = e^{-m} \sum_{x=1}^{\infty} \frac{xm^x}{x!} = e^{-m} \sum_{x=0}^{\infty} \frac{(x+1)m^{x+1}}{(x+1)!}$$

$$= e^{-m} m \sum_{x=0}^{\infty} \frac{m^x}{x!} = m,$$

$$\langle X^2 \rangle = e^{-m} \sum_{x=1}^{\infty} \frac{x^2 m^x}{x!} = e^{-m} \sum_{x=1}^{\infty} \frac{xm^x}{(x-1)!}$$

$$= e^{-m} \left[\sum_{x=2}^{\infty} \frac{m^{x-1}}{(x-2)!} + \sum_{x=1}^{\infty} \frac{m^{x-1}}{(x-1)!} \right]$$

$$= e^{-m} m[me^m + e^m] = m^2 + m,$$

$$V(X) = m.$$

The characteristic function

$$\phi(X;u) = \sum_{x=0}^{\infty} e^{iux} \frac{e^{-m} m^x}{x!} = e^{-m} \sum_{x=0}^{\infty} \frac{(me^{iu})^x}{x!}$$

$$= e^{-m} \exp(me^{iu}) = \exp\{m(e^{iu} - 1)\}.$$

3.3 Distributions of Several Random Variables

We consider a system which involves more than one random variable. Suppose that the random variables $X_1, X_2, \ldots, X_n$ are defined as functions on the same space, as happens, for example, if each X_i is the three-dimensional position vector $\mathbf{r}$ of a gas molecule in a container. In place of the relation (3,2.1) we now define the *joint distribution function* $F(X_1, X_2, \ldots X_n; x_1, x_2, \ldots x_n)$ by

$$F(X_1, X_2, \ldots X_n; x_1, x_2, \ldots x_n) = P\{s: X_1(s) \leqslant x_1, X_2(s) \leqslant x_2, \ldots X_n(s) \leqslant x_n\}, \quad (3,3.1)$$

where s is an element of the space and the probability P is taken for the values of s that satisfy the inequalities in (3,3.1). If the random variables are continuous, it will be possible to write

$$F(X_1, X_2, \ldots X_n; x_1, x_2, \ldots x_n) = \int_{-\infty}^{x_1} \mathrm{d}y_1 \int_{-\infty}^{x_2} \mathrm{d}y_2 \ldots \int_{-\infty}^{x_n} \mathrm{d}y_n f(X_1, X_2, \ldots X_n; y_1, y_2, \ldots y_n), \quad (3,3.2)$$

and $f(X_1, X_2, \ldots X_n; x_1, x_2, \ldots x_n)$ is called the *joint probability density function.*

The *joint moment-generating function* $\theta(X_1, X_2, \ldots X_n; k_1, k_2, \ldots k_n)$ and the *joint characteristic function* $\phi(X_1, X_2, \ldots X_n; u_1, u_2, \ldots u_n)$ are defined, respectively, by

$$\theta(X_1, X_2, \ldots X_n; k_1, k_2, \ldots k_n) = \langle \exp\{k_1 X_1 + k_2 X_2 + \ldots + k_n X_n\}\rangle,$$

$$\phi(X_1, X_2, \ldots X_n; u_1, u_2, \ldots u_n) = \langle \exp\{i(u_1 X_1 + u_2 X_2 + \ldots + u_n X_n)\}\rangle. \quad (3,3.3)$$

Mean value is now defined by

$$\langle g(X_1, X_2, \ldots X_n)\rangle = \int_{x_1=-\infty}^{\infty} \ldots \int_{x_n=-\infty}^{\infty} g(x_1, x_2, \ldots x_n) dF(X_1, X_2, \ldots X_n; x_1, x_2, \ldots x_n),$$

where the dF notation is to be interpreted as coming from obvious generalizations of (3,2.10) and (3,2.11). When the random variables are independent,

$$P\{s: X_1(s) \leqslant x_1, X_2(s) \leqslant x_2, \ldots X_n(s) \leqslant x_n\}$$
$$= P\{s: X_1(s) \leqslant x_1\} P\{s: X_2(s) \leqslant x_2\} \ldots P\{s: X_n(s) \leqslant x_n\}. \tag{3,3.4}$$

This implies that

$$F(X_1, X_2, \ldots X_n; x_1, x_2, \ldots x_n)$$
$$= F(X_1; x_1) F(X_2; x_2) \ldots F(X_n; x_n).$$

It follows that, if the variables are continuous, the necessary and sufficient condition for them to be independent is that the joint probability density function satisfies

$$f(X_1, X_2, \ldots X_n; x_1, x_2, \ldots x_n)$$
$$= f(X_1; x_1) f(X_2; x_2) \ldots f(X_n; x_n), \tag{3,3.5}$$

as may be deduced by writing down the probability of finding s in the element $dx_1 dx_2 \ldots dx_n$. Equation (3,3.4) also yields

$$\langle g_i(X_i) g_j(X_j) \rangle = \langle g_i(X_i) \rangle \langle g_j(X_j) \rangle = \langle g_j(X_j) \rangle \langle g_i(X_i) \rangle$$
$$= \langle g_j(X_j) g_i(X_i) \rangle, \tag{3,3.6}$$

which may obviously be extended to the product of three or more such functions. In particular

$$\phi(X_1, X_2, \ldots X_n; u_1, u_2, \ldots u_n) = \langle e^{iu_1 X_1} \rangle \langle e^{iu_2 X_2} \rangle \ldots \langle e^{iu_n X_n} \rangle,$$

which may be expressed as

$$\phi(X_1, X_2, \ldots X_n; u_1, u_2, \ldots u_n)$$
$$= \phi(X_1; u_1) \phi(X_2, u_2) \ldots \phi(X_n; u_n). \tag{3,3.7}$$

We similarly obtain

$$\theta(X_1, X_2, \ldots X_n) = \theta(X_1; k_1) \theta(X_2; k_2) \ldots \theta(X_n; k_n). \tag{3,3.8}$$

Let us take two independent random variables X_1 with mean m_1 and standard deviation σ_1, and X_2 with mean m_2 and standard deviation σ_2. Then

$$\langle X_1 + X_2 \rangle = \langle X_1 \rangle + \langle X_2 \rangle = m_1 + m_2,$$
$$\langle (X_1 + X_2)^2 \rangle = \langle X_1^2 \rangle + \langle X_2^2 \rangle + 2\langle X_1 \rangle \langle X_2 \rangle \qquad \text{(by (3,3.6))}$$
$$= \sigma_1^2 + m_1^2 + \sigma_2^2 + m_2^2 + 2m_1 m_2,$$
$$V(X_1 + X_2) = \sigma_1^2 + \sigma_2^2, \tag{3,3.9}$$

so that $X_1 + X_2$ is a random variable with mean $m_1 + m_2$ and standard deviation $(\sigma_1^2 + \sigma_2^2)^{1/2}$. Moreover by employing (3,3.6)

$$\langle \exp [iu(X_1 + X_2)] \rangle = \langle 1 + iu(X_1 + X_2) + \frac{(iu)^2}{2!}(X_1 + X_2)^2 + \ldots \rangle$$
$$= \langle e^{iuX_1} \rangle \langle e^{iuX_2} \rangle,$$

which may clearly be extended to give, for independent random variables,

$$\phi(X_1 + X_2 + \ldots X_t; u) = \phi(X_1; u)\phi(X_2; u) \ldots \phi(X_t; u). \quad (3,3.10)$$

Similarly for the moment-generating function

$$\theta(X_1 + X_2 + \ldots X_t; k) = \theta(X_1; k)\theta(X_2; k) \ldots \theta(X_t; k). \quad (3,3.11)$$

Equations (3,3.7) and (3,3.8) should not be confused with (3,3.10) amd (3,3.11). The former refer to the joint distribution of the independent random variables $X_1, X_2, \ldots X_n$, whereas the latter refer to the distribution of a single random variable $X_1 + X_2 + \ldots + X_t$ which is the sum of independent random variables.

We now consider the central limit theorem for a set of independent random variables $X_1, X_2, \ldots X_n$. An exact formulation and proof of this theorem has been given by Khinchin (1949, Appendix). We shall present it in a way that is adequate for our future applications. Suppose that the set is identically distributed, each member of the set having finite mean m and variance σ^2. According to (3,2.12) the moment-generating function of X_r is

$$1 + mk + \tfrac{1}{2}(\sigma^2 + m^2)k^2 + k^2\lambda(k),$$

where $\lambda(k)$ tends to zero with k and λ is independent of r since we have an identical distribution. For the variable $X_r - m$ the mean is zero and therefore

$$V(X_r - m) = \langle (X_r - m)^2 \rangle = \langle X_r^2 - m^2 \rangle$$
$$= V(X_r) = \sigma^2.$$

Hence the moment-generating function of $X_r - m$ is

$$1 + \tfrac{1}{2}\sigma^2 k^2 + k^2 \mu(k), \quad (3,3.12)$$

where $\mu(k)$ tends to zero with k and is independent of r. From the definition $\langle \exp(kY) \rangle$ of the moment-generating function of Y we see that the moment-generating function of lY is obtained from that of Y by replacing k by lk. Hence from (3,3.12) the moment-generating function of $(X_r - m)n^{-1/2}$, where n is a positive integer, is

$$1 + \frac{\sigma^2 k^2}{2n} + \frac{k^2}{n}\mu\left(\frac{k}{n^{1/2}}\right).$$

Let us now write

$$X_1 + X_2 + \ldots + X_n = X.$$

Then applying (3,3.11) to our last result we deduce that the moment-generating function of $(X - \langle X \rangle)n^{-1/2}$ is

$$\left[1 + \frac{\sigma^2 k^2}{2n} + \frac{k^2}{n}\mu\left(\frac{k}{n^{1/2}}\right)\right]^n. \qquad (3,3.13)$$

If we choose a positive quantity ϵ arbitrarily small, we can find an integer n such that $|\mu(kn^{-1/2})| < \epsilon$ and so express (3,3.13) as

$$\left[1 + \frac{(\sigma^2 + \eta)k^2}{2n}\right]^n,$$

where $|\eta| < 2\epsilon$. As n increases indefinitely, we obtain for the limit of (3,3.13)

$$\lim_{n\to\infty}\left[1 + \frac{(\sigma^2 + \eta)k^2}{2n}\right]^n = \lim_{n\to\infty}\left[1 + \frac{\sigma^2 k^2}{2n}\right]^n = \exp\left(\tfrac{1}{2}\sigma^2 k^2\right).$$

The characteristic function is therefore $\exp(-\frac{1}{2}\sigma^2 u^2)$. On account of the one-to-one correspondence between distribution functions and characteristic functions, this shows that in the limit the distribution of $[\sum_{r=1}^{n} X_r - \langle \sum_{r=1}^{n} X_r \rangle]n^{-1/2}$ becomes Gaussian with zero mean and variance σ^2. On dividing each X_r by σ and noting that by an extension of (3,3.9) the variance of $\Sigma_{r=1}^{n} X_r$ is $n\sigma^2$ we conclude that

$$P\left[\lim_{n\to\infty}\frac{\sum_{r=1}^{n} X_r - \sum_{r=1}^{n}\langle X_r \rangle}{[V(\sum_{r=1}^{n} X_r)]^{1/2}}; x\right] = \frac{1}{(2\pi)^{1/2}}\int_{-\infty}^{x}\exp\left[-\tfrac{1}{2}s^2\right]ds. \qquad (3,3.14)$$

This is a formulation of the *central limit theorem*, which is adequate for our applications to the theory of Brownian motion.

The *covariance* $\mathrm{Cov}(X_1, X_2)$ of any two random variables is defined by

$$\mathrm{Cov}(X_1, X_2) = \langle (X_1 - \langle X_1 \rangle)(X_2 - \langle X_2 \rangle) \rangle.$$

We see that

$$\mathrm{Cov}(X_2, X_1) = \mathrm{Cov}(X_1, X_2),$$
$$\mathrm{Cov}(X_1, X_2) = \langle X_1 X_2 \rangle - \langle X_1 \rangle \langle X_2 \rangle, \qquad (3,3.15)$$
$$\mathrm{Cov}(X_1, X_1) = V(X_1).$$

If $\mathrm{Cov}(X_1, X_2)$ vanishes, X_1 and X_2 are said to be *uncorrelated random variables.*

When X_1 and X_2 are independent, (3,3.6) gives

$$\mathrm{Cov}(X_1, X_2) = 0,$$

so two independent random variables and uncorrelated. Two uncorrelated random variables are not necessarily independent (Hunter, 1972). However, it will be proved in the next section that, if two Gaussian random variables are uncorrelated, they must be independent.

If X_1 and X_2 are uncorrelated, it follows from (3,3.15) that

$$\langle X_1 X_2 \rangle = \langle X_1 \rangle \langle X_2 \rangle. \qquad (3,3.16)$$

Then

$$\begin{aligned} & V(X_1 + X_2) - V(X_1) - V(X_2) \\ &= \langle (X_1 + X_2)^2 \rangle - \langle X_1 + X_2 \rangle^2 - \langle X_1^2 \rangle \\ &\quad + \langle X_1 \rangle^2 - \langle X_2^2 \rangle + \langle X_2 \rangle^2 \\ &= 2\langle X_1 \rangle \langle X_2 \rangle - (\langle X_1 \rangle + \langle X_2 \rangle)^2 + \langle X_1 \rangle^2 + \langle X_2 \rangle^2 = 0, \end{aligned}$$

so that

$$V(X_1 + X_2) = V(X_1) + V(X_2) \qquad (3,3.17)$$

for uncorrelated random variables. We had already established (3,3.16) and (3,3.17) for independent random variables in (3,3.6) and (3,3.9).

Let A and B be two random variables with finite second moments. If λ is any real number,

$$0 \leqslant \langle (A + \lambda B)^2 \rangle = \langle A^2 \rangle + 2\lambda \langle AB \rangle + \lambda^2 \langle B^2 \rangle \qquad (3,3.18)$$

and hence for all values of λ

$$0 \leqslant \langle B^2 \rangle \langle (A + \lambda B)^2 \rangle = (\lambda \langle B^2 \rangle + \langle AB \rangle)^2 + \langle A^2 \rangle \langle B^2 \rangle - \langle AB \rangle^2 .$$

Since λ can be chosen to make $(\lambda\langle B^2\rangle + \langle AB\rangle)^2$ vanish, we must have

$$\langle AB\rangle^2 \leqslant \langle A^2\rangle\langle B^2\rangle. \tag{3,3.19}$$

Taking any two random variables X_1, X_2 we put

$$X_1 - \langle X_1\rangle = A, \quad X_2 - \langle X_2\rangle = B$$

and deduce from (3,3.19) that

$$[\text{Cov}\,(X_1, X_2)]^2 \leqslant V(X_1)V(X_2). \tag{3,3.20}$$

We define the *covariance matrix* of X_1 and X_2 as

$$\begin{bmatrix} \text{Cov}\,(X_1, X_1) & \text{Cov}\,(X_1, X_2) \\ \text{Cov}\,(X_2, X_1) & \text{Cov}\,(X_2, X_2) \end{bmatrix}. \tag{3,3.21}$$

If we take any two non-vanishing real quantities p and q the quadratic form

$$\begin{aligned} &p^2\,\text{Cov}\,(X_1, X_1) + 2pq\,\text{Cov}\,(X_1, X_2) + q^2\,\text{Cov}\,(X_2, X_2) \\ &= p^2\langle A^2\rangle + 2pq\langle AB\rangle + q^2\langle B^2\rangle \\ &= p^2\{\langle A^2\rangle + \frac{2q}{p}\langle AB\rangle + \frac{q}{p}^2\langle B^2\rangle\} \geqslant 0, \end{aligned}$$

by (3,3.18). We may similarly to (3,3.21) define the covariance matrix of random variables $X_1, X_2, \ldots X_n$ as the $n \times n$ matrix whose ij-element is Cov (X_i, X_j).

Let X_1 and X_2 be two continuous random variables with probability density functions $f(X_1; x_1)$ and $f(X_2; x_2)$, respectively, and let $f(X_1, X_2; x_1, x_2)$ be their joint probability density function. Suppose that, given $X_1 = x_1$, $\Pi_2(X_1, X_2; x_1, x_2)dx_2$ is the probability that one finds X_2 in the interval $(x_2, x_2 + dx_2)$. Then Π_2 is called the *conditional probability density* for this situation. Now the probability that X_1 be in $(x_1, x_1 + dx_1)$ and that X_2 be in $(x_2, x_2 + dx_2)$ is expressible either as $f(X_1, X_2; x_1, x_2)dx_1dx_2$ or as $f(X_1; x_1)dx_1\Pi_2(X_1, X_2; x_1, x_2)dx_2$. We therefore conclude that

$$\Pi_2(X_1, X_2; x_1, x_2) = \frac{f(X_1, X_2; x_1, x_2)}{f(X_1; x_1)}. \tag{3,3.22}$$

3.4 Joint Normal Distributions

As an illustration of how joint distribution of random variables are treated we consider the important case of Gaussian or normal distributions. Let us for the moment consider n Gaussian variables X_1, $X_2, \ldots X_n$, let $\langle X_i \rangle = m_i$ and let C denote the covariant matrix defined towards the end of the previous section. We write $x - m$ for the column with elements $x_i - m_i$, $i = 1, 2, \ldots n$, and $(x - m)^T$ for the row with these elements. Assuming that the determinant of the covariant matrix does not vanish we define the *joint normal distribution* of $X_1, X_2, \ldots X_n$ as that with the probability density function

$$f(X_1, X_2, \ldots X_n; x_1, x_2, \ldots x_n) = \frac{\exp\{-\frac{1}{2}(x - m)^T C^{-1} (x - m)\}}{[(2\pi)^n \det C]^{1/2}}, \tag{3,4.1}$$

where C^{-1} is the reciprocal matrix of C (Cramér, 1951, Chaps 21 and 24).

We specialize to the case of $n = 2$, so that

$$f(X_1, X_2; x_1, x_2) = \frac{\exp\left\{-\frac{1}{2}\left[x_1 - m_1, x_2 - m_2\right] C^{-1} \begin{bmatrix} x_1 - m_1 \\ x_2 - m_2 \end{bmatrix}\right\}}{[(2\pi)^2 \det C]^{1/2}}, \tag{3,4.2}$$

and C is the matrix (3,3.21). We put

$$V(X_1) = \sigma^2, \quad V(X_2) = \sigma_2^2,$$
$$\mathrm{Cov}\,(X_1, X_2) = \mathrm{Cov}\,(X_2, X_1) = \rho\sigma_1\sigma_2, \tag{3,4.3}$$

so that, by (3,3.20),

$$\rho^2 \leqslant 1. \tag{3,4.4}$$

Hence

$$C = \begin{bmatrix} \sigma_1^2 & \rho\sigma_1\sigma_2 \\ \rho\sigma_1\sigma_2 & \sigma_2^2 \end{bmatrix}, \tag{3,4.5}$$

$$\det C = (1 - \rho^2)\sigma_1^2\sigma_2^2, \tag{3,4.6}$$

$$C^{-1} = \frac{1}{(1 - \rho^2)\sigma_1^2\sigma_2^2} \begin{bmatrix} \sigma_2^2 & -\rho\sigma_1\sigma_2 \\ -\rho\sigma_1\sigma_2 & \sigma_1^2 \end{bmatrix},$$

$$Q \equiv (x-m)^{\mathrm{T}} C^{-1}(x-m)$$

$$= \frac{1}{1-\rho^2}\left\{\left(\frac{x_1-m_1}{\sigma_1}\right)^2 - 2\rho\left(\frac{x_1-m_1}{\sigma_1}\right)\left(\frac{x_2-m_2}{\sigma_2}\right) + \left(\frac{x_2-m_2}{\sigma_2}\right)^2\right\} \tag{3,4.7}$$

and

$$f(X_1, X_2; x_1, x_2) = \frac{e^{-Q/2}}{2\pi(1-\rho^2)^{1/2}\sigma_1\sigma_2}. \tag{3,4.8}$$

We have so far just asserted that (3,4.2) is a joint normal distribution. If this assertion is true, it must lead to $\langle X_1\rangle = m_1, \langle X_2\rangle = m_2$ and eqn (3,4.3), and it must give the correct normal distribution for X_1 and for X_2. Let us verify that these results are consequences of (3.4.2). Writing

$$(1-\rho^2)Q = \left(\frac{x_2-m_2}{\sigma_2} - \frac{\rho(x_1-m_1)}{\sigma_1}\right)^2 + (1-\rho^2)\left(\frac{x_1-m_1}{\sigma_1}\right)^2 \tag{3,4.9}$$

we have from (3,4.7) and (3,4.8)

$$\langle X_1\rangle = \int_{-\infty}^{\infty} \frac{x_1\,dx_1}{2\pi\sigma_1\sigma_2(1-\rho^2)^{1/2}} \exp\left[-\frac{(x_1-m_1)^2}{2\sigma_1^2}\right] \times$$

$$\int_{-\infty}^{\infty} dx_2 \exp\left[\frac{-\left\{\dfrac{x_2-m_2}{\sigma_2} - \dfrac{\rho(x_1-m_1)}{\sigma_1}\right\}^2}{2(1-\rho^2)}\right]$$

$$= \int_{-\infty}^{\infty} \frac{x_1\,dx_1}{(2\pi\sigma_1^2)^{1/2}} \exp\left[-\frac{(x_1-m_1)^2}{2\sigma_1^2}\right] = m_1, \tag{3,4.10}$$

and similarly $\langle X_2\rangle = m_2$. The integrations here and later may be shortened by using results derived in Section 3.2. Next we obtain

$$V(X_1) = \langle (X_1-m_1)^2\rangle$$

$$= \int_{-\infty}^{\infty} \frac{(x_1-m_1)^2\,dx_1}{(2\pi\sigma_1^2)^{1/2}} \exp\left[-\frac{(x_1-m_1)^2}{2\sigma_1^2}\right] = \sigma_1^2,$$

so that $V(X_2) = \sigma_2^2$, and

$$\text{Cov}(X_1, X_2) = \langle\langle (X_1 - m_1)(X_2 - m_2) \rangle\rangle$$

$$= \int_{-\infty}^{\infty} \frac{(x_1 - m_1)dx_1}{2\pi\sigma_1\sigma_2(1-\rho^2)^{1/2}} \exp\left[-\frac{(x_1 - m_1)^2}{2\sigma_1^2}\right] \times$$

$$\int_{-\infty}^{\infty} dx_2 (x_2 - m_2) \exp\left[-\frac{\left\{\frac{x_2 - m_2}{\sigma_2} - \frac{\rho(x_1 - m_1)}{\sigma_1}\right\}^2}{2(1-\rho^2)}\right]$$

$$= \frac{\rho\sigma_2}{(2\pi)^{1/2}\sigma_1^2} \int_{-\infty}^{\infty} dx_1 (x_1 - m_1)^2 \exp\left[-\frac{(x_1 - m_1)^2}{2\sigma_1^2}\right]$$

$$= \rho\sigma_1\sigma_2 .$$

Finally, we obtain the probability density function of X_1 by integrating (3,4.8) with respect to x_2 from $-\infty$ to ∞. It is obvious from the derivation of (3,4.10) that we get

$$f(X_1; x_1) = \frac{\exp\left\{-\frac{(x_1 - m_1)^2}{2\sigma_1^2}\right\}}{(2\pi\sigma_1^2)^{1/2}}. \tag{3,4.11}$$

This will continue to be true, if we employ (3,4.1). Hence the joint normal distribution yields the normal distribution for X_i with mean m_i and variance σ_i^2. We shall assume that it also gives correct covariances.

When two Gaussian random variables X_1 and X_2 are uncorrelated, we see from (3,4.3) that ρ vanishes and from (3,4.7) and (3,4.8) that

$$f(X_1, X_2; x_1, x_2) = \frac{\exp\left\{-\frac{1}{2}\left(\frac{x_1 - m_1}{\sigma_1}\right)^2 - \frac{1}{2}\left(\frac{x_2 - m_2}{\sigma_2}\right)^2\right\}}{2\pi\sigma_1\sigma_2}$$

$$= f(X_1; x_1) f(X_2; x_2).$$

Thus the joint probability density function is the product of the probability density functions of X_1 and X_2, so that, as we saw when considering (3,3.5), X_1 and X_2 are independent. Hence, when two Gaussian variables are uncorrelated, they must also be independent.

The joint characteristic function of any two Gaussian random variables X_1 and X_2 is given by (3,3.3) as the mean value of $\exp[i(u_1 X_1 + u_2 X_2)]$, so on comparing with (3,4.10) we put

$$\phi(X_1, X_2; u_1, u_2)$$

$$= \frac{1}{2\pi\sigma_1\sigma_2(1-\rho^2)^{1/2}} \int_{-\infty}^{\infty} dx_1 e^{iu_1x_1} \exp\left[-\frac{(x_1 - m_1)^2}{2\sigma_1^2}\right]$$

$$\times \int_{-\infty}^{\infty} dx_2 e^{iu_2x_2} \exp\left[-\frac{\left\{\dfrac{x_2 - m_2}{\sigma^2} - \dfrac{\rho(x_1 - m_1)}{\sigma_1}\right\}^2}{2(1-\rho^2)}\right].$$

On employing (3,2.16) we deduce that

$$\phi(X_1, X_2; u_1, u_2)$$

$$= \exp\{i(m_1u_1 + m_2u_2) - \tfrac{1}{2}(\sigma_1^2u_1^2 + 2\rho\sigma_1\sigma_2u_1u_2 + \sigma_2^2u_2^2)\}. \tag{3,4.12}$$

If we denote by m the column with elements m_1, m_2, by u the column with elements u_1, u_2, by m^{T} and u^{T} the corresponding rows and use (3,4.5) for C, we may put the right-hand side equal to $\exp\{im^{\mathrm{T}}u - \frac{1}{2}u^{\mathrm{T}}Cu\}$. This result is generally true for n variables, that is,

$$\phi(X_1, X_2, \ldots X_n; u_1, u_2, \ldots u_n) = \exp\{im^{\mathrm{T}}u - \tfrac{1}{2}u^{\mathrm{T}}Cu\}.$$

When $u^{\mathrm{T}}Cu$ is positive definite, the joint normal distribution is said to be non-singular (Cramér, 1951, p. 311). A necessary condition for the positive definiteness is $\det C > 0$ (Littlewood, 1950, p. 45). This justifies, for non-singular distributions, the assumption made when we were defining a joint normal distribution.

Equation (3,3.22) gives the relation for the conditional probability density

$$\Pi_2(X_1, X_2; x_1, x_2) = \frac{f(X_1, X_2; x_1, x_2)}{f(X_1; x_1)}.$$

We apply this to the case of two Gaussian variables taking $f(X_1; x_1)$ from (3,4.11) and $f(X_1, X_2; x_1, x_2)$ from (3,4.8) and (3,4.9). Thus

$$\Pi_2(X_1, X_2; x_1, x_2) = \frac{(2\pi\sigma_1^2)^{1/2}}{2\pi(1-\rho^2)^{1/2}\sigma_1\sigma_2} \exp\left[-\frac{1}{2}\left(Q - \frac{(x_1 - m_1)^2}{\sigma_1^2}\right)\right],$$

so that

$$\Pi_2(X_1, X_2; x_1, x_2) = \frac{\exp\left[-\dfrac{1}{2}\left\{\dfrac{x_2 - m_2 - \dfrac{\rho\sigma_2}{\sigma_1}(x_1 - m_1)}{(1-\rho^2)^{1/2}\sigma_2}\right\}^2\right]}{[2\pi(1-\rho^2)\sigma_2^2]^{1/2}}. \tag{3,4.13}$$

This is a normal distribution $N(m, \sigma^2)$ with

$$m = m_2 + \frac{\rho\sigma_2(x_1 - m_1)}{\sigma_1},$$

$$\sigma = (1-\rho^2)^{1/2}\sigma_2.$$

Let us find the distribution of $X_1 + X_2$. According to (3,2.13) the characteristic function of $X_1 + X_2$ is $\langle \exp[iu(X_1 + X_2)]\rangle$. This is obtained by putting $u_1 = u_2 = u$ in (3,4.12), so the characteristic function is $\exp\{i(m_1 + m_2)u - \frac{1}{2}(\sigma_1^2 + 2\rho\sigma_1\sigma_2 + \sigma_2^2)u^2\}$. It follows from our discussion of (3,2.17) that $X_1 + X_2$ is a Gaussian random variable with mean $m_1 + m_2$ and variance $\sigma_1^2 + 2\rho\sigma_1\sigma_2 + \sigma_2^2$. If X_1 and X_2 are also uncorrelated, the $\sigma_1\sigma_2$ term in (3,4.12) vanishes and the variance becomes $\sigma_1^2 + \sigma_2^2$ in agreement with the result of (3,3.9) for independent variables. In the general case, since for any real constants c_1, c_2 the variables c_1X_1 and c_2X_2 are Gaussian, it follows that any linear combination of X_1 and X_2 is Gaussian. Clearly this may be extended to show that any linear combination of Gaussian random variables $X_1, X_2, \ldots X_n$ is itself Gaussian. Moreover, if $X_1, X_2, \ldots X_n$ are all centred, so is the linear combination.

We take centred Gaussian random variables $X_1, X_2, \ldots X_n$, real numbers $c_1, c_2, \ldots c_n$ and write

$$c_1X_1 + c_2X_2 + \ldots + c_nX_n = X. \tag{3,4.14}$$

This is a centred Gaussian random variable, so from (3,2.13) and (3,2.17)

$$\langle e^{iX}\rangle = \phi(X; 1) = \exp\{-\tfrac{1}{2}\langle X^2\rangle\},$$

that is,

$$\sum_{r=0}^{\infty} \frac{i^r}{r!}\langle X^r\rangle = \sum_{s=0}^{\infty} \frac{(-)^s}{2^s s!}\langle X^2\rangle^s. \tag{3,4.15}$$

From (3,4.14)

$$X^r = (c_1 X_1 + c_2 X_2 + \ldots + c_n X_n)^r$$

and using this relation we shall equate the multipliers of $c_1 c_2 \ldots c_n$ on both sides of (3,4.15). If $r \neq n$, $c_1 c_2 \ldots c_n$ does not occur in X^r. If $r = n$, it occurs $n!$ times and it is multiplied by $X_1 X_2 \ldots X_n$. Then the multiplier of $c_1 c_2 \ldots c_n$ in $\langle X^r \rangle$ is $\delta_{rn} n! \langle X_1 X_2 \ldots X_n \rangle$, where δ_{rn} is the *Kronecker delta* defined by

$$\delta_{rn} = \begin{cases} 1 & \text{for } r = n \\ 0 & \text{for } r \neq n. \end{cases} \tag{3,4.16}$$

We see from

$$X^2 = (c_1 X_1 + c_2 X_2 + \ldots + c_n X_n)^2$$

that we do not get $c_1 c_2 \ldots c_n$ in $\langle X^2 \rangle^s$ unless $n = 2s$. We get $c_a c_b (a < b)$ twice for X^2. We get $c_1 c_2 \ldots c_{2s}$ in $\langle X^2 \rangle^s$ only by taking two X_i's at a time, the multiplier of $c_a c_b$ being $2\langle X_a X_b \rangle$. Hence the coefficient of $c_1 c_2 \ldots c_{2s}$ in $\langle X^2 \rangle^s$ is

$$2^s s! \sum \langle X_{i,1} X_{j,1} \rangle \langle X_{i,2} X_{j,2} \rangle \ldots \langle X_{i,s} X_{j,s} \rangle,$$

where in the summation $i, 1 < i, 2 < \ldots < i, s; i, 1 < j, 1, i, 2 < j, 2 \ldots i, s < j, s$. On comparing coefficients on both sides of (3,4.15) we deduce that

$$\langle X_1 X_2 \ldots X_{2t-1} \rangle = 0, \tag{3,4.17}$$

$$\langle X_1 X_2 \ldots X_{2t} \rangle = \sum \langle X_{i,1} X_{j,1} \rangle \langle X_{i,2} X_{j,2} \rangle \ldots \langle X_{i,t} X_{j,t} \rangle \tag{3,4.18}$$

with the above restrictions on the i, p's and j, q's. These relations are frequently used in Brownian motion calculations.

4

Random Processes and Correlations Functions

4.1 Random Processes

In rotational Brownian motion a body is tossed around by random couples arising from the agitated motion of the particles in the heat bath. The angular velocity of the body and the moments of the couples are stochastic variables that depend on the time; they are random functions of the time. We define a *random process* or a *stochastic process* as a set of random variables $X(t)$ that is defined for a range of values of t that belong to a set $\mathbf{T}$. For definiteness we shall take t to denote the time. X may also depend on other variables like the angles which specify the orientation of the body.

It is reasonable to approximate the random process by $X(t_1)$, $X(t_2), \ldots X(t_n)$ with $t_1, t_2, \ldots t_n$ $(t_1 < t_2 < \ldots < t_n)$ all lying close to one another and extending over $\mathbf{T}$. We should not confuse $X(t_1), X(t_2), \ldots X(t_n)$ with $X_1, X_2, \ldots X_n$ used in the previous chapter to denote an arbitrary set of stochastic variables, though in many respects the discussion of both proceeds along similar lines. The joint probability law for $X(t_1), X(t_2), \ldots X(t_n)$ may be given by the joint distribution function F:

$$F[X(t_1), X(t_2), \ldots X(t_n); x_1, x_2, \ldots x_n] =$$
$$P[X(t_1) \leqslant x_1, X(t_2) \leqslant x_2, \ldots X(t_n) \leqslant x_n], \qquad (4,1.1)$$

or by the joint characteristic function ϕ:

$$\phi[X(t_1), X(t_2), \ldots X(t_n); u_1, u_2, \ldots u_n] =$$
$$\langle \exp \{i(u_1 X(t_1) + u_2 X(t_2) + \ldots + u_n X(t_n))\}\rangle. \qquad (4,1.2)$$

A *stationary random process* or *steady-state random process* is one whose joint distribution function satisfies

$$F[X(t_1+\tau), X(t_2+\tau), \ldots X(t_n+\tau); x_1, x_2, \ldots x_n] =$$
$$F(X(t_1), X(t_2), \ldots X(t_n); x_1, x_2, \ldots x_n] \quad (4,1.3)$$

for values of all $t_i + \tau$ lying in **T**. This is sometimes expressed by saying that F is invariant under time translations.

We shall henceforth assume that $X(t)$ is continuous, so that the distributions of $X(t_1), X(t_2), \ldots X(t_n)$ may be specified in terms of probability density functions as explained in Section 3.2. We can then describe the stochastic process as follows (Ming Chen Wang and Uhlenbeck, 1945): let $w_1(x_1, t_1)dx_1$ be the probability of finding $X(t_1)$ in the range (x_1, x_1+dx_1), $w_2(x_1, t_1; x_2, t_2)dx_1dx_2$ the probability of finding $X(t_1)$ in (x_1, x_1+dx_1) and $X(t_2)$ in $(x_2, x_2+dx_2), \ldots w_n(x_1, t_1; x_2, t_2; \ldots x_n, t_n)dx_1dx_2 \ldots dx_n$ the joint probability of finding $X(t_1)$ in $(x_1, x_1+dx_1), X(t_2)$ in $(x_2, x_2+dx_2) \ldots X(t_n)$ in (x_n, x_n+dx_n). The $w_1, w_2, \ldots w_n$ are probability density functions and they are defined for any set of times $t_1, t_2, \ldots t_n$. We can express the F of (4,1.1) in terms of w_n, in analogy with (3,3.2). Each w_i is greater than or equal to zero, and it is symmetric in the sets of variables $x_1, t_1; x_2, t_2; \ldots$. We see from the definitions of $w_1, w_2, \ldots w_n$ that

$$w_1(x_1, t_1) = \int w_2(x_1, t_1; x_2, t_2)dx_2, \quad (4,1.4)$$

and more generally

$$w_k(x_1, t_1; \ldots x_k, t_k) =$$
$$\int \ldots \int dx_{k+1} \ldots dx_n w_n(x_1, t_1; \ldots x_n, t_n).$$

If successive values of $X(t)$ are independent of one another,

$$w_2(x_1, t_1; x_2, t_2)dx_1dx_2 = w_1(x_1, t_1)dx_1 w_1(x_2, t_2)dx_2,$$

so that

$$w_2(x_1, t_1; x_2, t_2) = w_1(x_1, t_1)w_1(x_2, t_2), \quad (4,1.5)$$
$$w_n(x_1, t_1; x_2, t_2; \ldots x_n, t_n) =$$
$$w_1(x_1, t_1)w_2(x_2, t_2) \ldots w_n(x_n, t_n).$$

Then, in analogy with (3,3.7), eqn (4,1.2) is expressible as

$$\phi[X(t_1), X(t_2), \ldots X(t_n); u_1, u_2, \ldots u_n] =$$
$$\phi[X(t_1), u_1]\phi[X(t_2)u_2] \ldots \phi[X(t_n), u_n].$$

To specify the joint probability distribution for this case it is therefore sufficient to know the characteristic function of $\phi[X(t), u]$ for all values of t that belong to the set **T**.

When the successive values $X(t_1), X(t_2), \ldots X(t_n)$ of $X(t)$ are not independent of one another but are distributed as Gaussian random variables, the process is called a *Gaussian stochastic process* or a *Gaussian random process.* Then $w_1, w_2, \ldots w_n$ are all Gaussian probability densities. This means that

$$w_k(x_1, t_1; x_2, t_2; \ldots x_k, t_k) = \frac{\exp\{-\frac{1}{2}(x-m)^{\mathrm{T}} C^{-1}(x-m)\}}{[(2\pi)^k \det C]^{1/2}}, \tag{4,1.6}$$

where we employ the notation of (3,4.1) with $m_i = \langle X(t_i)\rangle$ and C the covariant $k \times k$ matrix with ij-element equal to $\mathrm{Cov}\,(X(t_i), X(t_j))$.

We shall next apply to stochastic processes the concept of conditional probability density introduced in Section 3.3. Let us denote by $\Pi_2(x_1, t_1; x_2, t_2)dx_2$ the conditional probability that given $X(t_1) = x_1$ one finds $X(t_2)$ in $(x_2, x_2 + dx_2)$ at a later time t_2. From the definitions of w_1, w_2, Π_2,

$$w_2(x_1, t_1; x_2, t_2)dx_1 dx_2 = w_1(x_1, t_1)dx_1 \; \Pi_2(x_1, t_1; x_2, t_2)dx_2,$$

so that

$$w_2(x_1, t_1; x_2, t_2) = w_1(x_1, t_1) \; \Pi_2(x_1, t_1; x_2, t_2), \tag{4,1.7}$$

in analogy with (3,3.22). Since $w_1 \geqslant 0$ and $w_2 \geqslant 0$, it follows that $\Pi_2 \geqslant 0$. From (4,1.7) and (4,1.4)

$$\begin{aligned} w_1(x_1, t_1) \int \Pi_2(x_1, t_1; x_2, t_2)dx_2 &= \int w_2(x_1, t_1; x_2, t_2)dx_2 \\ &= w_1(x_1, t_1), \end{aligned}$$

so

$$\int \Pi_2(x_1, t_1; x_2, t_2)dx_2 = 1. \tag{4,1.8}$$

Then from (4,1.7) and the symmetry of w_2 in x_1, t_1 and x_2, t_2

$$\int w_1(x_1, t_1)\Pi_2(x_1, t_1; x_2, t_2)dx_1 = \int w_2(x_1, t_1; x_2, t_2)dx_1$$

$$= \int w_2(x_2, t_2; x_1, t_1)dx_1,$$

and employing (4,1.4) we conclude that

$$\int w_1(x_1, t_1)\Pi_2(x_1, t_1; x_2, t_2)dx_1 = w_1(x_2, t_2). \quad (4,1.9)$$

In the case of Brownian motion the probability of finding $X(t_2)$ in $(x_2, x_2 + dx_2)$ cannot depend on the value of $X(t)$ at a previous time t_1, when $t_2 - t_1$ is very great. Hence in this limiting case we may accept (4,1.5). Combining this with (4,1.7) gives

$$\lim_{t_2 - t_1 \to \infty} \Pi_2(x_1, t_1; x_2, t_2) = w_1(x_2, t_2) \quad (4,1.10)$$

for Brownian motion.

4.2 Markov Processes

A *Markov process* is a random process for which the relation

$$\Pi_n(x_1, t_1; x_2, t_2; \ldots x_{n-1}, t_{n-1}; x_n, t_n) = \Pi_2(x_{n-1}, t_{n-1}; x_n, t_n) \quad (4,2.1)$$

$$(t_1 < t_2 < \ldots < t_{n-1} < t_n)$$

holds for the conditional probability densities just discussed in the previous section. This relation states that Π_n is independent of what happened before t_{n-1}. The probability that $X(t_n)$ be in $(x_n, x_n + dx_n)$ depends only on t_n, x_n and $X(t_{n-1})$, and so at time t_n the stochastic process has no memory of what happened before time t_{n-1}.

Let us consider for a Markov process the expression

$$dx_2 \int \Pi_2(x_1, t_1; x, t)\Pi_2(x, t; x_2, t_2)dx,$$

the integral being taken over the values of x that occur when t satisfies $t_1 < t < t_2$. Since for a Markov process all the conditional probability densities reduce to Π_2's, the above expression is equal to the probability of finding $X(t_2)$ in $(x_2, x_2 + dx_2)$, if $X(t_1) = x_1$. Equating this probability to $\Pi_2(x_1, t_1; x_2, t_2)dx_2$ we deduce the *Chapman–Kolmogorov equation* (Kolmogorov, 1931, eqn (3))

$$\Pi_2(x_1, t_1; x_2, t_2) = \int \Pi_2(x_1, t_1; x, t)\, \Pi_2(x, t; x_2, t_2)dx. \tag{4,2.2}$$

In order that a stochastic process be Markovian the conditional probability density must satisfy $\Pi_2 \geqslant 0$ and eqns (4,1.7), (4,1.8) and (4,2.2).

Markov processes occur in Brownian motion. Since, for example, a polar molecule in a liquid solution undergoes something of the order of 10^{21} random collisions per second with the particles of the environment, we can make each $t_i - t_{i-1}$ macroscopically very small even though during it very many collisions occur. These numerous impacts destroy all correlation between what happens during the time interval (t_{i-1}, t_i) and what has happened before t_{i-1}. The same is true for any Brownian particle, whether it be a polar molecule or not and whether we are examining its translational or rotational motion. Hence any stochastic process associated with the random motion of a Brownian particle is a Markov process. The process may be a linear or angular coordinate that specifies the position or orientation of the particle, it may be a component of its linear or angular velocity, and it may be the random driving force or couple exerted by the heat bath on the particle.

When a process X is Markovian the probability $w_3(x_1, t_1; x_2, t_2; x_3, t_3)dx_1 dx_2 dx_3$ of finding $X(t_1)$ in $(x_1, x_1 + dx_1)$, $X(t_2)$ in $(x_2, x_2 + dx_2)$ and $X(t_3)$ in $(x_3, x_3 + dx_3)$ is expressible as

$$w_2(x_1, t_1; x_2, t_2)dx_1 dx_2\, \Pi_2(x_2, t_2; x_3, t_3)dx_3.$$

We therefore deduce that

$$w_3(x_1, t_1; x_2, t_2; x_3, t_3) = w_2(x_1, t_1; x_2, t_2)\, \Pi_2(x_2, t_2; x_3, t_3). \tag{4,2.3}$$

This extension of (4,1.7) holds only for a Markov process.

4.3 The Wiener Process

A particular type of Gaussian Markov process that will be found important for describing the random couples exerted by a heat bath on a polar molecule is the Wiener process. A *Wiener process* is a set of Gaussian random variables $W(t)$ defined for $t \geqslant 0$ and having the following properties:

(i) the ensemble average $\langle W(t) \rangle = 0$,
(ii) $W(0)$ has a prescribed value, which we take to be zero: $W(0) = 0$,
(iii) the increments $W(t') - W(t'')$ for $t', t'' \geqslant 0$ are stationary and independent: $W(t' + \tau) - W(t'' + \tau) = W(t') - W(t'')$ for $t' + \tau$, $t'' + \tau \geqslant 0$; $W(t_i) - W(t_j)$, $W(t_k) - W(t_l)$ are independent for $t_i > t_j \geqslant t_k > t_l \geqslant 0$.

The Wiener process itself is not stationary; indeed the property $W(0) = 0$ shows that (4,1.3) is not obeyed. Since $W(t)$ is Gaussian and centred, so also are the increments, as was shown in Section 3.4.

Let us examine the variance of $W(t)$. Since its mean value is zero, its variance is equal to its second moment. For any $t_1, t_2 \geqslant 0$

$$\begin{aligned} V(W(t_1 + t_2)) &= \langle \{W(t_1 + t_2)\}^2 \rangle \\ &= \langle \{W(t_1 + t_2) - W(t_1) + W(t_1) - W(0)\}^2 \rangle \\ &= \langle \{W(t_1 + t_2) - W(t_1)\}^2 \rangle + \langle \{W(t_1) - W(0)\}^2 \rangle \end{aligned}$$

since $W(t_1 + t_2) - W(t_1)$ and $W(t_1) - W(0)$ are independent and have zero mean. On account of the stationary property of the increments we can express the last equation as

$$V(W(t_1 + t_2)) = V(W(t_1)) + V(W(t_2)). \tag{4,3.1}$$

If we write

$$V(W(t)) = c(t), \tag{4,3.2}$$

where $t > 0$, $c(t) \geqslant 0$, we can express (4,3.1) as

$$c(t_1 + t_2) = c(t_1) + c(t_2). \tag{4,3.3}$$

Hence for r a positive integer $c(r) = rc(1)$ and it immediately follows that this is true for r the quotient of two positive integers. It is also true for a positive irrational number, if we take this as the intersection of two limiting sequences of rational numbers, $c(t)$ being continuous because $W(t)$ is. The function $c(t)$ is therefore proportional to t and we may express (4,3.2) as

$$V(W(t)) = \sigma^2 t$$

where σ^2 is a positive quantity to be determined by the specific problem that happens to be under investigation. More generally

$$\begin{aligned} V(W(t) - W(s)) &= V(W(t-s) - W(0)) \\ &= V(W(t-s)), \end{aligned}$$

which we write as

$$V(W(t) - W(s)) = \sigma^2 |t-s| \tag{4,3.4}$$

to ensure that we have a positive time interval and that the variance is non-negative. Equation (3,2.17) gives the characteristic function for a Wiener process

$$\phi[W(t) - W(s); u] = \exp\{-\tfrac{1}{2}\sigma^2 |t-s| u^2\}.$$

For two members of a Wiener process $W(s)$ and $W(t)$ with $0 \leqslant s \leqslant t$ we have from the defining properties (i), (ii), (iii),

$$\begin{aligned} \langle W(s)W(t)\rangle &= \langle W(s)(W(s) + W(t) - W(s))\rangle \\ &= \langle W(s)W(s)\rangle + \langle (W(s) - W(0))(W(t) - W(s))\rangle \\ &= V(W(s)) = \sigma^2 s, \end{aligned}$$

by (4,3.4). Similarly, when $0 \leqslant t \leqslant s$,

$$\langle W(s)W(t)\rangle = \langle W(t)W(s)\rangle = \sigma^2 t.$$

To comprise both cases we write

$$\langle W(t)W(s)\rangle = \sigma^2 \min(s, t), \ (s, t \geqslant 0) \tag{4,3.5}$$

where min (s, t) denotes the lesser of s and t.

Let us take $t_1 < t_2$ and $t_1' < t_2'$ and examine the value of $\langle (W(t_2) - W(t_1))(W(t_2') - W(t_1'))\rangle$. For the case of $t_1 < t_1' < t_2 < t_2'$ we have from (4,3.5)

$$\begin{aligned} &\langle (W(t_2) - W(t_1))(W(t_2') - W(t_1'))\rangle \\ &= \langle W(t_2)W(t_2')\rangle + \langle W(t_1)W(t_1')\rangle - \langle W(t_2)W(t_1')\rangle - \langle W(t_1)W(t_2')\rangle \\ &= \sigma^2 (t_2 + t_1 - t_1' - t_1) = \sigma^2 (t_2 - t_1'), \end{aligned}$$

which, as we see from Fig. 4.1, is σ^2 multiplied by the length of the intersection of the intervals $t_2 - t_1$ and $t'_2 - t'_1$. By selecting different positions of t'_1 and t'_2 relative to t_1 and t_2 we may verify that this is generally true; that is,

$$\langle (W(t_2) - W(t_1))(W(t'_2) - W(t'_1)) \rangle = \sigma^2 [(t_2 - t_1) \cap (t'_2 - t'_1)]. \quad (4,3.6)$$

t_1 t'_1 t_2 t'_2

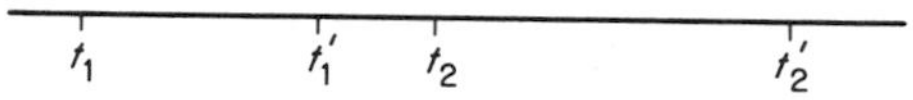

Fig. 4.1. The values of the time in the argument of the Wiener process $W(t)$.

If we now write

$$t_1 = t, t_2 = t + dt, t'_1 = t', t'_2 = t' + dt',$$
$$W(t + dt) - W(t) = dW(t),$$
$$W(t' + dt') - W(t') = dW(t'),$$

we deduce from (4,3.6)

$$\langle dW(t) dW(t') \rangle = \sigma^2 (dt \cap dt'), \quad (4,3.7)$$

and in particular for $t' = t$

$$\langle (dW(t))^2 \rangle = \sigma^2 dt, \quad (4,3.8)$$

the time differential being taken positive.

We choose an arbitrary non-stochastic function $f(t, t')$ continuous in t and t' and consider the integral $\int\int f(t, t')(dW(t)/dt)(dW(t')/dt')dtdt'$ taken over the ranges where W is defined. Since f is non-stochastic,

$$\int\int f(t, t') \left\langle \frac{dW(t)}{dt} \frac{dW(t')}{dt'} \right\rangle dtdt'$$

$$= \left\langle \int\int f(t, t') \frac{dW(t)}{dt} \frac{dW(t')}{dt'} dtdt' \right\rangle$$

$$= \left\langle \int\int f(t, t') dW(t) dW(t') \right\rangle$$

$$= \int\int f(t, t') \langle dW(t) dW(t') \rangle$$

$$= \sigma^2 \int\int f(t, t')(dt \cap dt'), \text{ by (4,3.7)},$$

$$= \sigma^2 \int f(t, t) dt,$$

the double integration reducing to a single integration since there is no contribution unless dt and dt' coincide. Hence on introducing the Dirac delta function that appeared in (2,2.5)

$$\iint f(t, t') \left\langle \frac{dW(t)}{dt} \frac{dW(t')}{dt'} \right\rangle dt dt'$$

$$= \iint f(t, t') \sigma^2 \delta(t - t') dt dt',$$

from which we conclude that

$$\left\langle \frac{dW(t)}{dt} \frac{dW(t')}{dt'} \right\rangle = \sigma^2 \delta(t - t'). \qquad (4,3.9)$$

When introducing the notion of random variables in Section 3.2 we pointed out that such a variable could be multi-dimensional. Let us suppose that $W(t)$ is a vector in q-dimensional Euclidean space having components $W_1(t), W_2(t), \ldots, W_q(t)$, each $W_i(t)$ being a Wiener process. We shall be concerned only with the cases of $q = 1, 2$ or 3. Let us further assume that the W's with different subscripts are uncorrelated, so that being Gaussian they are independent and therefore

$$\langle W_i(t) W_j(s) \rangle = \langle W_i(t) \rangle \langle W_j(s) \rangle = 0 \quad (i \neq j).$$

Moreover, by (4,3.5),

$$\langle W_i(t) W_i(s) \rangle = \sigma_i^2 \min(s, t). \qquad (4,3.10)$$

We write $W(t)$ as a column vector and $W^{\mathrm{T}}(t)$ as the row that is its transpose, so that

$$W(t)W^{\mathrm{T}}(s) = \begin{bmatrix} W_1(t)W_1(s) & W_1(t)W_2(s) & \ldots & W_1(t)W_q(s) \\ W_2(t)W_1(s) & W_2(t)W_2(s) & \ldots & W_2(t)W_q(s) \\ \cdot & \cdot & \cdot & \cdot \\ \cdot & \cdot & \cdot & \cdot \\ \cdot & \cdot & \cdot & \cdot \\ W_q(t)W_1(s) & W_q(t)W_2(s) & \ldots & W_q(t)W_q(s) \end{bmatrix}$$

and

$$\langle W(t)W^{\mathrm{T}}(s)\rangle = \min(s,t)\begin{bmatrix} \sigma_1^2 & 0 & \dots & 0 \\ 0 & \sigma_2^2 & \dots & 0 \\ \cdot & \cdot & & \\ \cdot & \cdot & & \\ 0 & 0 & \dots & \sigma_q^2 \end{bmatrix}. \qquad (4,3.11)$$

If the system under investigation is invariant under rotations in the space, as happens, for example, in the case of a spherical Brownian particle in three-dimensional space, we can put $\sigma_1 = \sigma_2 = \ldots = \sigma_q$. Then

$$\langle W(t)W^{\mathrm{T}}(s)\rangle = \sigma^2 \min(s,t)\mathbf{I},$$

where $\mathbf{I}$ is the $q \times q$ unit matrix. Moreover it is obvious that (4,3.6), (4,3.8) and (4,3.9) will generalize to

$$\langle (W_i(t_2) - W_i(t_1))(W_j(t_2') - W_j(t_1'))\rangle = \delta_{ij}\sigma_i^2[(t_2 - t_1) \cap (t_2' - t_1')],$$

$$\langle dW_i(t)dW_j(t)\rangle = \delta_{ij}\sigma_i^2\,dt, \qquad (4,3.12)$$

$$\left\langle \frac{dW_i(t)}{dt}\,\frac{dW_j(t')}{dt'}\right\rangle = \delta_{ij}\sigma_i^2\,\delta(t - t'), \qquad (4,3.13)$$

where δ_{ij} is the Kronecker delta defined in (3,4.16) and dt is taken positive.

4.4 Correlation Functions and Times

The results of many of our calculations on stochastic processes will be expressed in terms of correlation functions. These are always related to random processes. We have in fact already used them in Section 2.2 and in the previous section of the present chapter, but we shall now introduce them formally. When we speak of correlation functions, we shall in future mean time-correlation functions as distinguished from space-correlation functions which are encountered, for example, in the theory of turbulence (Batchelor, 1953). We should point out that time-correlation functions have applications not only in Brownian motion studies but also in a wide range of fields of physics including Raman scattering, neutron scattering, Rayleigh scattering, Brillouin scattering, absorption of radiation, fluorescence, magnetic resonance absorption, nuclear magnetic lineshape (Gordon, 1968; Berne and Harp, 1970; Evans 1977).

Considering some physical system we take two continuous variables A and B of the system, such as the linear velocity of the centre of mass of a molecule, its angular velocity about the centre of mass, a unit vector through the centre of mass of a rotating molecule and rotating with it, a spherical harmonic whose angles are related to such a vector. A and B will in general depend on position and velocity variables or position and conjugate momentum variables of the system, which we denote generically by $u(t)$, and also explicitly on the time t. For brevity we shall write the values of A and B at time t simply as $A(t)$ and $B(t)$. We suppose that the physical system has attained a steady state. The *time-correlation function* of A and B is defined as the ensemble average $\langle A^*(0)B(t)\rangle$ of $A^*(0)$ and $B(t)$ in the absence of an external field as described is Section 2.2 and taken at time zero, the asterisk denoting complex conjugate. Thus

$$\langle A^*(0)B(t)\rangle = \int du(0)f(0)A^*(0)B(t), \tag{4,4.1}$$

where $f(t)$ is the probability density function of the system at time t in the absence of an external field. For any time s we have

$$f(s)du(s) = f(0)du(0),$$

since the probability that the steady state system be found with the u's in the interval $(u, u + du)$ is independent of the time. Hence from (4,4.1)

$$\begin{aligned}\langle A^*(s)B(t+s)\rangle &= \int du(0)f(0)A^*(s)B(t+s)\\ &= \int du(s)f(s)A^*(s)B(t+s),\end{aligned}$$

which on change of integration variable from s to 0 gives

$$\langle A^*(s)B(t+s)\rangle = \langle A^*(0)B(t)\rangle, \tag{4,4.2}$$

where both ensemble averages are taken at time zero. It is evident from this calculation that the ensemble averages may be taken at any time.

When A and B are both functions of only the orientation of a Brownian particle, the correlation function is known as an *orientational correlation function.* If A and B are both angular velocities, the correlation function is an *angular velocity correlation function.* When A and B are spatial vectors, we shall understand by $\langle A^*(0)B(t)\rangle$ the ensemble average of their scalar product $(A^*(0)\cdot B(t))$. If A and B are different variables, $\langle A^*(0)B(t)\rangle$ is also called the *cross-correlation function* of A and B. When A and B are identical, we have $\langle A^*(0)A(t)\rangle$, which is the *autocorrelation function* of A. We deduce from (4,4.2) that

$$\langle A^*(0)A(-t)\rangle = \langle A^*(t)A(0)\rangle.$$

If $A^*(t)$ commutes with $A(0)$,

$$\langle A^*(0)A(-t)\rangle = \langle A^*(0)A(t)\rangle^*, \qquad (4,4.3)$$

which shows that the autocorrelation function for time $-t$ is the complex conjugate of the autocorrelation function for time t. When in our subsequent calculations for particular problems real expressions for autocorrelation functions are given in terms of t, it must be understood that t is to be interpreted for negative values as $|t|$. The ensemble average $\langle A^*(0)A(0)\rangle$ must be real, since it is the mean value of the real quantity $|A(0)|^2$. The *normalized autocorrelation function* of A is defined as the dimensionless quantity $\langle A^*(0)A(t)\rangle/\langle A^*(0)A(0)\rangle$.

To link the above with results derived earlier, eqn (2,2.26)

$$\langle(n(0)\cdot n(t))\rangle = e^{-t/\tau_D} \qquad (t \geqslant 0)$$

shows that in the Debye theory the autocorrelation function of a unit vector in the direction of the dipole axis of a spherical molecule is e^{-t/τ_D}. Since

$$\langle(\mathbf{n}(0)\cdot\mathbf{n}(0))\rangle = \langle 1\rangle = 1,$$

the autocorrelation function is also the normalized autocorrelation function. In the previous section of the present chapter it was pointed out that the Wiener process $W(t)$ is not stationary. However since the increment $W(t+dt) - W(t)$ is stationary, $dW(t)$ and therefore $dW(t)/dt$ is a stationary process. Since $\delta(\lambda)$ is an even function of λ, we deduce, on putting $t = 0$, $t' = t$, in eqn (4,3.9),

$$\left\langle \frac{dW(t)}{dt}\,\frac{dW(t')}{dt'}\right\rangle = \sigma^2\,\delta(t-t'),$$

that the autocorrelation function of $dW(t)/dt$ is $\sigma^2\,\delta(t)$. In future applications to rotational Brownian motion we shall need correlation functions for components of angular velocity, for vectors in two and three dimensions, and also for spherical harmonics $Y_{jm}(\beta(t), \alpha(t))$. The $\beta(t)$, $\alpha(t)$ are Euler angles, which will be defined in Section 7.3 and which will be seen to be stochastic dynamical variables.

Correlation functions have a special connection with spectral densities, to which reference was made in Section 2.2 in the course of establishing the Kubo relation. We take a stationary stochastic process $X(t)$, a real function of the time, and define $\tilde{X}(\omega, \tau)$ by

$$\tilde{X}(\omega, \tau) = \int_{-\tau/2}^{\tau/2} X(t)e^{-i\omega t}dt. \tag{4,4.4}$$

On comparing this with (2,2.1) we see that we may take over results from Section 2.2 by making the replacements

$$\mathbf{M}(t) \to X(t),\ \tilde{\mathbf{M}}(\omega, \tau) \to \tilde{X}(\omega, \tau).$$

Thus we define the *spectral density*, or *power spectrum*, $X(\omega)$ of $X(t)$ by

$$X(\omega) = \lim_{\tau \to \infty} \frac{|\tilde{X}(\omega, \tau)|^2}{\tau}.$$

Equations (2,2.6) and (2,2.4) give

$$\frac{1}{2\pi}\int_{-\infty}^{\infty} e^{i\omega t}X(\omega)d\omega = \overline{X(t')X(t' + t)}$$

$$= \langle X(t')X(t' + t)\rangle,$$

by the ergodic theorem. Hence (4,4.2) yields

$$\langle X(0)X(t)\rangle = \frac{1}{2\pi}\int_{-\infty}^{\infty} e^{i\omega t}X(\omega)d\omega. \tag{4,4.5}$$

If $\int_{-\infty}^{\infty} |\langle X(0)X(t)\rangle| dt$ is finite, we can, as was stated in Section 3.2, invert (4,4.5) and obtain

$$X(\omega) = \int_{-\infty}^{\infty} e^{-i\omega t}\langle X(0)X(t)\rangle dt, \tag{4,4.6}$$

so that the spectral density and the autocorrelation function are Fourier transforms of each other. This is the *Wiener–Khinchin theorem* (Wiener, 1930; Khinchin, 1934). Berne and Pecora (1976, p. 18) have explained how in light scattering experiments the spectral density of the electric field of the scattered light is sometimes measured, and how consequently the spectrum of the scattered light is determined by the autocorrelation functions of the electric field at the detector.

Related to the autocorrelation function is the correlation time. We define the *correlation time* τ_A for the random variable $A(t)$ as the integral with respect to t from 0 to ∞ of its normalized autocorrelation function, if the integral exists;

$$\tau_A = \int_0^\infty \frac{\langle A^*(0)A(t)\rangle}{\langle A^*(0)A(0)\rangle}\, dt. \qquad (4,4.7)$$

The integral clearly exists, if $\int_{-\infty}^{\infty} |\langle A^*(0)A(t)\rangle| dt$ is finite. In our future calculations $\langle A^*(0)A(t)\rangle$ will always be real, the integral will exist and so there will be a real correlation time. We may regard τ_A as the limit as $\omega \to 0$ of the Fourier transform of the normalized autocorrelation function taken only over positive values of t, that is,

$$\tau_A = \lim_{\omega \to 0} \int_0^\infty \frac{\langle A^*(0)A(t)\rangle}{\langle A^*(0)A(0)\rangle}\, e^{-i\omega t} dt. \qquad (4,4.8)$$

If the normalized autocorrelation function is plotted as a function of t, then τ_A is the area in the positive half-plane between the curve and the t-axis.

5
Stochastic Differential Equations

5.1 The Langevin Equation

At the end of Chapter 2 it was pointed out that equations like (2,5.1) and (2,5.5) which attempted to include inertial effects in rotational Brownian motion studies suffered the defect that their $I\ddot{\theta}$ terms do not exist. The use of such equations for translational Brownian motion goes back to Langevin (1908) who employed the equation

$$m \frac{d^2 x}{dt^2} = -6\pi\eta a \frac{dx}{dt} + X, \tag{5,1.1}$$

where m is the mass and a the radius of a spherical Brownian particle, η the viscosity of the surrounding fluid and X the random force in the x-direction arising from the heat bath. Assuming that the mean kinetic energy of the particle is $\frac{1}{2}kT$ and that the mean value of Xx is zero Langevin deduced that

$$\langle (x(t)-x(0))^2 \rangle = \frac{kTt}{3\pi\eta a}, \tag{5,1.2}$$

so that

$$\frac{\sqrt{\langle (x(t)-x(0))^2 \rangle}}{t} = \left(\frac{kT}{3\pi\eta a}\right)^{1/2} t^{-1/2}. \tag{5,1.3}$$

The left-hand side is the root-mean-square of the average velocity over time t. Equation (5,1.3) shows that, as had already been noted by Einstein (1906), the mean velocity becomes infinite for indefinitely small values of t. Hence the acceleration term in (5,1.1) does not exist. Precisely the same kind of difficulty occurs for rotational Brownian motion, if x is interpreted as an angular coordinate and X as a random couple.

A way out of the difficulty was provided by Uhlenbeck and Ornstein (1930). They made several assumptions about the random force X in (5,1.1) and after lengthy calculations obtained Gaussian probability distribution functions for the displacement x and the velocity dx/dt. In place of (5,1.2) they got

$$\langle (x(t) - x(0))^2 \rangle = \frac{2kT(Bt - 1 + e^{-Bt})}{mB^2}, \tag{5,1.4}$$

where we have written B for $6\pi\eta a/m$. For small values of Bt eqn (5,1.4) gives

$$\frac{(\langle (x(t) - x(0))^2 \rangle)^{1/2}}{t} = \left(\frac{kT}{m}\right)^{1/2},$$

so that the mean velocity has then a finite value and the difficulty in (5,1.3) disappears.

A differential equation in which one or more coefficients is random is called a *stochastic differential equation.* The solution of such an equation will be a random function. Equation (5,1.1) is an example of a stochastic differential equation, since X is a random variable. Many of our calculations will be concerned with the solutions of such equations.

In the next section we shall consider the solution of a one-dimensional Langevin equation. We shall derive in a different way results of Uhlenbeck and Ornstein for the linear velocity, and we shall deduce similar results for angular velocity that will be required for our later discussion of rotational Brownian motion.

5.2 The Ornstein–Uhlenbeck Process

We consider the Langevin equation

$$\frac{du(t)}{dt} = -Bu(t) + A(t). \tag{5,2.1}$$

In this, $u(t)$ is a stochastic variable which for translational Brownian motion we may identify with the dx/dt of (5,1.1). For rotational

Brownian motion we shall identify $u(t)$ with the angular velocity $\omega(t)$. Then for a Brownian particle with moment of inertia I about an axis through its centre of mass about which it rotates there is, according to (5,2.1), a random driving couple $IA(t)$ and a retarding frictional couple $-IB\omega(t)$ proportional to the angular velocity, but no external couple. As has already been pointed out, (5,2.1) as it stands is meaningless because $u(t)$, interpreted either as linear or angular velocity, is discontinuous and consequently $du(t)/dt$ does not exist. To ascribe a meaning to it we write it as (Doob, 1942)

$$du(t) = -Bu(t)dt + dU(t) \tag{5,2.2}$$

with $dU(t) = A(t)dt$. We then interpret (5,2.2) as signifying that for $f(t)$ any non-stochastic continuous function of t and for all $a, b\ (> a)$

$$\int_{t=a}^{b} f(t)du(t) = -B\int_{a}^{b} f(t)u(t)dt + \int_{t=a}^{b} f(t)dU(t). \tag{5,2.3}$$

In particular $f(t) \equiv 1$ gives

$$u(b) - u(a) = -B\int_{a}^{b} u(t)dt + U(b) - U(a). \tag{5,2.4}$$

Let us consider the implications of this equation for rotational Brownian motion when $u(t)$ is identified with $\omega(t)$. We take a plane in the body through the axis of rotation and another plane through the axis but in a fixed direction in space, and we denote the angle between these two planes by $\theta(t)$. Then (5,2.4) yields

$$I\omega(b) - I\omega(a) = -IB[\theta(b) - \theta(a)] + IU(b) - IU(a).$$

We may interpret this by saying that the change in angular momentum of the Brownian particle in the time interval $b - a$ is equal to $IB[\theta(b) - \theta(a)]$ due to the frictional drag subtracted from $IU(b) - IU(a)$ arising from the random rotational motion of the environment.

Putting $a = 0$ and $b = t$ we shall now discuss $U(t)$; a more precise discussion is to be found in Arnold (1974, Chap. 8). We take t to be very long and divide it into intervals, writing

$$U(t) - U(0) = \sum_{k=1}^{n} (U(t_k) - U(t_{k-1})) \tag{5,2.5}$$

with $t_0 = 0 < t_1 < t_2 \ldots < t_n = t$. For Brownian motion the increments

$$U(t_1) - U(0), U(t_2) - U(t_1), \ldots U(t) - U(t_{n-1}) \quad (5,2.6)$$

are independent and $U(t)$ is a Markov process, as was explained in Section 4.2. We suppose that the random motion of the heat bath has attained a steady state. Then the increments (5,2.6) are also stationary and identically distributed. On account of the randomness of the motion $\langle U(t_k) \rangle = 0$, and we choose our time origin so that $U(0) = 0$. Hence each increment of (5,2.6) has zero mean and applying the central limit theorem (3,3.14) to the right-hand side of (5,2.5) we deduce that $U(t)/[V(U(t))]^{1/2}$ is Gaussian with zero mean and unit variance. Thus $U(t)$ is a Gaussian random variable with zero mean. In addition it has all the other requirements specified in Section 4.3 for a Wiener process. If therefore, notwithstanding the nonexistence of du/dt, we wish for convenience to write down a one-dimensional Langevin equation for the motion of a Brownian particle subject to no external forces, we see from (5,2.2) that it will assume the form

$$\frac{du(t)}{dt} = -Bu(t) + \frac{dW(t)}{dt}, \quad (5,2.7)$$

where $W(t)$ is a Wiener process.

To solve (5,2.7) with the initial condition that for zero time $u(t)$ is equal to a constant $u(0)$ we write down (5,2.3) replacing $f(t)$ by e^{Bt}, $U(t)$ by $W(t)$, a by 0, b by t, and use s as the integration variable:

$$\int_{s=0}^{t} e^{Bs} du(s) = -B \int_0^t e^{Bs} u(s) ds + \int_{s=0}^{t} e^{Bs} dW(s). \quad (5,2.8)$$

On integrating the left-hand side by parts and dividing across by e^{Bt} we obtain from (5,2.8)

$$u(t) = u(0)e^{-Bt} + \int_{s=0}^{t} e^{-B(t-s)} dW(s). \quad (5,2.9)$$

Now expressing $dW(s)$ as $W(s+ds) - W(s)$ we see that $dW(s)$ is Gaussian and centred, and so taking ensemble averages we deduce from (5,2.9) that

$$\langle u(t) \rangle = u(0)e^{-Bt}. \quad (5,2.10)$$

The integral in (5,2.9) being the limit of a sum of centred Gaussian random variables is itself a centred Gaussian random variable, as we saw in Section 3.4. Hence $u(t)$ is a Gaussian random variable with mean $u(0)e^{-Bt}$. To find its variance we deduce from (5,2.9) and (5,2.10) that

$$\begin{aligned}
&\langle (u(t) - \langle u(t) \rangle)^2 \rangle \\
&= \left\langle \int_0^t e^{-B(t-t_1)} \frac{dW(t_1)}{dt_1} dt_1 \int_0^t e^{-B(t-t_2)} \frac{dW(t_2)}{dt_2} dt_2 \right\rangle \\
&= \sigma^2 e^{-2Bt} \int_0^t dt_1 \int_0^t dt_2 e^{B(t_1+t_2)} \delta(t_1 - t_2), \quad \text{by (4,3.9)}, \\
&= \sigma^2 e^{-2Bt} \int_0^t e^{2Bt_1} dt_1,
\end{aligned}$$

that is,

$$\langle (u(t) - \langle u(t) \rangle)^2 \rangle = \frac{\sigma^2 (1 - e^{-2Bt})}{2B}. \tag{5,2.11}$$

The variance is $\sigma^2(1 - e^{-2Bt})/(2B)$. If therefore we know σ, both the mean and variance of $u(t)$ are determined and its probability distribution function will be specified as in (3,2.4) by (5,2.10) and (5,2.11).

The value of σ^2 is obtainable from the condition that, as t tends to infinity the system approaches a state of statistical equilibrium. Then the mean energy associated with the variable $u(t)$ is $\frac{1}{2}kT$. As $t \to \infty$, $\langle u(t) \rangle \to 0$ according to (5,2.10), and $\langle (u(t))^2 \rangle \to \sigma^2/2B$ according to (5,2.11). If $u(t)$ is the linear velocity, we have

$$\frac{1}{2}kT = \frac{1}{2}m \lim_{t \to \infty} \langle (u(t))^2 \rangle = \frac{1}{2}m \frac{\sigma^2}{2B}$$

$$\sigma^2 = \frac{2BkT}{m}. \tag{5,2.12}$$

Then (5,2.10)–(5,2.12) yield

$$\langle (u(t))^2 \rangle = (u(0))^2 e^{-2Bt} + \frac{kT(1 - e^{-2Bt})}{m}. \tag{5,2.13}$$

Equations (5,2.10) and (5,2.13) give the Gaussian distribution in Uhlenbeck and Ornstein (1930). If $u(t)$ is the angular velocity, (5,2.12) is obviously replaced by

$$\sigma^2 = \frac{2BkT}{I}. \qquad (5,2.14)$$

The derivation of the result that $u(t)$ is a Gaussian random process was based on three assumptions:

(i) $u(t)$ satisfies the Langevin equation (5,2.1);
(ii) $A(t)$ is a centred Gaussian random variable;
(iii) the initial value of $u(t)$ is a constant.

For the present purpose it was not necessary to assume that $A(t)$ is the time derivative of a Wiener process. Indeed, if we allow $A(t)$ to remain in the Langevin equation, (5,2.9) becomes

$$u(t) = u(0)e^{-Bt} + \int_{s=0}^{t} e^{-B(t-s)}A(s)ds$$

and the integral is a centred Gaussian random variable. It is not generally true that a random process associated with Brownian motion is Gaussian. This will be seen clearly in Section 11.1 when we calculate components of angular velocity of an asymmetric body that is undergoing rotational Brownian motion. It will, however, be a Markov process, as was explained in Section 4.2.

In future investigations we shall be very much concerned with the steady state solution, or stationary solution, of the Langevin equation (5,2.7). Since the Brownian motion has been in operation for a very long time, we replace the lower limits of integration in (5,2.8) by $y = -\infty$, where y is now the integration variable, and obtain

$$u(t) = \int_{y=-\infty}^{t} e^{-B(t-y)}dW(y). \qquad (5,2.15)$$

We see that $u(t)$ is a centred Gaussian random variable. Then for $t \geqslant s$

$$\begin{aligned}\langle u(t)u(s)\rangle &= \left\langle \int_{-\infty}^{t} dt_1 \int_{-\infty}^{s} dt_2 e^{-B(t-t_1)} \frac{dW(t_1)}{dt_1} e^{-B(s-t_2)} \frac{dW(t_2)}{dt_2} \right\rangle \\ &= \sigma^2 e^{-B(t+s)} \int_{-\infty}^{t} dt_1 \int_{-\infty}^{s} dt_2 e^{B(t_1+t_2)} \delta(t_1 - t_2),\end{aligned}$$

by (4,3.9). There is a contribution to the double integral only for $t_1 = t_2 \leqslant s$, so the upper integration limit of t_1 is s. Hence we deduce

$$\langle u(t)u(s)\rangle = \sigma^2 e^{-B(t+s)} \int_{-\infty}^{s} dt_1 \int_{-\infty}^{s} dt_2 e^{B(t_1+t_2)}\delta(t_1 - t_2)$$

$$= \sigma^2 e^{-B(t+s)} \int_{-\infty}^{s} e^{2Bt_1}\, dt_1$$

$$= \frac{\sigma^2 e^{-B(t-s)}}{2B}.$$

For $t \leqslant s$ we have similarly

$$\langle u(t)u(s)\rangle = \langle u(s)u(t)\rangle = \frac{\sigma^2 e^{-B(s-t)}}{2B}.$$

Both cases are covered by the relation

$$\langle u(t)u(s)\rangle = \frac{\sigma^2 e^{-B|t-s|}}{2B}. \tag{5,2.16}$$

This gives for $t = s$

$$\langle (u(t))^2\rangle = \frac{\sigma^2}{2B}, \tag{5,2.17}$$

which is now the variance. It is independent of t and again leads to (5,2.12) or (5,2.14). We have therefore determined the distribution of $u(t)$ for the steady state case.

When dealing with rotational Brownian motion we often write the Langevin equation as

$$I\,\frac{d\omega(t)}{dt} = -\zeta\omega(t) + N(t), \tag{5,2.18}$$

where we have put $N(t)$ for the random couple $I(dW(t)/dt)$ and ζ for IB. Since $N(t)$ is expressible as $I \lim_{dt\to 0} [W(t+dt) - W(t)]/dt$, it is stationary, Gaussian and centred. According to (5,2.14)–(5,2.16) we have for the steady state

$$\omega(t) = \int_{y=-\infty}^{t} e^{-B(t-y)} dW(y) \tag{5,2.19}$$

$$\langle \omega(t)\omega(s)\rangle = (kT/I)e^{-B|t-s|}. \tag{5,2.20}$$

Equation (4,3.9) is now equivalent to

$$\langle N(t)N(t')\rangle = 2\zeta kT\delta(t-t'), \tag{5,2.21}$$

which shows that $N(t)$ and $N(t')$ are uncorrelated when $t' \neq t$. Since the correlation function $\langle N(0)N(t)\rangle$ of $N(t)$ is an even function of t,

$$\int_0^\infty \langle N(0)N(t)\rangle dt = \tfrac{1}{2}\int_{-\infty}^{\infty} \langle N(0)N(t)\rangle dt$$

$$= \zeta kT \int_{-\infty}^{\infty} \delta(t)dt,$$

by (5,2.21). We therefore have the relation between the friction constant ζ and the random couple $N(t)$

$$\zeta = \frac{1}{kT}\int_0^\infty \langle N(0)N(t)\rangle dt.$$

It is not suprising that there exists a relation between ζ and $N(t)$, since both the random rotational couple and the frictional couple owe their origins to the Brownian motion of the particles in the heat bath.

According to the Wiener–Khinchin theorem as expressed by (4,4.6) the spectral density of $N(t)$ is $\int_{-\infty}^{\infty} \langle N(0)N(t)\rangle e^{-i\omega t}dt$. On substituting the value of $\langle N(0)N(t)\rangle$ from (5,2.21) we obtain the spectral density $2\zeta kT$, which is independent of ω. A stationary Gaussian random process, which has zero mean and whose spectral density is constant for all values of ω is called *white noise*. It is so called in analogy with white light, whose spectrum is constant in the visible range of frequencies. Thus the random process $N(t)$, where $-\infty < t < \infty$, is white noise.

Suppose that we have a process $X(t)$ defined for $-\infty < t < \infty$ such that for $t_1 < t_2 \ldots < t_n$ the random variables $X(t_1), X(t_2), \ldots X(t_n)$ have a joint normal distribution as explained in Sections 3.4 and 4.1, and such that for every value of t each $X(t)$ has the same mean and has the same variance σ_0^2. This process is called an *Ornstein–Uhlenbeck process*, if in addition

$$\mathrm{Cov}\,(X(t_i), X(t_j)) = \sigma_0^2 e^{-\lambda|t_i - t_j|} \qquad (5,2.22)$$

for all t_i and t_j, λ being a positive constant. The steady state solution (5,2.15) of the Langevin equation has for all values of t zero mean and variance $\sigma_0^2 = \sigma^2/(2B)$. Each $u(t_1), u(t_2), \ldots u(t_n)$ is Gaussian, so taken together they have a joint normal distribution. Then

(5,2.16) shows that $u(t)$ is an Ornstein–Uhlenbeck process. In particular this is so when $u(t)$ is the angular velocity $\omega(t)$.

We shall now establish *Doob's theorem* that the only stationary Markov processes such that for distinct s and v any pair $X(s)$ and $X(v)$ have a non-singular joint normal distribution are those in which $X(t_1), X(t_2), \ldots X(t_n)$ for $t_1 < t_2 \ldots < t_n$ are either mutually independent or constitute an Ornstein–Uhlenbeck process (Doob, 1942).

To simplify the proof of this theorem we reinterpret $X(t)$ as $[X(t) - \langle X(t)\rangle][V(X(t))]^{-1/2}$, which has zero mean and variance unity. Since $X(t_1), X(t_2), \ldots X(t_n)$ have a joint normal distribution, $X(t_1)$ has a Gaussian distribution so that in the notation of Section 4.1

$$w_1(x_1, t_1) = \frac{\exp(-\frac{1}{2}x_1^2)}{(2\pi)^{1/2}}. \tag{5,2.23}$$

Then the conditional probability density

$$\Pi_2(x_1, t_1; x_2, t_2) = \frac{\exp\left[-\dfrac{(x_2 - \rho_1 x_1)^2}{2(1-\rho_1^2)}\right]}{[2\pi(1-\rho_1^2)]^{1/2}}, \tag{5,2.24}$$

as we deduce from (3,4.13) by putting $m_1 = m_2 = 0$, $\sigma_1 = \sigma_2 = 1$, $\rho = \langle X(t_1)X(t_2)\rangle = \rho_1$, say. The non-singular property of the joint distribution of $X(t_1)$ and $X(t_2)$ means that the quadratic form $u_1^2 + 2\rho_1 u_1 u_2 + u_2^2$, obtained from the quadratic form in (3,4.12), is positive definite. Thus for all non-vanishing values of u_1 and u_2

$$(u_1 + \rho_1 u_2)^2 + (1 - \rho_1^2)u_2^2 > 0.$$

We must therefore have $\rho_1^2 < 1$, and so $|\rho_1| < 1$. From (4,1.7), (5,2.23) and (5,2.24)

$$w_2(x_1, t_1; x_2, t_2) = \frac{\exp\left[-\frac{1}{2}x_1^2 - \dfrac{(x_2 - \rho_1 x_1)^2}{2(1-\rho_1^2)}\right]}{2\pi(1-\rho_1^2)^{1/2}}. \tag{5,2.25}$$

Since the process is Markovian, we have from (4,2.3)

$$w_3(x_1, t_1; x_2, t_2; x_3, t_3) = w_2(x_1, t_1; x_2, t_2)\Pi_2(x_2, t_2; x_3, t_3). \tag{5,2.26}$$

Then from (5,2.24)

$$\Pi_2(x_2, t_2; x_3, t_3) = \frac{\exp\left[-\dfrac{(x_3 - \rho_2 x_2)^2}{2(1-\rho_2)^2}\right]}{[2\pi(1-\rho_2^2)]^{1/2}} \tag{5,2.27}$$

with $\rho_2 = \langle X(t_2)X(t_3)\rangle$ and $|\rho_2| < 1$. From (5,2.25)–(5,2.27) we obtain

$$w_3(x_1, t_1; x_2, t_2; x_3, t_2) =$$

$$\frac{\exp\left[-\frac{1}{2}x_1^2 - \dfrac{(x_2-\rho_1 x_1)^2}{2(1-\rho_1^2)} - \dfrac{(x_3-\rho_2 x_2)^2}{2(1-\rho_2^2)}\right]}{[(2\pi)^3(1-\rho_1^2)(1-\rho_2^2)]^{1/2}}. \tag{5,2.28}$$

Let us show that w_3 gives a joint normal distribution. If this is so, then, according to (4,1.6), it must be possible to express w_3 by

$$w_3(x_1, t_1; x_2, t_2; x_3, t_3) = \frac{\exp\{-\frac{1}{2}x^{\mathrm{T}}C^{-1}x\}}{[(2\pi)^3 \det C]^{1/2}}. \tag{5,2.29}$$

In order that the exponents in (5,2.28) and (5,2.29) may agree we must have

$$C^{-1} = \begin{bmatrix} \dfrac{1}{1-\rho_1^2} & -\dfrac{\rho_1}{1-\rho_1^2} & 0 \\ -\dfrac{\rho_1}{1-\rho_1^2} & \dfrac{1}{1-\rho_1^2} + \dfrac{\rho_2^2}{1-\rho_2^2} & -\dfrac{\rho_2}{1-\rho_2^2} \\ 0 & -\dfrac{\rho_2}{1-\rho_2^2} & \dfrac{1}{1-\rho_2^2} \end{bmatrix},$$

and hence

$$\det C^{-1} = (1-\rho_1^2)^{-1}(1-\rho_2^2)^{-1},$$

$$\det C = (1-\rho_1^2)(1-\rho_2^2).$$

The denominators on the right-hand sides of (5,2.28) and (5,2.29) will therefore be equal, so w_3 does give a joint normal distribution. Then on inverting C^{-1} we find that $C_{11} = C_{22} = C_{33} = 1$, $C_{12} = \rho_1$ and $C_{23} = \rho_2$ as expected, and that $C_{13} = \rho_1\rho_2$.

The last relation leads to Doob's theorem. By definition

$$\begin{aligned}\rho_1 &= \langle X(t_1)X(t_2)\rangle = \langle X(t_2)X(t_1)\rangle \\ &= \langle X(t_2 - t_1)X(0)\rangle, \text{ by stationarity,} \\ &= \theta(t_2 - t_1),\end{aligned}$$

say, and similarly

$$\begin{aligned}\rho_2 &= \langle X(t_2)X(t_3)\rangle = \langle X(t_3)X(t_2)\rangle \\ &= \theta(t_3 - t_2), \\ C_{13} &= \langle X(t_1)X(t_3)\rangle = \langle X(t_3)X(t_1)\rangle \\ &= \theta(t_3 - t_1),\end{aligned}$$

so that $\theta(t)$ is an even function of t. Then, since

$$\begin{aligned}|\theta(t_2 - t_1)| &= |\rho_1| < 1, \quad |\theta(t_3 - t_2)| = |\rho_2| < 1, \\ |\theta(t_3 - t_1)| &= |C_{13}| = |\rho_1 \rho_2| < 1,\end{aligned}$$

it follows that $|\theta(t)| < 1$. The relation $C_{13} = \rho_1 \rho_2$ gives

$$\theta(t_3 - t_1) = \theta(t_3 - t_2)\theta(t_2 - t_1),$$

which we express as

$$\theta(v + w) = \theta(v)\theta(w),$$

where v and w are positive. Hence we deduce that

$$\ln |\theta(v + w)| = \ln |\theta(v)| + \ln |\theta(w)|. \tag{5,2.30}$$

In the discussion of (4,3.3), which we now write as

$$c(v + w) = c(v) + c(w), \tag{5,2.31}$$

we saw that $c(t)$ is proportional to t for positive t. However, we now do not require the constant of proportionality to be positive because $c(t)$ is no longer a variance. On comparing (5,2.31) with (5,2.30) we deduce that, since $\theta(t)$ is an even function of t,

$$\ln |\theta(t)| = -\lambda |t|,$$

where λ is a positive constant because $|\theta(t)| < 1$. Hence

$$\theta(t) = e^{-\lambda |t|}, \tag{5,2.32}$$

since $\theta(0)$ being the variance of $X(t)$ is unity. If $\lambda = \infty$, $X(t_1)$, $X(t_2), \ldots X(t_n)$ are uncorrelated, and therefore independent,

Gaussian random variables. When $\lambda \neq \infty$ and we return to the interpretation of $X(t)$ as having non-zero mean and variance σ_0^2, we deduce from (5,2.32) that

$$\langle (X(t_i) - \langle X(t_i)\rangle)(X(t_j) - \langle X(t_j)\rangle)\rangle = \sigma_0^2 e^{-\lambda|t_i - t_j|},$$

which is the condition (5,2.22) for an Ornstein–Uhlenbeck process. Doob's theorem is therefore established.

In future studies of steady state rotational Brownian motion we shall frequently make use of the correlation function of $\omega(t)$ as given by (5,2.20), and it might be asked whether we could obtain this relation directly from Doob's theorem without having to study the solution of the Langevin equation. However, without this equation we could not know that $\omega(t)$ is Gaussian and therefore that $\omega(s)$ and $\omega(v)$ have a non-singular joint normal distribution. We are therefore not entitled to invoke Doob's theorem. Moreover, even if (5,2.22) is accepted, we have no way of relating λ to the friction constant B.

5.3 Generalizations of the Langevin Equation

Let us first consider the extension of the Langevin equation to more than one dimension. For later applications we keep in mind rotational Brownian motion; the discussion of translational motion follows along the same lines. We take the Langevin equation

$$I_i \frac{d\omega_i}{dt} = -\zeta_i + N_i(t), \qquad (5,3.1)$$

where the subscript i refers to rectangular coordinates in Euclidean space. These axes will be drawn through the centre of mass of the rotating body and in the directions of the principal axes of inertia. I_i is the appropriate moment of inertia, ζ_i is the frictional constant and N_i the component of the random driving couple. The case of $i = 1$ only has already been met in the rotational Brownian motion of a sphere when the axis of rotation always points in a fixed direction. Indeed the mathematical problem is the same for a body of any shape, provided that the frictional couple is proportional to the angular velocity. As will be discussed in Chapter 10, the case of $i = 1$ and 2 applies to the rotational motion of a linear molecule, that is, one whose shape is approximated by a thin rod or needle. The case of $i = 1, 2$ and 3 is relevant for the rotational motion of a sphere whose axis is free to orientate in three-dimensional space. The rotational Brownian motion of a body with no special symmetry will require separate examination.

To link (5,3.1) with the last section we put

$$\zeta_i = I_i B_i, \quad N_i(t) = I_i \frac{dW_i(t)}{dt}. \tag{5,3.2}$$

The stationary solution of (5,3.1) is, from (5,2.19),

$$\omega_i(t) = \int_{y=-\infty}^{t} e^{-B_i(t-y)} dW_i(y), \tag{5,3.3}$$

a centred Gaussian random variable. We assume that dW_i and dW_j are uncorrelated for $j \neq i$. Then eqn (4,3.13) for the correlation function of $dW_i(t)/dt$ and $dW_j(t)/dt$ is expressible as

$$\langle N_i(t) N_j(t') \rangle = \delta_{ij} I_i^2 \sigma_i^2 \delta(t - t'), \tag{5,3.4}$$

and (5,3.3) leads to the relation

$$\langle \omega_i(t) \omega_j(s) \rangle = \frac{\delta_{ij} \sigma_i^2 e^{-B_i|t-s|}}{2B_i}. \tag{5,3.5}$$

The equipartition of energy yields for the ith direction

$$\tfrac{1}{2} kT = \tfrac{1}{2} I_i \langle (\omega_i(t))^2 \rangle = \tfrac{1}{2} I_i \frac{\sigma_i^2}{2B_i},$$

so that

$$\sigma_i^2 = \frac{2B_i kT}{I_i}. \tag{5,3.6}$$

Then from (5,3.2), (5,3.4) and (5,3.5)

$$\langle N_i(t) N_j(t') \rangle = 2\delta_{ij} \zeta_i kT \delta(t - t'),$$

$$\zeta_i = \frac{1}{kT} \int_0^{\infty} \langle N_i(0) N_i(t) \rangle dt, \tag{5,3.7}$$

$$\langle \omega_i(t) \omega_j(s) \rangle = \delta_{ij} (kT/I_i) e^{-B_i|t-s|}, \tag{5,3.8}$$

and the process $N_i(t)$ for $-\infty < t < \infty$ is white noise.

Let us again turn our attention to the one-dimensional Langevin equation

$$\frac{du}{dt} = -Bu + \frac{dW(t)}{dt}. \tag{5,3.9}$$

The right-hand side gives the force or couple which the environment exerts on the Brownian particle. Since the mean value of the last term in (5,3.9) vanishes, the mean value of the force or couple is $-mB\langle u\rangle$ or $-IB\langle u\rangle$ proportional to the mean velocity or angular velocity. The $dW(t)/dt$ term in (5,3.9) gives the residual fluctuating force or couple $N(t)$ exerted by the heat bath, when the frictional part has been subtracted. It is a Gaussian, Markovian, white noise term and it is uncorrelated with its value at any other instant of time. This is, of course, an idealization made for mathematical convenience. Physical processes are smooth and are never strictly Markovian.

The Langevin equation arose originally from the hydrodynamical model of a spherical Brownian particle immersed in a fluid consisting of particles whose linear dimensions are small compared with the radius of the Brownian particle. However, the Langevin equation may have wider applications. Thus Ford *et al.* (1965) constructed a mechanical model of a heat bath from a chain of coupled harmonic oscillators. On making a specific limiting process, a certain time function $\gamma(t)$ that occurs in their calculations becomes equal to a constant. This allows their equation of motion for the Brownian particle under consideration to be identified with the Langevin equation. Kirkwood (1946) deduced the Langevin equation from statistical mechanics and related the frictional constant B to intermolecular forces. He showed that B does not depend on space variables, that it does not depend on the momentum, if the Brownian particle has a mass large relative to those of the environment, and that B may not depend very much on the momentum when the masses are comparable, if we are dealing with condensed systems. In the light of such considerations it seems not unreasonable to tentatively apply the Langevin equation to physical situations in which the conditions of the hydrodynamical model are not fulfilled. This is in fact commonly done even when the surrounding fluid consists of the same molecules as the Brownian particle (Evans and Evans, 1978; Reid *et al.*, 1978).

A much studied modification of the Langevin equation due to Kubo (1966) is

$$\frac{du(t)}{dt} = -\int_0^t K(t-s)u(s)ds + F(t), \qquad (5,3.10)$$

where $F(t)$ is stochastic, but not necessarily Gaussian or Markovian, and satisfies

$$\langle F(t)\rangle = 0, \langle u(0)F(t)\rangle = 0. \qquad (5,3.11)$$

The reason for the latter condition is that $F(t)$ depends only on fluctuations in the environment and not on the initial value of the velocity of the Brownian particle. The condition is satisfied also in (5,3.9) when $F(t)$ is interpreted as $dW(t)/dt$. The integral in the *Kubo equation* (5,3.10) is called a *memory term,* since it describes frictional effects due to all times between 0 and t. A memory term also occurs in the above theory of Ford *et al.*, when $\gamma(t)$ is not constant. One deduces from (5,3.10) and (5,3.11) that

$$\frac{d}{dt}\langle u(0)u(t)\rangle = -\int_0^t K(t-s)\langle u(0)u(s)\rangle ds,$$

an integral equation for the autocorrelation function of $u(t)$. We shall not use the Kubo equation in our calculations on Brownian motion.

5.4 Averaging Method Solution of Stochastic Differential Equations

In the course of calculations on the theory of rotational Brownian motion we shall frequently encounter stochastic equations of the type

$$\frac{dx(t)}{dt} = \epsilon O(t)x(t), \qquad (5,4.1)$$

and a discussion of this equation will shorten future calculations considerably. In the equation $x(t)$ is a random variable, ϵ a small parameter and $O(t)$ a *stochastic operator,* that is, an operator that involves random variables. The variable $x(t)$ may also be a stochastic operator, and we must be careful to preserve the order in which such operators occur. We assume that $O(t)$ is independent of ϵ.

The solution of stochastic differential equations has been discussed at length by Van Kampen (1976). In the present section we follow the method of Ford (1975), an exposition of which is to be found in Ford *et al.* (1976, Section 1).

The solution of (5,4.1) will be slowly varying and will consist of a non-stochastic mean $\langle x(t)\rangle$ about which there will be small fluctuations, as sketched in Fig. 5.1. We express the solution as a power series in ϵ:

$$x(t) = \langle x(t)\rangle + \epsilon F^{(1)}(t)\langle x(t)\rangle + \epsilon^2 F^{(2)}(t)\langle x(t)\rangle + \ldots, \qquad (5,4.2)$$

where $F^{(n)}(t)$ is a stochastic operator that clearly satisfies

$$\langle F^{(n)}(t)\rangle = 0. \qquad (5,4.3)$$

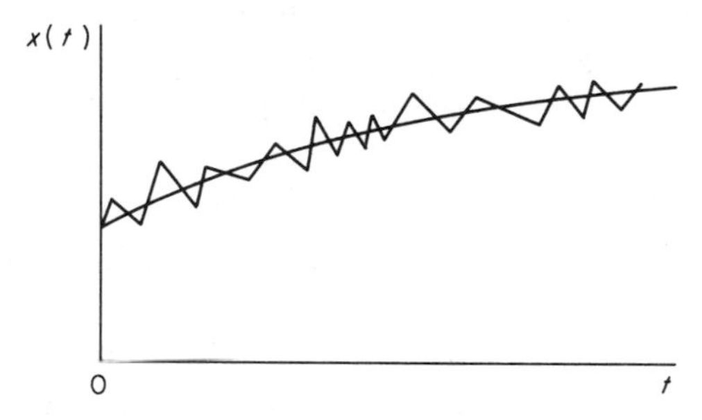

Fig. 5.1. The fluctuation of $x(t)$ about $\langle x(t)\rangle$, represented by the smooth curve.

The mean $\langle x(t)\rangle$. will itself satisfy some non-stochastic differential equation

$$\frac{d\langle x(t)\rangle}{dt} = \epsilon\Omega^{(1)}(t)\langle x(t)\rangle + \epsilon^2\Omega^{(2)}(t)\langle x(t)\rangle + \ldots, \qquad (5,4.4)$$

where $\Omega^{(n)}(t)$ being non-stochastic satisfies

$$\langle \Omega^{(n)}(t)\rangle = \Omega^{(n)}(t). \qquad (5,4.5)$$

Our objective will be to find $\langle x(t)\rangle$, and in order to do this we must derive expressions for $\Omega^{(1)}(t)$, $\Omega^{(2)}(t)$, etc.

On substituting (5,4.2) into (5,4.1), using (5,4.4) and equating terms with the same power of ϵ we deduce that

$$\Omega^{(1)}(t) + \dot{F}^{(1)}(t) = O(t) \qquad (5,4.6)$$

$$\Omega^{(2)}(t) + \dot{F}^{(2)}(t) = O(t)F^{(1)}(t) - F^{(1)}(t)\Omega^{(1)}(t) \qquad (5,4.7)$$

$$\Omega^{(3)}(t)+\dot{F}^{(3)}(t)=O(t)F^{(2)}(t)-F^{(1)}(t)\Omega^{(2)}-F^{(2)}(t)\Omega^{(1)}(t) \quad (5,4.8)$$

. . . . . .

$$\Omega^{(n)}(t)+\dot{F}^{(n)}(t) = O(t)F^{(n-1)}(t)-\sum_{j=1}^{n-1} F^{(j)}(t)\Omega^{(n-j)}(t). \quad (5,4.9)$$

In solving these equations it is useful to note that from (5,4.3), (5,4.5) and (5,4.9)

$$\langle F^{(j)}(t)\Omega^{(n-j)}(t)\rangle = \langle F^{(j)}(t)\rangle\Omega^{(n-j)}(t) = 0$$

$$\Omega^{(n)}(t) = \langle O(t)F^{(n-1)}(t)\rangle. \quad (5,4.10)$$

From (5,4.6) we deduce that

$$\Omega^{(1)}(t) = \langle O(t)\rangle, \quad (5,4.11)$$

and therefore that

$$F^{(1)}(t) = \int_0^t [O(t_1)-\langle O(t_1)\rangle]\,dt_1. \quad (5,4.12)$$

Then from (5,4.10) and (5,4.12)

$$\Omega^{(2)}(t) = \int_0^t [\langle O(t)O(t_1)\rangle-\langle O(t)\rangle\langle O(t_1)\rangle]\,dt_1,$$

so that

$$\Omega^{(2)}(t_1) = \int_0^t [\langle O(t_1)O(t_2)\rangle-\langle O(t_1)\rangle\langle O(t_2)\rangle]\,dt_2. \quad (5,4.13)$$

From (5,4.7), (5,4.12) and (5,4.13) we find on integration that

$$F^{(2)}(t)=\int_0^t dt_1\int_0^{t_1} dt_2\,[O(t_1)O(t_2)-O(t_1)\langle O(t_2)\rangle-O(t_2)\langle O(t_1)\rangle$$
$$+\langle O(t_2)\rangle\langle O(t_1)\rangle-\langle O(t_1)O(t_2)\rangle+\langle O(t_1)\rangle\langle O(t_2)\rangle]. \quad (5,4.14)$$

Then on substituting (5,4.14) into (5,4.10) we deduce that

$$\Omega^{(3)}(t_1)=\int_0^{t_1} dt_2\int_0^{t_2} dt_3\,[\langle O(t_1)O(t_2)O(t_3)\rangle-\langle O(t_1)O(t_2)\rangle\langle O(t_3)\rangle$$
$$-\langle O(t_1)O(t_3)\rangle\langle O(t_2)\rangle-\langle O(t_1)\rangle\langle O(t_2)O(t_3)\rangle$$
$$+\langle O(t_1)\rangle\langle O(t_2)\rangle\langle O(t_3)\rangle+\langle O(t_1)\rangle\langle O(t_3)\rangle\langle O(t_2)\rangle].$$

Continuing in this way we obtain

$$\Omega^{(4)}(t_1) = \int_0^{t_1} dt_2 \int_0^{t_2} dt_3 \int_0^{t_3} dt_4 \times$$

$$\begin{aligned}
&[\langle O(t_1)O(t_2)O(t_3)O(t_4)\rangle - \langle O(t_1)O(t_2)O(t_3)\rangle\langle O(t_4)\rangle \\
&- \langle O(t_1)O(t_2)O(t_4)\rangle\langle O(t_3)\rangle + \langle O(t_1)O(t_2)\rangle\langle O(t_4)\rangle\langle O(t_3)\rangle \\
&- \langle O(t_1)O(t_2)\rangle\langle O(t_3)O(t_4)\rangle + \langle O(t_1)O(t_2)\rangle\langle O(t_3)\rangle\langle O(t_4)\rangle \\
&- \langle O(t_1)\rangle\langle O(t_2)O(t_3)O(t_4)\rangle + \langle O(t_1)\rangle\langle O(t_2)O(t_3)\rangle\langle O(t_4)\rangle \\
&+ \langle O(t_1)\rangle\langle O(t_2)O(t_4)\rangle\langle O(t_3)\rangle - \langle O(t_1)\rangle\langle O(t_2)\rangle\langle O(t_4)\rangle\langle O(t_3)\rangle \\
&- \langle O(t_1)\rangle\langle O(t_2)\rangle\langle O(t_3)\rangle\langle O(t_4)\rangle + \langle O(t_1)\rangle\langle O(t_2)\rangle\langle O(t_3)O(t_4)\rangle \\
&- \langle O(t_1)O(t_3)\rangle\langle O(t_2)O(t_4)\rangle + \langle O(t_1)\rangle\langle O(t_3)\rangle\langle O(t_2)O(t_4)\rangle \\
&+ \langle O(t_1)O(t_3)\rangle\langle O(t_2)\rangle\langle O(t_4)\rangle - \langle O(t_1)\rangle\langle O(t_3)\rangle\langle O(t_2)\rangle\langle O(t_4)\rangle \\
&- \langle O(t_1)O(t_4)\rangle\langle O(t_2)O(t_3)\rangle + \langle O(t_1)\rangle\langle O(t_4)\rangle\langle O(t_2)O(t_3)\rangle \\
&+ \langle O(t_1)O(t_4)\rangle\langle O(t_2)\rangle\langle O(t_3)\rangle - \langle O(t_1)\rangle\langle O(t_4)\rangle\langle O(t_2)\rangle\langle O(t_3)\rangle \\
&- \langle O(t_1)O(t_3)O(t_4)\rangle\langle O(t_2)\rangle + \langle O(t_1)O(t_3)\rangle\langle O(t_4)\rangle\langle O(t_2)\rangle \\
&+ \langle O(t_1)O(t_4)\rangle\langle O(t_3)\rangle\langle O(t_2)\rangle - \langle O(t_1)\rangle\langle O(t_4)\rangle\langle O(t_3)\rangle\langle O(t_2)\rangle \\
&+ \langle O(t_1)\rangle\langle O(t_3)O(t_4)\rangle\langle O(t_2)\rangle - \langle O(t_1)\rangle\langle O(t_3)\rangle\langle O(t_4)\rangle\langle O(t_2)\rangle] .
\end{aligned}$$

A general expression for $\Omega^{(n)}(t_1)$ is to be found in eqn (21) of Ford *et al.* (1976).

The foregoing results hold for any stochastic operator $O(t)$. In the case of a Brownian particle that rotates about an axis pointing in a fixed direction, or of one that is linear or spherical in shape and is free to rotate in three-dimensional space, it will be proved that $O(t)$ has the following properties:

(i) $O(t)$ is a centred Gaussian stochastic operator. This means that, if $O(t)$ is represented by a matrix, as explained in Appendix A, all its elements are centred Gaussian random variables.

(ii) $\langle O(t_1)O(t_2)\ldots O(t_{2n-1})\rangle = 0$ for any $2n-1$ instants of time.

If $O(t)$ were not an operator but just a commuting random variable, the second property would result from eqn (3,4.17).

On using these two properties the expressions for $\Omega^{(n)}(t_1)$ reduce considerably. It is readily deduced that all $\Omega^{(2n-1)}(t_1)$ vanish and that

$$\Omega^{(2)}(t_1) = \int_0^{t_1} \langle O(t_1)O(t_2)\rangle dt_2, \tag{5,4.15}$$

$$\begin{aligned}\Omega^{(4)}(t_1) = \int_0^{t_1} dt_2 \int_0^{t_2} dt_3 \int_0^{t_3} dt_4 [&\langle O(t_1)O(t_2)O(t_3)O(t_4)\rangle \\ &- \langle O(t_1)O(t_2)\rangle\langle O(t_3)O(t_4)\rangle - \langle O(t_1)O(t_3)\rangle\langle O(t_2)O(t_4)\rangle \\ &- \langle O(t_1)O(t_4)\rangle\langle O(t_2)O(t_3)\rangle],\end{aligned} \tag{5,4.16}$$

$$\begin{aligned}\Omega^{(6)}(t_1) = \int_0^{t_1} dt_2 \int_0^{t_2} dt_3 \int_0^{t_3} dt_4 \int_0^{t_4} dt_5 \int_0^{t_5} dt_6 \times \\
&[\langle O(t_1)O(t_2)O(t_3)O(t_4)O(t_5)O(t_6)\rangle \\
&- \langle O(t_1)O(t_2)\rangle\langle O(t_3)O(t_4)O(t_5)O(t_6)\rangle \\
&- \langle O(t_1)O(t_3)\rangle\langle O(t_2)O(t_4)O(t_5)O(t_6)\rangle \\
&- \langle O(t_1)O(t_4)\rangle\langle O(t_2)O(t_3)O(t_5)O(t_6)\rangle \\
&- \langle O(t_1)O(t_5)\rangle\langle O(t_2)O(t_3)O(t_4)O(t_6)\rangle \\
&- \langle O(t_1)O(t_6)\rangle\langle O(t_2)O(t_3)O(t_4)O(t_5)\rangle \\
&- \langle O(t_1)O(t_2)O(t_3)O(t_4)\rangle\langle O(t_5)O(t_6)\rangle \\
&- \langle O(t_1)O(t_2)O(t_3)O(t_5)\rangle\langle O(t_4)O(t_6)\rangle \\
&- \langle O(t_1)O(t_2)O(t_4)O(t_5)\rangle\langle O(t_3)O(t_6)\rangle \\
&- \langle O(t_1)O(t_3)O(t_4)O(t_5)\rangle\langle O(t_2)O(t_6)\rangle \\
&- \langle O(t_1)O(t_2)O(t_3)O(t_6)\rangle\langle O(t_4)O(t_5)\rangle \\
&- \langle O(t_1)O(t_2)O(t_4)O(t_6)\rangle\langle O(t_3)O(t_5)\rangle \\
&- \langle O(t_1)O(t_3)O(t_4)O(t_6)\rangle\langle O(t_2)O(t_5)\rangle \\
&- \langle O(t_1)O(t_2)O(t_5)O(t_6)\rangle\langle O(t_3)O(t_4)\rangle \\
&- \langle O(t_1)O(t_3)O(t_5)O(t_6)\rangle\langle O(t_2)O(t_4)\rangle \\
&- \langle O(t_1)O(t_4)O(t_5)O(t_6)\rangle\langle O(t_2)O(t_3)\rangle \\
&+ 2\langle O(t_1)O(t_2)\rangle\langle O(t_3)O(t_4)\rangle\langle O(t_5)O(t_6)\rangle \\
&+ 2\langle O(t_1)O(t_2)\rangle\langle O(t_3)O(t_5)\rangle\langle O(t_4)O(t_6)\rangle \\
&+ 2\langle O(t_1)O(t_2)\rangle\langle O(t_3)O(t_6)\rangle\langle O(t_4)O(t_5)\rangle\end{aligned}$$

$$
\begin{aligned}
&+ 2\langle O(t_1)O(t_3)\rangle\langle O(t_2)O(t_4)\rangle\langle O(t_5)O(t_6)\rangle \\
&+ 2\langle O(t_1)O(t_3)\rangle\langle O(t_2)O(t_5)\rangle\langle O(t_4)O(t_6)\rangle \\
&+ 2\langle O(t_1)O(t_3)\rangle\langle O(t_2)O(t_6)\rangle\langle O(t_4)O(t_5)\rangle \\
&+ 2\langle O(t_1)O(t_4)\rangle\langle O(t_2)O(t_3)\rangle\langle O(t_5)O(t_6)\rangle \\
&+ 2\langle O(t_1)O(t_4)\rangle\langle O(t_2)O(t_5)\rangle\langle O(t_3)O(t_6)\rangle \\
&+ 2\langle O(t_1)O(t_4)\rangle\langle O(t_2)O(t_6)\rangle\langle O(t_3)O(t_5)\rangle \\
&+ 2\langle O(t_1)O(t_5)\rangle\langle O(t_2)O(t_3)\rangle\langle O(t_4)O(t_6)\rangle \\
&+ 2\langle O(t_1)O(t_5)\rangle\langle O(t_2)O(t_4)\rangle\langle O(t_3)O(t_6)\rangle \\
&+ 2\langle O(t_1)O(t_5)\rangle\langle O(t_2)O(t_6)\rangle\langle O(t_3)O(t_4)\rangle \\
&+ 2\langle O(t_1)O(t_6)\rangle\langle O(t_2)O(t_3)\rangle\langle O(t_4)O(t_5)\rangle \\
&+ 2\langle O(t_1)O(t_6)\rangle\langle O(t_2)O(t_4)\rangle\langle O(t_3)O(t_5)\rangle \\
&+ 2\langle O(t_1)O(t_6)\rangle\langle O(t_2)O(t_5)\rangle\langle O(t_3)O(t_4)\rangle] \,.
\end{aligned}
\tag{5,4.17}
$$

When the values of $\Omega^{(2)}(t)$, $\Omega^{(4)}(t)$, $\Omega^{(6)}(t)$, etc., are substituted into (5,4.4), it may be possible to solve this equation for $\langle x(t)\rangle$. Considering matrix elements with respect to a basis, that does not of course depend on stochastic variables, we see from eqn (A.9) of Appendix A that

$$
\begin{aligned}
\langle x_{mm'}(t)\rangle &= \left\langle \int f_m^* x(t) f_{m'} dq \right\rangle \\
&= \int f_m^* \langle x(t)\rangle f_{m'} dq \,,
\end{aligned}
$$

so that

$$
\langle x_{mm'}(t)\rangle = \langle x(t)\rangle_{mm'} . \tag{5,4.18}
$$

Hence the solution of (5,4.4) will give the mean value of an element of the matrix representative of $x(t)$.

6
Partial Differential Equations for Stochastic Processes

6.1 Diffusion Processes

In the Debye theory, as presented in Chapter 1, the investigation of Brownian motion was based on the solution of partial differential equations for the probability density function. We shall see in Chapters 8, 9 and 10 that the earliest systematic studies of rotational Brownian motion with inclusion of inertial effects were made by solving such equations. These equations are quite deterministic; they have no random coefficients and the probability density function itself is nonstochastic, as we already know. In some cases we shall treat Brownian motion problems by means of one or more stochastic equations, and alternatively by means of a partial differential equation. When we do this, we find that the two methods lead to identical results for physical problems in which we are interested. The reason for this is a theorem for a stationary Markov process $X(t)$, which we suppose to have p components $X_1(t)$, $X_2(t), \ldots X_p(t)$, that to a set of stochastic equations

$$dX_i(t) = f_i(X(t), t)dt + \sum_{l=1}^{m} G_{il}(X(t), t)dW_l(t) \qquad (i = 1, 2, \ldots p), \tag{6,1.1}$$

where $W(t)$ is an m-dimensional Wiener process, there corresponds a second order partial differential equation for the conditional probability density Π_2 defined in Section 3.3 and applied to stochastic processes in Section 4.1.

We first of all restrict our attention to the case where both $X(t)$ and $W(t)$ are one-dimensional, so that (6,1.1) reduces to

$$dX(t) = f(X(t), t)dt + G(X(t), t)dW(t). \tag{6,1.2}$$

Let us suppose that $X(t)$ has a specified value x and let us consider the stochastic behaviour of $X(t)$ in the time interval $(t, t + \delta t)$. We see from (6,1.2) that

$$dx = f(x, t)dt + G(x, t)dW(t),$$

$$(dx)^2 = (f(x,t))^2(dt)^2 + 2f(x,t)G(x,t)dtdW(t) + (G(x,t))^2(dW(t))^2.$$

On taking ensemble averages and employing the relations

$$\langle dW(t)\rangle = \langle W(t + \delta t)\rangle - \langle W(t)\rangle = 0$$

and

$$\langle (dW(t))^2\rangle = \sigma^2 dt,$$

as given by (4,3.8), we deduce that, neglecting terms of order higher than the first in dt,

$$\langle dx\rangle = f(x, t)dt,$$

$$\langle (dx)^2\rangle = \sigma^2(G(x, t))^2 dt,$$

$$0 = \langle (dx)^3\rangle = \langle (dx)^4\rangle, \text{ etc.}$$

Let us write $X(t + \delta t) = y$, so that $y - x = \delta x$ and to first order in δt

$$\langle y - x\rangle = f(x, t)\delta t, \tag{6,1.3}$$

$$\langle (y - x)^2\rangle = \sigma^2(G(x, t))^2 \delta t, \tag{6,1.4}$$

$$0 = \langle (y - x)^3\rangle = \langle (y - x)^4\rangle, \text{ etc.} \tag{6,1.5}$$

Since we have taken $X(t)$ to be not only a Markov but also a stationary process, the conditional probability density $\Pi_2(x_1, t_1; x_2, t_2)$ depends on the time only through the positive interval $t_2 - t_1$, and we have

$$\Pi_2(x_1, t_1; x_2, t_2) = \Pi_2(x_1, 0; x_2, t_2 - t_1).$$

We write this $\Pi(x_1 | x_2, t_2 - t_1)$ omitting the subscript 2 of Π, since Π_3, Π_4, etc., will no longer appear in our calculations. We may therefore define Π by saying that, if X is equal to u at any instant, the probability of finding X in the interval $(v, v + dv)$ at an instant which is later by a time τ is $\Pi(u|v, \tau)dv$. In terms of Π the Chapman–Kolmogorov equation (4,2.2) becomes

$$\Pi(x_1 | x_2, t) = \int \Pi(x_1 | z, s)\Pi(z | x_2, t - s)dz, \tag{6,1.6}$$

the integral being taken over the values of z that occur when s goes from 0 to t. If now we consider the case of $x_1 = x, x_2 = y = x + \delta x$, then t is replaced by $t + \delta t$ and, since we have a Markov process, s can take only the value t in $\Pi(x, z, s)$, so that (6,1.6) becomes

$$\Pi(x|y, t + \delta t) = \int \Pi(x|z, t)\Pi(z|y, \delta t)dz. \tag{6,1.7}$$

Moreover from the definition of $\Pi(x|y, \delta t)$ we see that

$$\int (y - x)\Pi(x|y, \delta t)dy = \langle y - x \rangle,$$

$$\int (y - x)^2 \Pi(x|y, \delta t)dy = \langle (y - x)^2 \rangle, \text{ etc.}$$

Hence from (6,1.3)–(6,1.5)

$$\lim_{\delta t \to 0} \frac{1}{\delta t} \int (y - x)\Pi(x|y, \delta t)dy = f(x, t), \tag{6,1.8}$$

$$\lim_{\delta t \to 0} \frac{1}{\delta t} \int (y - x)^2 \Pi(x|y, \delta t)dy = \sigma^2 (G(x, t))^2, \tag{6,1.9}$$

$$0 = \lim_{\delta t \to 0} \frac{1}{\delta t} \int (y - x)^3 \Pi(x|y, \delta t)dy, \text{ etc.} \tag{6,1.10}$$

Equations (6,1.8) and (6,1.9) are simplified forms of two of the three conditions that the Markov process $X(t)$ be a *diffusion process*, a simplified form of the third condition being

$$\lim_{\delta t \to 0} \frac{1}{\delta t} \int_{|y-x| > \epsilon} \Pi(x|y, t)dy = 0$$

for every $\epsilon > 0$. A precise definition of a diffusion process, together with complete proofs of theorems and full discussion of material presented in the present section, have been given by Gikhman and Skorokhod (1969, Chap. 8).

We now seek a differential equation for $\Pi(x|y, t)$. We choose a continuous arbitrary function $\phi(y)$ with continuous first and second derivatives and such that $\phi(y)$ and $d\phi/dy$ vanish outside a prescribed finite interval. We consider $\int dy \phi(y) \partial \Pi(x|y, t)/\partial t$ taken over a range which includes the finite interval, so that both $\phi(y)$ and $d\phi/dy$

vanish at the end points of the range. Writing

$$\frac{\partial \Pi(x|y,t)}{\partial t} = \lim_{\delta t \to 0} \frac{\Pi(x|y,t+\delta t) - \Pi(x|y,t)}{\delta t}$$

and employing (6,1.7) we get

$$\frac{\partial \Pi(x|y,t)}{\partial t} = \lim_{\delta t \to 0} \frac{1}{\delta t}\left\{\int \Pi(x|z,t)\Pi(z|y,\delta t)dz - \Pi(x|y,t)\right\},$$

so that

$$\int dy\phi(y)\frac{\partial \Pi(x(y,t)}{\partial t} = \lim_{\delta t \to 0} \frac{1}{\delta t}\left\{\int dy\phi(y)\int \Pi(x|z,t)\Pi(z|y,\delta t)dz \right.$$

$$\left. - \int dz\phi(z)\Pi(x|z,t)\right\}, \qquad (6,1.11)$$

where we have changed the integration variable in the last integral from y to z. We expand $\phi(y)$ as a Taylor series in $(y-z)$:

$$\phi(y) = \phi(z) + (y-z)\frac{d\phi(z)}{dz} + \frac{1}{2}(y-z)^2\frac{d^2\phi(z)}{dz^2}$$

$$+ \frac{1}{6}(y-z)^3\frac{d^3\phi(z)}{dz^3} + \dots. \qquad (6,1.12)$$

On inverting the order of the y and z integrations in (6,1.11)

$$\int dy\phi(y)\frac{\partial \Pi(x|y,t)}{\partial t}$$

$$= \lim_{\delta t \to 0} \frac{1}{\delta t}\int dz\Pi(x|z,t)\left[\int dy\Pi(z|y,\delta t)\phi(y) - \phi(z)\right]. \qquad (6,1.13)$$

When we substitute the value of $\phi(y)$ from (6,1.12), the first term $\phi(z)$ contributes to the integral with respect to y the quantity $\phi(z)\int\Pi(z|y, \delta t)dy$. This is just $\phi(z)$, since the relation $\int\Pi(z|y, \delta t)dy = 1$ clearly comes from (4,1.8), namely, $\int\Pi_2(x_1, t_1; x_2, t_2)dx_2 = 1$. Thus the first term in the Taylor series produces a contribution which cancels that arising from the $-\phi(z)$ in (6,1.13), and therefore

$$\int dy\phi(y)\frac{\partial \Pi(x|y,t)}{\partial t}$$

$$= \lim_{\delta t \to 0} \frac{1}{\delta t} \int dz \Pi(x|z,t) \int dy \Pi(z|y,\delta t) \left\{ (y-z)\frac{d\phi(z)}{dz} \right.$$

$$\left. + \frac{1}{2}(y-z)^2 \frac{d^2\phi(z)}{dz^2} + \frac{1}{6}(y-z)^3 \frac{d^3\phi(z)}{dz^3} + \ldots \right\}.$$

When we substitute from (6,1.8)–(6,1.10) we deduce that

$$\int dy \phi(y) \frac{\partial \Pi(x|y,t)}{\partial t}$$

$$= \int dz \Pi(x|z,t) \left\{ f(z,t)\frac{d\phi(z)}{dz} + \frac{1}{2}\sigma^2 (G(z,t))^2 \frac{d^2\phi(z)}{dz^2} \right\}$$

$$= \int dy \Pi(x|y,t) \left\{ f(y,t)\frac{d\phi(y)}{dy} + \frac{1}{2}\sigma^2 (G(y,t))^2 \frac{d^2\phi(y)}{dy^2} \right\}, \tag{6,1.14}$$

on change of integration variable. Integrating by parts we find that

$$\int dy \Pi(x|y,t) f(y,t) \frac{d\phi(y)}{dy}$$

$$= f(y,t)\Pi(x|y,t)\phi(y) - \int \phi(y) \frac{\partial}{\partial y}(f(y,t)\Pi(x|y,t))dy$$

$$= - \int \phi(y) \frac{\partial}{\partial y}(f(y,t)\Pi(x|y,t))dy,$$

since $\phi(y)$ vanishes at the limits of integration. Since $d\phi(y)/dy$ also vanishes at the limits, successive integration by parts gives

$$\int dy \Pi(x|y,t)(G(y,t))^2 \frac{d^2\phi(y)}{dy^2}$$

$$= \int \phi(y) \frac{\partial^2}{\partial y^2}[(G(y,t))^2 \Pi(x|y,t)].$$

Substituting these results in (6,1.14) yields

$$\int dy \phi(y) \left\{ \frac{\partial \Pi(x|y,t)}{\partial t} + \frac{\partial}{\partial y}(f(y,t)\Pi(x|y,t)) \right.$$

$$\left. - \frac{1}{2}\sigma^2 \frac{\partial^2}{\partial y^2}[(G(y,t))^2 \Pi(x|y,t)] \right\} = 0.$$

Since $\phi(y)$ is an arbitrary function, the quantity that multiplies it in the integrand must vanish. On replacing x by x_0 and y by $\dot{x}$ we deduce that

$$\frac{\partial \Pi(x_0|x,t)}{\partial t} + \frac{\partial}{\partial x}(f(x,t)\Pi(x_0|x,t)) - \frac{1}{2}\sigma^2\frac{\partial^2}{\partial x^2}[(G(x,t))^2\Pi(x_0|x,t)] = 0. \qquad (6,1.15)$$

The condition (4,1.10) for Brownian motion,

$$\lim_{t_2 - t_1 \to \infty} \Pi_2(x_1, t_1; x_2, t_2) = w_1(x_2, t_2),$$

may be expressed as

$$\lim_{t \to \infty} \Pi(x_0|x,t) = w_1(x,t).$$

Since w_2, w_3, etc., will no longer be required for our discussion, we suppress the suffix 1 and write w for the probability density function, as we had done in Chapter 1. Hence for steady state Brownian motion eqn (6,1.15) gives

$$\frac{\partial w(x,t)}{\partial t} + \frac{\partial}{\partial x}[f(x,t)w(x,t)] - \frac{1}{2}\sigma^2\frac{\partial^2}{\partial x^2}[(G(x,t))^2 w(x,t)] = 0. \qquad (6,1.16)$$

We examine what this equation becomes, when we return from (6,1.2) to the set of equations (6,1.1). We put $X_i(t) = x_i$ so that

$$dx_i = f_i(x,t)dt + \sum_{l=1}^{m} G_{il}(x,t)dW_l(t),$$

where

$$f_i(x,t) \equiv f_i(x_1, x_2, \ldots x_p, t);\; G_{il}(x,t) \equiv G_{il}(x_1, x_2, \ldots x_p, t).$$

Then

$$dx_i dx_j = \left(f_i(x,t)dt + \sum_{l=1}^{m} G_{il}(x,t)dW_l(t)\right) \times \left(f_j(x,t)dt + \sum_{k=1}^{m} G_{jk}(x,t)dW_k(t)\right),$$

so that up to first order in dt

$$\langle dx_i \rangle = f_i(x, t)dt, \qquad (6,1.17)$$

$$\langle dx_i dx_j \rangle = \sum_{l=1}^{m} \sum_{k=1}^{m} G_{il}(x, t) G_{jk}(x, t) \delta_{lk} \sigma_l^2 dt,$$

by (4,3.12). We denote by G the matrix whose ilth element is $G_{il}(x, t)$ and by G^{T} its transpose, so that $G_{jk} = G_{kj}^{\mathrm{T}}$ and

$$\langle dx_i dx_j \rangle = \sum_{l=1}^{m} \sigma_l^2 G_{il} G_{lj}^{\mathrm{T}} dt. \qquad (6,1.18)$$

To generalize (6,1.7) one must interpret $\Pi(x|z, t)$ as $\Pi(x_1, x_2, \ldots x_p | z_1, z_2, \ldots z_p, t)$ and dz as the volume element $dz_1 dz_2 \ldots dz_p$. From (6,1.17) and (6,1.18), eqns (6,1.8)–(6,1.10) become

$$\lim_{\delta t \to 0} \frac{1}{\delta t} \int (y_i - x_i) \Pi(x | y, \delta t) dy = f_i(x, t),$$

$$\lim_{\delta t \to 0} \frac{1}{\delta t} \int (y_i - x_i)(y_j - x_j) \Pi(x | y, \delta t) dy = \sum_{l=1}^{m} \sigma_l^2 G_{il} G_{lj}^{\mathrm{T}},$$

$$0 = \lim_{\delta t \to 0} \frac{1}{\delta t} \int (y_i - x_i)(y_j - x_j)(y_k - x_k) \Pi(x | y, \delta t) dy, \text{ etc.}$$

We choose an arbitrary function $\phi(y_1, y_2, \ldots y_p)$ continuous in the variables $y_1, y_2, \ldots y_p$ and such that its first and second partial derivatives are continuous, and we suppose that $\phi(y) \equiv \phi(y_1, y_2, \ldots y_p)$ and its first derivatives vanish outside a finite p-dimensional region. We construct the Taylor expansion

$$\phi(y) = \phi(z) + \sum_{i=1}^{p} (y_i - z_i) \frac{\partial \phi(z)}{\partial z_i}$$

$$+ \frac{1}{2} \sum_{i,j=1}^{p} (y_i - z_i)(y_j - z_j) \frac{\partial^2 \phi(z)}{\partial z_i \partial z_j} + \ldots .$$

On integrating $\int \phi(y_1, y_2, \ldots y_p)(\partial \Pi(x|y, t)/\partial t) dy_1, dy_2 \ldots dy_p$ over a volume whose boundary is outside the region where ϕ and $\partial \phi / \partial y_i$ are non-vanishing and proceeding as before for the case of y and t we deduce without difficulty the equation

$$\frac{\partial w(x,t)}{\partial t} + \sum_{i=1}^{p} \frac{\partial}{\partial x_i}(f_i(x,t)w(x,t))$$

$$-\frac{1}{2}\sum_{l=1}^{m} \sigma_l^2 \sum_{i,j=1}^{p} \frac{\partial^2}{\partial x_i \partial x_j}(G_{il}G_{lj}^{\mathrm{T}} w(x,t)) = 0, \qquad (6,1.19)$$

where $w(x, t)$ denotes the probability density function $w(x_1, x_2, \ldots x_p, t)$.

This is the partial differential equation corresponding to (6,1.1) that we set out to find. If $\sigma_1^2 = \sigma_2^2 = \ldots = \sigma_m^2 = \sigma^2$, the summation over l gives the ij-element of GG^{T} and (6,1.19) becomes

$$\frac{\partial w(x,t)}{\partial t} + \sum_{i=1}^{p} \frac{\partial}{\partial x_i}(f_i(x,t)w(x,t))$$

$$-\frac{1}{2}\sigma^2 \sum_{i,j=1}^{p} \frac{\partial^2}{\partial x_i \partial x_j}[(GG^{\mathrm{T}})_{ij} w(x,t)] = 0. \qquad (6,1.20)$$

Equation (6,1.20) with $w(x,t)$ replaced by $\Pi(x_0|x,t)$ is *Kolmogorov's forward equation* (Kolmogorov, 1931, eqn (133)).

In the next section we shall apply the above equations to linear and rotational Brownian motion when inertial effects are neglected. Account will be taken of inertial effects in the subsequent section.

6.2 The Smoluchowski Equation

As a first application of the results of the previous section we consider the translational Brownian motion of a spherical particle of mass m, which is subject to an external force $\mathbf{F}$. The stochastic differential equation is

$$m\ddot{\mathbf{r}} = -\zeta\dot{\mathbf{r}} + \mathbf{F} + m\frac{d\mathbf{W}(t)}{dt}. \qquad (6,2.1)$$

We suppose that the components of $\mathbf{W}(t)$ are mutually independent and that $\mathbf{W}(t)$ is independent of $\mathbf{F}$. Then, on account of the spherical symmetry of the problem when $\mathbf{F}$ vanishes, we may say that σ_i^2 for each component $W_i(t)$ is equal to the same quantity σ^2.

Let us first investigate the problem when inertial effects are neglected. This means that the $m\ddot{\mathbf{r}}$ term in (6,2.1) is put equal to zero and that the equation may be expressed as

$$dx_j = (1/\zeta)F_j dt + (m/\zeta)dW_j(t) \qquad (j = 1, 2, 3),$$

where (x_1, x_2, x_3) are the components of $\mathbf{r}$ referred to Cartesian axes. In the notation of (6,1.1)

$$p = m = 3, \quad f_i = \frac{F_i}{\zeta}, \quad G_{il} = \frac{\delta_{il}}{B},$$

where δ_{il} is the Kronecker delta and $B = \zeta/m$. The element of volume is $dx_1 dx_2 dx_3$, so we may employ (6,1.20), which becomes

$$\frac{\partial w(\mathbf{r}, t)}{\partial t} + \frac{1}{\zeta}\,\mathrm{div}\,(\mathbf{F}w) - \frac{\sigma^2}{2B^2}\nabla^2 w = 0. \tag{6,2.2}$$

We shall now determine the value of σ^2 from this equation. If V is the potential energy giving rise to a time-independent $\mathbf{F}$, so that $\mathbf{F} = -\,\mathrm{grad}\, V$, the energy of the particle is $\frac{1}{2}mv^2 + V$, where $\mathbf{v} = \dot{\mathbf{r}}$. The steady state value of w comes from the Maxwell–Boltzmann distribution:

$$w = \text{constant} \exp\left\{-\frac{\frac{1}{2}mv^2 + V}{kT}\right\}. \tag{6,2.3}$$

From this we obtain

$$\frac{\partial w}{\partial t} = 0, \quad \frac{\partial w}{\partial x_i} = -\frac{1}{kT}\frac{\partial V}{\partial x_i}w = \frac{1}{kT}F_i w,$$

$$\nabla^2 w = \frac{1}{kT}\,\mathrm{div}\,(\mathbf{F}w).$$

On substitution into (6,2.2) we deduce that

$$\sigma^2 = \frac{2B^2 kT}{\zeta} = \frac{2BkT}{m}$$

in agreement with (5,2.12). With this value of σ^2 eqn (6,2.2) becomes

$$\frac{\partial w(\mathbf{r}, t)}{\partial t} = \frac{kT}{\zeta}\nabla^2 w(\mathbf{r}, t) - \frac{1}{\zeta}\,\mathrm{div}\,(\mathbf{F}w(\mathbf{r}, t)). \tag{6,2.4}$$

This is Smoluchowski's equation (von Smoluchowski, 1915). It is a differential equation in configuration space because w does not depend on the velocity.

In the special case of free Brownian motion **F** vanishes and (6,2.4) reduces to

$$\frac{\partial w(\mathbf{r}, t)}{\partial t} = \frac{kT}{\zeta} \nabla^2 w(\mathbf{r}, t).$$

This equation was derived by Einstein (1905). It has the same structure as the diffusion equation of heat conduction with diffusion coefficient D given by the *Einstein relation*

$$D = \frac{kT}{\zeta}. \qquad (6,2.5)$$

For rotational Brownian motion of a polar spherical molecule, when the axis of rotation points in a fixed direction, (6,2.1) is replaced by

$$I\ddot{\theta} = -\zeta\dot{\theta} - \mu F \sin\theta + I\frac{dW(t)}{dt}, \qquad (6,2.6)$$

which is (1,1.2) supplemented by a white noise driving couple. Again we ignore inertial effects, and so replace (6,2.6) by

$$d\theta = -\frac{\mu F \sin\theta}{\zeta} dt + \frac{I}{\zeta} dW(t). \qquad (6,2.7)$$

Since there is only one spatial coordinate, the volume element is $d\theta$. We may employ (6,1.2) and (6,1.16), identifying x with θ and obtaining

$$\frac{\partial w(\theta, t)}{\partial t} + \frac{\partial}{\partial \theta}\left[-\frac{\mu F \sin\theta w}{\zeta}\right] - \frac{\sigma^2 I^2}{2\zeta^2}\frac{\partial^2 w}{\partial \theta^2} = 0. \qquad (6,2.8)$$

For the steady state

$$w = \text{constant} \exp\left\{\frac{-\frac{1}{2}I\dot{\theta}^2 + \mu F \cos\theta}{kT}\right\}, \qquad (6,2.9)$$

as in (1,1.7), and on substitution into (6,2.8) this gives $\sigma^2 = 2\zeta kT/I^2$, which confirms (5,2.14). Thus (6,2.8) becomes

$$\frac{\partial w(\theta, t)}{\partial t} = \frac{kT}{\zeta}\frac{\partial^2 w(\theta, t)}{\partial \theta^2} + \frac{\mu F}{\zeta}\frac{\partial}{\partial \theta}(\sin\theta w(\theta, t)), \qquad (6,2.10)$$

which agrees with (1,1.8). In the absence of an external field (6,2.10) reduces to

$$\frac{\partial w(\theta, t)}{\partial t} = \frac{kT}{\zeta}\frac{\partial^2 w(\theta, t)}{\partial \theta^2}.$$

This is a diffusion equation with diffusion coefficient still given by (6,2.5) but with ζ now having the physical significance that the frictional couple is ζ times the angular velocity.

It is seen from the above that, when inertial effects are ignored, we obtain a partial differential equation in configuration space. We agree to call such an equation a *Smoluchowski equation*, no longer restricting this designation to (6,2.4).

6.3 The Fokker–Planck Equation

We return to eqn (6,2.1) for a spherical particle, retaining the inertial term, and write it in differential form:

$$d\mathbf{v} = \frac{1}{m}(-\zeta\mathbf{v} + \mathbf{F})dt + d\mathbf{W}(t).$$

We have

$$dx_l = v_l dt, \quad dv_l = \frac{1}{m}(-\zeta v_l + F_l)dt + dW_l(t), \qquad (6,3.1)$$

and we take a six-vector with components $(x_1, x_2, x_3, v_1, v_2, v_3)$, so that $p = 6$ in (6,1.1). Comparing (6,3.1) with (6,1.1) we see that

$$f_i(x, t) = v_i, \quad G_{ik} = 0 \quad (i = 1, 2, 3),$$

$$f_i(x, t) = \frac{1}{m}(-\zeta v_j + F_j) \quad (i = 4, 5, 6; j = i - 3),$$

$$G_{ik} = \delta_{i-3,k} = \delta_{jk}.$$

Hence eqn (6,1.20) for the probability density function $w(\mathbf{r}, \mathbf{v}, t)$ becomes

$$\frac{\partial w(\mathbf{r}, \mathbf{v}, t)}{\partial t} + \sum_{i=1}^{3} \frac{\partial}{\partial x_i}(v_i w)$$
$$+ \sum_{j=1}^{3} \frac{\partial}{\partial v_j}\left[\frac{F_j - \zeta v_j}{m} w\right] - \frac{1}{2}\sigma^2 \sum_{j=1}^{3} \frac{\partial^2 w}{\partial v_j^2} = 0. \qquad (6,3.2)$$

If we can put $F_i = -\partial V/\partial x_i$, eqn (6,3.2) is expressible as

$$\frac{\partial w(\mathbf{r}, \mathbf{v}, t)}{\partial t} + (\mathbf{v} \cdot \operatorname{grad} w) - \frac{1}{m}\left(\operatorname{grad} V \cdot \frac{\partial w}{\partial \mathbf{v}}\right) = \frac{\zeta}{m}\left(\frac{\partial}{\partial \mathbf{v}} \cdot (\mathbf{v}w)\right) + \frac{1}{2}\sigma^2 \nabla_v^2 w, \tag{6,3.3}$$

where

$$\nabla_v^2 = \frac{\partial^2}{\partial v_1^2} + \frac{\partial^2}{\partial v_2^2} + \frac{\partial^2}{\partial v_3^2}.$$

Substituting the steady state solution (6,2.3) into (6,3.3) we obtain again the value of σ^2 given in (5,2.12). We then write (6,3.3) as

$$\frac{\partial w(\mathbf{r}, \mathbf{v}, t)}{\partial t} + (\mathbf{v} \cdot \operatorname{grad} w) - \frac{1}{m}\left(\operatorname{grad} V \cdot \frac{\partial w}{\partial \mathbf{v}}\right) = \frac{\zeta}{m}\left(\frac{\partial}{\partial \mathbf{v}} \cdot \left\{\mathbf{v}w + \frac{kT}{m}\frac{\partial w}{\partial \mathbf{v}}\right\}\right). \tag{6,3.4}$$

This is a differential equation in coordinate and velocity space, which is equivalent to the phase space of coordinates and momenta. For one-dimensional motion (6,3.4) becomes

$$\frac{\partial w(x, v, t)}{\partial t} + v\frac{\partial w}{\partial x} - \frac{1}{m}\frac{\partial V}{\partial x}\frac{\partial w}{\partial v} = \frac{\zeta}{m}\frac{\partial}{\partial v}\left(vw + \frac{kT}{m}\frac{\partial w}{\partial v}\right), \tag{6,3.5}$$

an equation due to Kramers (1940, eqn (9)).

In non-stochastic mechanics the left-hand side of (6,3.4) is just

$$\frac{\partial w}{\partial t} + \sum_{i=1}^{3} \frac{\partial w}{\partial x_i}\dot{x}_i + \sum_{i=1}^{3} \frac{\partial w}{\partial v_i}\dot{v}_i,$$

which is the total derivative of $w(\mathbf{r}, \mathbf{v}, t)$ with respect to the time. Liouville's theorem states that this total derivative is zero (Landau and Lifschitz, 1958, Section 3). We may therefore regard (6,3.4) for Brownian motion as a *generalized Liouville equation*, the effect of the random motion of the particles in the heat bath on the probability density function being described by the terms on the right hand side of (6,3.4).

If w is independent of the spatial coordinates, grad w vanishes and (6,3.4) becomes

$$\frac{\partial w(\mathbf{v}, t)}{\partial t} - \frac{1}{m}\left(\operatorname{grad} V \cdot \frac{\partial w}{\partial v}\right)$$

$$= \frac{\zeta}{m}\left(\frac{\partial}{\partial \mathbf{v}} \cdot \left\{\mathbf{v}w + \frac{kT}{m}\frac{\partial w}{\partial \mathbf{v}}\right\}\right). \qquad (6,3.6)$$

If no external force is present, this simplifies to

$$\frac{\partial w(\mathbf{v}, t)}{\partial t} = \frac{\zeta}{m}\left(\frac{\partial}{\partial \mathbf{v}} \cdot \left\{\mathbf{v}w + \frac{kT}{m}\frac{\partial w}{\partial \mathbf{v}}\right\}\right),$$

an equation in velocity space. This is, strictly speaking, the *Fokker–Planck equation* (Fokker, 1914; Planck, 1917). The generalized Liouville equation is sometimes called the *Fokker–Planck–Kramers equation.* Such distinctions have been eroded in recent literature, and we shall henceforth understand by a Fokker–Planck equation a partial differential equation of the type (6,1.19) in velocity or momentum variables and possibly also in linear or angular coordinates.

To construct an example of a Fokker–Planck equation in angular coordinates and velocities we take eqn (6,2.6) for a polar spherical molecule whose axis of rotation points in a fixed direction:

$$I\ddot{\theta} = -\zeta\dot{\theta} - \mu F \sin\theta + I\frac{dW(t)}{dt}.$$

We put $\theta = x_1, \theta = x_2$ and then

$$dx_1 = \dot{\theta}dt,$$

$$dx_2 = -\frac{\zeta\dot{\theta} + \mu F \sin\theta}{I}dt + dW(t),$$

$$f_1 = \dot{\theta}, \quad f_2 = -\frac{\zeta\dot{\theta} + \mu F \sin\theta}{I},$$

$$G_{1i} = 0, \quad G_{2i} = \delta_{i1}.$$

On substituting into (6,1.20) and using the steady state solution (6,2.9) to obtain the same value of σ^2 as before we deduce

$$\frac{\partial w(\theta, \dot{\theta}, t)}{\partial t} + \dot{\theta}\frac{\partial w}{\partial \theta} - \frac{\mu F \sin\theta}{I}\frac{\partial w}{\partial \dot{\theta}}$$

$$= \frac{\zeta}{I}\frac{\partial}{\partial \dot{\theta}}\left(\dot{\theta}w + \frac{kT}{I}\frac{\partial w}{\partial \dot{\theta}}\right). \qquad (6,3.7)$$

This could be got from (6,3.5) by replacing x by θ, v by $\dot{\theta}$, m by I and the external force $-\partial V/\partial x$ by the couple $-\mu F \sin\theta$.

It should be noted that a Smoluchowski equation is not just a special case of a Fokker–Planck equation; for example, if w is independent of v, (6,3.5) becomes

$$\frac{\partial w(x,t)}{\partial t} + v\frac{\partial w(x,t)}{\partial x} = \frac{\zeta w(x,t)}{m}, \tag{6,3.8}$$

whereas the one-dimensional version of (6,2.4) is

$$\frac{\partial w(x,t)}{\partial t} = \frac{kT}{\zeta}\frac{\partial^2 w(x,t)}{\partial x^2} - \frac{1}{\zeta}\frac{\partial}{\partial x}(Fw(x,t)). \tag{6,3.9}$$

The reason for this disagreement is that (6,3.9) corresponds to the first order stochastic differential equation

$$\zeta\frac{dx}{dt} = F + m\frac{dW(t)}{dt},$$

while (6,3.8) corresponds to the second order equation

$$m\frac{d^2x}{dt^2} + \zeta\frac{dx}{dt} = F + m\frac{dW(t)}{dt},$$

and it may not be permissible to deduce the solution of a first order equation from the limiting value of the solution of a second order equation when the coefficient of the second derivative tends to zero. Consequently the partial differential equations may differ from one another. This difficulty does not occur in the transition from (6,3.4) to (6,3.6), since both are related to the same stochastic differential equation (6,2.1).

The derivation of the Fokker–Planck equation in this section has been based on probability concepts developed in earlier chapters. The position and velocity vectors of a Brownian particle are stochastic variables but, when the system under consideration contains many such particles, it is possible to establish a deterministic partial differential equation for the probability density function $w(\mathbf{r}, \mathbf{v}, t)$. The situation is much the same as in quantum mechanics where the position and momentum of an electron are random variables, and nevertheless the solution of the Schrödinger equation provides a deterministic probability density function. We now refer to alternative approaches to the derivation of the Fokker–Planck equation.

Wang Chang and Uhlenbeck (1970, Chap, 5) have investigated this problem by means of standard scattering theory. They established (6,3.4) with $-(1/m)$ grad V replaced by $-\omega_0^2\mathbf{r} + \delta_{i1}E_0 \cos \omega t$ as approximately true under certain conditions. They assumed that $m \gg M$, the mass of the particles constituting the heat bath, and that both the velocity v of the Brownian particle and the velocity V of a heat bath particle are always close to their equipartition values. Hence

$$\tfrac{1}{2}mv^2 \doteqdot \tfrac{3}{2}kT \doteqdot \tfrac{1}{2}MV^2,$$

which shows that v/V is of order $(M/m)^{1/2}$. With this condition the frictional force on the Brownian particle will on the average be proportional to v. They obtained for the frictional constant

$$\zeta = \frac{16\pi^{1/2}}{3} NM \left(\frac{M}{2kT}\right)^{5/2} \int_0^{\pi} d\theta \sin\theta\,(1-\cos\theta) \times$$

$$\int_0^{\infty} dV V^5 \exp\left\{-\frac{MV^2}{2kT}\right\} I(V,\theta),$$

where $I(V, \theta)$ is the differential cross-section for scattering of the heat bath particles and N is their number per unit volume.

It is possible to obtain a Fokker–Planck equation for a Brownian particle, on which a constant weak external field acts, from a microscopic theory which takes account of the dynamics of both the Brownian particle and of the particles of the fluid that is in thermal motion (Lebowitz and Rubin, 1963); Rêsibois and Davis, 1964; Lebowitz and Rêsibois, 1965). On starting from a generalized Liouville equation for the joint probability density of the Brownian particle and fluid, the probability density w for the Brownian particle is deduced by integration over the fluid particles. It is found to the first order of approximation in $(M/m)^{1/2}$ that w obeys the Fokker–Planck equation. When a weak periodic electric field $E_0 e^{i\omega t}$ is acting and one considers a steady state, the ζ in the Fokker–Planck equation is no longer constant but is a frequency-dependent $\zeta(\omega)$ satisfying

$$\zeta(\omega) = \frac{1}{3kT} \lim_{t_1 \to \infty} \int_0^{\infty} e^{-t/t_1} e^{-i\omega t} \langle(\mathscr{F}(0) \cdot \mathscr{F}(t))\rangle dt, \tag{6,3.10}$$

where $\mathscr{F}(t)$ is the residual microscopic fluctuating force exerted on the Brownian particle by the surrounding fluid.

Such a residual force for the one-dimensional Langevin theory was encountered in Section 5.3, where it was found to be equal to $N(t)$. Let us therefore compare (6,3.10) with (5,3.7), which holds for both rotational and translational motion and which for the present case of spherical symmetry is just

$$\zeta_i = \zeta = \frac{1}{3kT} \int_0^\infty \langle (\mathbf{N}(0) \cdot \mathbf{N}(t)) \rangle dt.$$

If we identify $\mathscr{F}(t)$ in (6,3.10) with $\mathbf{N}(t)$ and apply (5,3.4) and (5,3.6) to translational motion, then

$$\begin{aligned}
\zeta(\omega) &= \frac{1}{3kT} \lim_{t_1 \to \infty} \int_0^\infty e^{-t/t_1} e^{-i\omega t} \sum_{i=1}^{3} 2\zeta_i kT \delta(t) dt \\
&= 2\zeta \lim_{t_1 \to \infty} \int_0^\infty e^{-t/t_1} e^{-i\omega t} \delta(t) dt \\
&= \zeta \lim_{t_1 \to \infty} \int_{-a}^\infty e^{-t/t_1} e^{-i\omega t} \delta(t) dt \quad (a > 0) \\
&= \zeta.
\end{aligned}$$

The above theory replaces $N(t)$ by $\mathscr{F}(t)$, which is physically more acceptable then the Markovian white noise force.

7

Equations of Rotational Brownian Motion of a Rigid Body

7.1 The Euler–Langevin Equations

We now begin to prepare the way for the solution of problems arising from the rotational motion of rigid Brownian particles of a specified shape. Our objective will be to obtain correlation functions, introduced in Section 4.4, with a view to employing them in the interpretation of experiments. To calculate correlation functions one may use probability density functions that are solutions of Fokker–Planck equations. Alternatively one may deduce correlation functions for angular velocity components from a study of one or more stochastic differential equations, and then use these to derive orientational correlation functions that are of interest in a particular problem. Since we have a method, described in Section 6.1, for proceeding from stochastic differential equations to Fokker–Planck equations, we shall first consider how the former arise.

Let us examine the Brownian motion of a body that is subject to no external force or couple. To set up stochastic differential equations for the rotating body, we take coordinate axes labelled 1, 2, 3 through the centre of mass of the body and in the directions of the principal axes of inertia. The coordinate axes are therefore fixed in the body. We denote by I_1, I_2, I_3 the principal moments of inertia and by $\omega_1, \omega_2, \omega_3$ the components of angular velocity of the body. The components of angular momentum of the body are $I_1\omega_1, I_2\omega_2, I_3\omega_3$ and the time rate of change of its first component is

$$I_1\dot{\omega}_1 - (I_2 - I_3)\omega_2\omega_3,$$

as we may deduce from the expression $\dot{\mathbf{V}} + (\boldsymbol{\omega} \times \mathbf{V})$ for the rate of change of a vector **V** whose components are referred to rotating axes (Synge and Griffith, 1959, p. 311).

The body is subject to a white noise driving couple and to a frictional couple. We write the former

$$I_1 \frac{dW_1(t)}{dt}, \quad I_2 \frac{dW_2(t)}{dt}, \quad I_3 \frac{dW_3(t)}{dt}$$

and we suppose that W_1, W_2, W_3 are mutually independent. Retaining the hydrodynamical picture of Brownian motion we may say that the frictional couple about the first axis is a linear combination of ω_1, ω_2, ω_3. When the body has an irregular shape, there is no *a priori* reason why this frictional couple should be proportional to ω_1, but to avoid mathematical complexity we shall assume that it is. Making similar assumptions for the other two axes we have a frictional couple with components $-I_1 B_1 \omega_1, -I_2 B_2 \omega_2, -I_3 B_3 \omega_3$, where B_1, B_2, B_3 are positive. Collecting our results we have the equations of motion

$$\begin{aligned} I_1 \dot{\omega}_1 - (I_2 - I_3)\omega_2 \omega_3 &= -I_1 B_1 \omega_1 + I_1 \frac{dW_1(t)}{dt}, \\ I_2 \dot{\omega}_2 - (I_3 - I_1)\omega_3 \omega_1 &= -I_2 B_2 \omega_2 + I_2 \frac{dW_2(t)}{dt}, \\ I_3 \dot{\omega}_3 - (I_1 - I_2)\omega_1 \omega_2 &= -I_3 B_3 \omega_3 + I_3 \frac{dW_3(t)}{dt}. \end{aligned} \qquad (7,1.1)$$

The expressions on the left-hand side are familiar from Euler's treatment of the dynamics of a rotating body, those on the right appear in Langevin equations, so we call (7,1.1) the *Euler–Langevin equations.* They are applicable to bodies for which the above assumption about the frictional couple is a reasonable one.

If the second terms on the left-hand sides of (7,1.1) vanish, the equations decouple into three independent Langevin equations and may therefore be investigated by the methods of Section 5.3. The three components of angular velocity for a steady state will be centred Gaussian random variables given by (5,3.3) and the correlation functions for the components of angular velocity will be given by (5,3.8). Equations (7,1.1) decouple in a trivial manner for $\omega_2 = \omega_3 = 0$. Then the body can rotate only about the first principal axis of inertia, and consequently this axis will point in a fixed direction in space. The rotational motion is one-dimensional and for brevity we shall refer to a body rotating in this way as a disc, even though it may not be laminar in shape. In this nomenclature the sphere in the two-dimensional Debye theory of Section 1.1 is a disc. Equations (7,1.1) will obviously decouple when the Brownian particle is a sphere, since then $I_1 = I_2 = I_3$. The equations will also decouple if $I_1 = I_2$, $\omega_3 = 0$. This gives a linear rotator, or needle, the third axis being the line of symmetry. When (7,1.1) do not decouple, they are nonlinear equations and the components of angular velocity are not in general Gaussian random variables. However, if the Brownian particle is a solid revolution about the third axis so that $I_1 = I_2$, then ω_3 will satisfy a Langevin equation and the steady state value of ω_3 will be a centred Gaussian random variable, as we saw in Section 5.2.

7.2 The Fokker–Planck Equation for a Rotating Brownian Particle

The rotational motion of a Brownian particle may be regarded as a succession of infinitesimal rotations $\boldsymbol{\omega}(t)\delta t$ about the instantaneous axis of rotation. In the time interval δt the body turns about the first coordinate axis through an angle $\omega_1(t)\delta t$, and during a positive time interval $t - t_0$ it turns through an angle $\theta_1(t)$ given by $\theta_1(t) = \int_{t_0}^{t} \omega_1(u)du$. We similarly define $\theta_2(t)$ and $\theta_3(t)$. Since $\omega_1(t)$, $\omega_2(t)$, $\omega_3(t)$ are stochastic, so also are $\theta_1(t)$, $\theta_2(t)$, $\theta_3(t)$. We wish to establish a partial differential equation for the probability density function $w(\theta_1, \theta_2, \theta_3, \omega_1, \omega_2, \omega_3, t)$, or more briefly $w(\boldsymbol{\theta}, \boldsymbol{\omega}, t)$.

The stochastic differential equations for this problem are

$$d\theta_i = \omega_i dt, \tag{7,2.1}$$

and (7,1.1), which we express as

$$d\omega_i = \left(\frac{I_j - I_k}{I_i}\omega_j\omega_k - B_i\omega_i\right)dt + dW_i(t), \tag{7,2.2}$$

where i, j, k is a cyclic permutation of 1, 2, 3. We write $\theta_1 = x_1$, $\theta_2 = x_2$, $\theta_3 = x_3$, $\omega_1 = x_4$, $\omega_2 = x_5$, $\omega_3 = x_6$ and on comparing (6,1.1) with (7,2.1) and (7,2.2) find that

$$f_1 = \omega_1, \quad f_2 = \omega_2, \quad f_3 = \omega_3$$

$$f_4 = \frac{I_2 - I_3}{I_1}\omega_2\omega_3 - B_1\omega_1, \text{ etc.}$$

$$G_{1l} = G_{2l} = G_{3l} = 0,$$

$$G_{4l} = \delta_{1l}, \quad G_{5l} = \delta_{2l}, \quad G_{6l} = \delta_{3l}.$$

Hence (6,1.19) yields

$$\frac{\partial w(\boldsymbol{\theta}, \boldsymbol{\omega}, t)}{\partial t} + \frac{\partial}{\partial \theta_1}(\omega_1 w) + \frac{\partial}{\partial \theta_2}(\omega_2 w) + \frac{\partial}{\partial \theta_3}(\omega_3 w)$$

$$+ \frac{\partial}{\partial \omega_1}\left[\left(\frac{I_2 - I_3}{I_1}\omega_2\omega_3 - B_1\omega_1\right)w\right] + \frac{\partial}{\partial \omega_2}\left[\left(\frac{I_3 - I_1}{I_2}\omega_3\omega_1 - B_2\omega_2\right)w\right]$$

$$+ \frac{\partial}{\partial \omega_3}\left[\left(\frac{I_1 - I_2}{I_3}\omega_1\omega_2 - B_3\omega_3\right)w\right] - \frac{1}{2}\left(\sigma_1^2\frac{\partial^2 w}{\partial \omega_1^2} + \sigma_2^2\frac{\partial^2 w}{\partial \omega_2^2} + \sigma_3^2\frac{\partial^2 w}{\partial \omega_3^2}\right)$$

$$= 0 \qquad (7,2.3)$$

with $\sigma_1^2, \sigma_2^2, \sigma_3^2$ still to be determined.

To determine these we substitute into (7,2.3) the steady state value of w. Since the energy of the body is just its kinetic energy, $\frac{1}{2}(I_1\omega_1^2 + I_2\omega_2^2 + I_3\omega_3^2)$,

$$w = \text{constant}\exp\left\{-\frac{I_1\omega_1^2 + I_2\omega_2^2 + I_3\omega_3^2}{2kT}\right\}$$

for the steady state. Then

$$\frac{\partial w}{\partial t} = 0, \quad \frac{\partial w}{\partial \theta_i} = 0, \quad \frac{\partial w}{\partial w_i} = -\frac{I_i\omega_i}{kT}w$$

$$\frac{\partial^2 w}{\partial \omega_i^2} = -\frac{I_i}{kT}w + \left(\frac{I_i\omega_i}{kT}\right)^2 w,$$

and substitution into (7,2.3) yields

$$\sum_{i=1}^{3}(I_i\omega_i^2 - kT)(2kTB_i - \sigma_i^2 I_i) = 0.$$

Since this is true for all values of $\omega_1, \omega_2, \omega_3$, we have $\sigma_i^2 = 2B_ikT/I_i$, as in (5,3.6). Thus the Fokker–Planck equation is

$$\frac{\partial w(\boldsymbol{\theta}, \boldsymbol{\omega}, t)}{\partial t} + \sum_{i=1}^{3} \omega_i \frac{\partial w}{\partial \theta_i} + \sum_{i=1}^{3} \frac{\partial}{\partial \omega_i} \left[\left(\frac{I_j - I_k}{I_i} \omega_j \omega_k - B_i \omega_i \right) w \right]$$

$$- kT \sum_{i=1}^{3} \frac{B_i}{I_i} \frac{\partial^2 w}{\partial \omega_i^2} = 0. \qquad (7,2.4)$$

An equation equivalent to this was derived from first principles by Hubbard (1972, eqn (2.19)). Usually it will not be easy to solve (7,2.4). We shall therefore employ alternative forms of the Fokker–Planck equation that are more suited to the investigation of special models of polar molecules.

7.3 Rotation Operators for a Rigid Body

When investigating rotational Brownian motion problems it will often be found extremely helpful to work with rotation operators. In Appendix B some standard results from elementary quantum mechanics are recalled, and rotation operators are related to quantum-mechanical angular momentum operators. We shall now start afresh and later link up with Appendix B. In the course of this section we shall also need results for the representation of linear operators that may be found in Appendix A.

Let us take a set of rectangular axes Ox, Oy, Oz and a function $\psi(x, y, z)$. Let us rotate the axes to Ox', Oy', Oz' and consider ψ for the new coordinates x', y', z' of the point whose coordinates were previously x, y, z. We then write

$$\psi(x', y', z') = R\psi(x, y, z), \qquad (7,3.1)$$

and we call R the rotation operator for the specified rotation of axes and for the function ψ. When ψ has several components ψ_1, ψ_2, etc., for example if $\psi(x, y, z)$ is the column vector $\begin{bmatrix} x \\ y \\ z \end{bmatrix}$, eqn (7,3.1) stands for a set of equations

$$\psi_i(x', y', z') = \sum_k R_{ik} \psi_k(x, y, z) \qquad (7,3.2)$$

and R will be a matrix. If the coordinate axes are rotated through an angle χ about an axis in the direction of the unit vector $\mathbf{n}$, eqn (B.20) of Appendix B gives

$$R\psi(x, y, z) = \exp\,[-i\chi(\mathbf{n}\cdot\mathbf{J})]\,\psi(x, y, z). \qquad (7,3.3)$$

The operator $\mathbf{J}$ with self-adjoint components J_x, J_y, J_z is defined as the angular momentum operator divided by $\hbar$. When the axes Ox, Oy, Oz are fixed in space, the components obey (B.16):

$$[J_y, J_z] = iJ_x, \quad [J_z, J_x] = iJ_y, \quad [J_x, J_y] = iJ_z. \quad (7,3.4)$$

If the coordinate system Ox, Oy, Oz is rotating, (7,3.4) is altered to

$$[J_y, J_z] = -iJ_x, \quad [J_z, J_x] = -iJ_y, \quad [J_x, J_y] = -iJ_z. \qquad (7,3.5)$$

The reason for this rather unexpected result is that the direction cosines of the rotating axes with respect to the fixed axes do not commute with the J_x, J_y, J_z of the fixed axes. Consequently (7,3.4) do not preserve their structure, when they are projected onto the rotating axes (Van Vleck, 1951).

At this stage we introduce the time explicitly into (7,3.1) and (7,3.3). Suppose that we have rotating Cartesian axes, the co-ordinates of a fixed point at time zero being $x(0)$, $y(0)$, $z(0)$ and those at time t being $x(t)$, $y(t)$, $z(t)$. Then R will be time dependent and we write

$$R(t)\,\psi(x(0), y(0), z(0)) = \Phi(t).$$

Next applying (7,3.3) to the situation where x, y, z are the co-ordinates at time t and the axes are then rotating with angular velocity $\boldsymbol{\omega}(t)$ during a short interval of time δt we deduce that

$$\begin{aligned}\Phi(t+\delta t) &= \exp\,[-i(\mathbf{J}\cdot\boldsymbol{\omega}(t))\delta t]\,\Phi(t)\\ &= [\mathbf{I} - i(\mathbf{J}\cdot\boldsymbol{\omega}(t))\delta t + \ldots]\,R(t)\,\psi(x(0), y(0), z(0)),\end{aligned}$$

where $\mathbf{I}$ is the identity operator. Hence

$$\Phi(t+\delta t) - \Phi(t) = [-i(\mathbf{J}\cdot\boldsymbol{\omega}(t))\delta t + \ldots]\,R(t)\,\psi(x(0), y(0), z(0)), \qquad (7,3.6)$$

but on the other hand

$$\begin{aligned}\Phi(t+\delta t) - \Phi(t) &= [R(t+\delta t) - R(t)]\,\psi(x(0), y(0), z(0))\\ &= \left[\frac{dR(t)}{dt}\delta t + \ldots\right]\psi(x(0), y(0), z(0)). \qquad (7,3.7)\end{aligned}$$

On comparing (7,3.6) and (7,3.7) we conclude that

$$\frac{dR(t)}{dt} = -i(\mathbf{J} \cdot \boldsymbol{\omega}(t))R(t). \tag{7,3.8}$$

Since in deriving this result we have considered the motion only during an infinitesimal time interval, eqn (7,3.8) holds even if the axis of rotation changes its direction with time. If it does, we must remember that, since the components of **J** in (7,3.8) are referred to the axes at time t and since these axes are rotating, these components obey (7,3.5).

We now consider the kinematics of a rigid body that rotates about its centre of mass O. Taking coordinate axes through O and fixed in the body we let Ox, Oy, Oz be their initial positions. If Oz' is in the position of Oz after a rotation of the body, we let β be the angle between Oz and Oz', and α the angle between the planes zOx and zOz', as shown in Fig. 7.1. The most general rotation of the body is

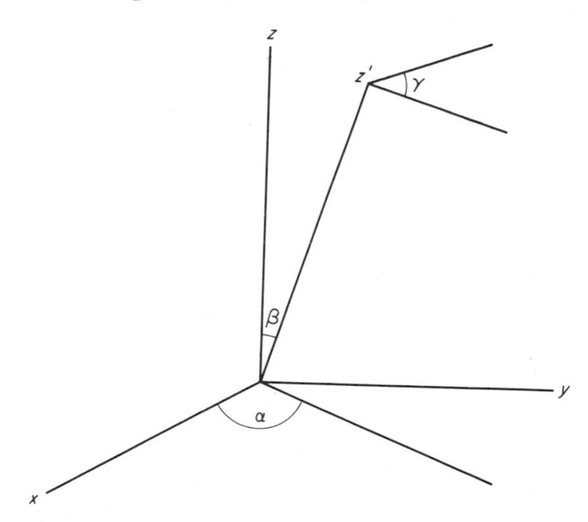

Fig. 7.1 The Euler angles α, β, γ for the orientation of a rigid body.

then obtained by rotating the plane zOz' about Oz' through an angle γ, say. We call α, β, γ the *Euler angles* (Synge and Griffith, 1959, p. 259). We see from Fig. 7.1 that we may arrive at the new orientation of the body by rotating it about Oz through angle γ, then about Oy through angle β and finally about Oz through angle α. These axes of rotation are fixed with respect to the laboratory.

According to (7,3.3) the rotation operators for the three successive rotations are, respectively, $\exp[-i\gamma J_z]$, $\exp[-i\beta J_y]$ and $\exp[-i\alpha J_z]$, so the rotation operator R for the complete rotation is given by

$$R = \exp[-i\alpha J_z]\exp[-i\beta J_y]\exp[-i\gamma J_z]. \qquad (7,3.9)$$

Since, as was pointed out in writing down (7,3.1), R depends not only on the prescribed rotation of axes but also on the function or functions ψ involved in the transformation, the representation of the operators J_y and J_z in (7,3.9) will depend on ψ.

A set of functions ψ with which we shall often be concerned are the spherical harmonics $Y_{jm}(\theta, \phi)$ defined in eqn (B.10) by

$$Y_{jm}(\theta, \phi) = \left[\frac{2j+1}{4\pi}\frac{(j-m)!}{(j+m)!}\right]^{1/2} \frac{e^{im\phi}(-\sin\theta)^m}{2^j j!}\left[\frac{d}{d(\cos\theta)}\right]^{j+m}(\cos^2\theta - 1)^j, \qquad (7,3.10)$$

where θ, ϕ are the usual spherical polar angular coordinates and m can take the values $-j, -j+1, \ldots j-1, j$. The harmonics satisfy the relation (B.11):

$$\int_0^{2\pi} d\phi \int_0^{\pi} \sin\theta d\theta\, Y^*_{jm}(\theta, \phi) Y_{j'm'}(\theta, \phi) = \delta_{jj'}\delta_{mm'}. \qquad (7,3.11)$$

We select a point whose spherical polar coordinates with respect to Ox, Oy, Oz are (r, θ, ϕ), so that the Cartesian coordinates of the point are given by

$$x = r\sin\theta\cos\phi, \quad y = r\sin\theta\sin\phi, \quad z = r\cos\theta.$$

We now rotate the coordinate axes to Ox', Oy', Oz', the rotation being specified by (7,3.9). The angles α, β, γ are quite independent of θ and ϕ. Let the spherical polar coordinates of the selected point with respect to Ox', Oy', Oz' be (r, θ', ϕ'). Then applying (7,3.1) we have

$$Y_{jm}(\theta', \phi') = RY_{jm}(\theta, \phi), \qquad (7,3.12)$$

where R is defined by (7,3.9) with a $(2j+1)$-dimensional representation of J_x, J_y, J_z. We agree to write (7,3.12) as

$$Y_{jm}(\theta', \phi') = \sum_{m'=-j}^{j} D^j_{m'm}(\alpha, \beta, \gamma) Y_{jm'}(\theta, \phi). \qquad (7,3.13)$$

By employing (7,3.9) and (7,3.11) it will follow that

$$D^j_{m'm}(\alpha, \beta, \gamma) = \int_0^{2\pi} d\phi \int_0^{\pi} d\theta \sin\theta\, Y^*_{jm'}(\theta, \phi) \exp[-i\alpha J_z] \times$$

$$\exp[-i\beta J_y] \exp[-i\gamma J_z] Y_{jm}(\theta, \phi). \quad (7,3.14)$$

On referring to eqn (A.9) and recalling that dq is the volume element, which in the present case is $\sin\theta d\theta d\phi$, we see that $D^j_{m'm}$ is the $m'm$-element of the matrix representative of R in the representation with basis elements $Y_{js}(\theta, \phi)$, where $s = -j, -j+1, \ldots j-1, j$. Since, however, θ and ϕ are integration variables in (7,3.14), $D^j_{m'm}$ is independent of them. Thus we can take as basis any set Y_{js} having the same two polar angles for arguments. We saw in Appendix B that Y_{jm} is an eigenfunction of J_z with eigenvalue m. Hence in the present representation J_z is a diagonal matrix with diagonal elements $-j, -j+1, \ldots j-1, j$. We work in this representation and then

$$\exp[-i\gamma J_z] Y_{jm}(\theta, \phi) = \exp[-im\gamma] Y_{jm}(\theta, \phi).$$

Moreover, since J_z is a self-adjoint operator so that $J_z^+ = J_z$, we can by (A.2) replace $Y^*_{jm'}(\theta, \phi) \exp[-i\alpha J_z]$ in (7,3.14) by $\{\exp[i\alpha J_z] Y_{jm'}(\theta, \phi)\}^*$, that is, $\exp[-im'\alpha] Y^*_{jm'}(\theta, \phi)$. Hence

$$D^j_{m'm}(\alpha, \beta, \gamma) = \exp[-i(m'\alpha + m\gamma)] \int_0^{2\pi} d\phi$$

$$\int_0^{\pi} d\theta \sin\theta\, Y^*_{jm'}(\theta, \phi) \exp[-i\beta J_y] Y_{jm}(\theta, \phi),$$

and it may be shown (Rose, 1957, pp. 228–234) that the double integral is the real quantity

$$[(j+m)!(j-m)!(j+m')!(j-m')!]^{1/2}$$

$$\times \sum_s (-)^s \frac{(\cos\frac{1}{2}\beta)^{2j+m-m'-2s}(-\sin\frac{1}{2}\beta)^{m'-m+2s}}{(j-m'-s)!(j+m-s)!(s+m'-m)!s!},$$

where the sum is taken over all values of s for which the arguments of the factorials are non-negative. We see from (7,3.12) and (7,3.13) that for the column vector with elements $Y_{js}(\theta, \phi)$, R is the transpose of the matrix whose $m'm$-element is $D^j_{m'm}(\alpha, \beta, \gamma)$.

The function $D^j_{m'm}$ has the following mathematical properties:

$$\int_0^{2\pi} d\gamma \int_0^{\pi} d\beta \sin\beta \int_0^{2\pi} d\alpha D^j_{mk}(\alpha,\beta,\gamma) D^{j'}_{m'k'}(\alpha,\beta,\gamma)$$

$$= \frac{8\pi^2}{2j+1}\,\delta_{jj'}\delta_{mm'}\delta_{kk'},$$

$$\sum_{m'} D^j_{mm'}(\alpha,\beta,\gamma) D^{j*}_{m''m'}(\alpha,\beta,\gamma) = \delta_{mm''},$$

$$\tag{7,3.15}$$

$$D^{j*}_{m0}(\alpha,\beta,\gamma) = \left(\frac{4\pi}{2j+1}\right)^{1/2} Y_{jm}(\beta,\alpha),$$

$$D^j_{00}(\alpha,\beta,\gamma) = P_j(\cos\beta),$$

where P_j is the Legendre polynomial of order j (Edmonds, 1968, Chap. 4). The above properties include (7,3.11). The physical interpretation of the integers j, m, m' is the following: $D^j_{m'm}$ is an eigenfunction of the square J^2 of the rotation operator $\mathbf{J}$ with eigenvalue $j(j+1)$, of its third component with respect to the laboratory frame J_z with eigenvalue m' and of its third component with respect to body frame $J_{z'}$ with eigenvalue m (Brink and Satchler, 1975, p. 25).

7.4 The Stochastic Rotation Operator and Correlation Functions

In the previous section we considered three-dimensional rotations of a coordinate frame of reference. In eqn (7,3.1) we defined the rotation operator R which, as we saw, depends not only on the three-dimensional rotation but also on the function ψ that undergoes the rotation transformation.

We apply the concept of rotation operator to the Brownian motion of a rotating body. We take the centre of mass O as the origin and a laboratory frame of coordinate axes Ox, Oy, Oz pointing in fixed directions in space, and we take a body frame $O1$, $O2$, $O3$ of axes fixed with respect to the body. We denote by $\alpha(t)$, $\beta(t)$, $\gamma(t)$ the Euler angles of the body frame at time t referred to the laboratory frame, the spherical polar angles of axis 3 being $\beta(t)$, $\alpha(t)$. We denote by $\alpha'(t)$, $\beta'(t)$, $\gamma'(t)$ the Euler angles of the body frame at time t referred to the body frame at time zero, so that $\beta'(t)$ is the angle between the directions of axis 3 at times t and zero,

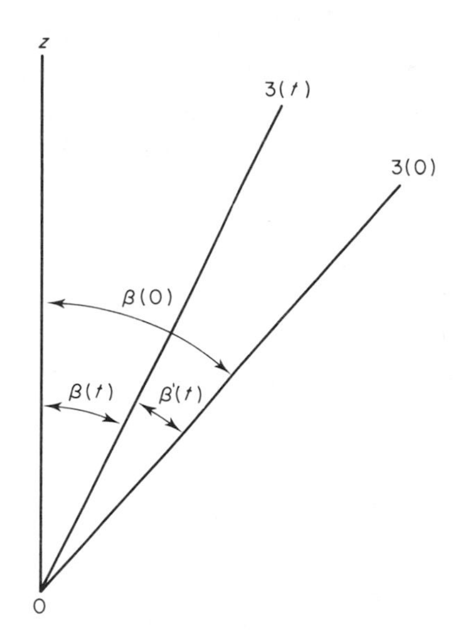

Fig. 7.2 The angles between the laboratory axis Oz, the third body axis $O3(t)$ at time t and the third body axis $O3(0)$ at time zero. In general these three axes are not coplanar.

as shown in Fig. 7.2. We then deduce from (7,3.9) that the rotation operator which brings the body frame at time zero to the body frame at time t is a time dependent $R(t)$ given by

$$R(t) = \exp[-i\alpha'(t)J_3] \exp[-i\beta'(t)J_2] \exp[-i\gamma'(t)J_3], \tag{7,4.1}$$

where J_1, J_2, J_3 are the components of $\mathbf{J}$ in the directions of the rotating 1, 2, 3-axes at time zero. According to (7,3.5) these components satisfy

$$J_k J_l - J_l J_k = -i \sum_n \epsilon_{kln} J_n, \tag{7,4.2}$$

where ϵ_{kln} is the totally antisymmetric *Levi Civita symbol*:

$$\epsilon_{kln} = \begin{cases} 1 \text{ for } k, l, n \text{ an even permutation of } 1, 2, 3 \\ -1 \text{ for } k, l, n \text{ an odd permutation of } 1, 2, 3 \\ 0 \text{ when two suffixes are equal.} \end{cases} \tag{7,4.3}$$

Since $\alpha'(t)$, $\beta'(t)$, $\gamma'(t)$ are random variables for the Brownian motion of the rotating body, $R(t)$ is a stochastic operator – the *stochastic rotation operator*. Moreover, as was discussed in Section 4.2, $\alpha(t)$, $\beta(t)$, $\gamma(t)$, $\alpha'(t)$, $\beta'(t)$, $\gamma'(t)$ are all Markov processes.

We now derive some properties of $R(t)$. Since $\alpha'(0) = \beta'(0) = \gamma'(0) = 0$, it follows from (7,4.1) that

$$R(0) = \mathbf{I}, \tag{7,4.4}$$

the identity operator. Since $\mathbf{J}$ is a self-adjoint operator and since $(AB)^+ = B^+A^+$ from (A.4), we deduce that

$$\begin{aligned} R^+(t) &= (\exp[-i\gamma'(t)J_3])^+(\exp[-i\beta'(t)J_2])^+(\exp[-i\alpha'(t)J_3])^+ \\ &= \exp[i\gamma'(t)J_3]\exp[i\beta'(t)J_2]\exp[i\alpha'(t)J_3]. \end{aligned} \tag{7,4.5}$$

Now

$$\begin{aligned} &\exp[i\alpha'(t)J_3]\exp[-i\alpha'(t)J_3] \\ &= [\mathbf{I} + i\alpha'(t)J_3 - \tfrac{1}{2}(\alpha'(t))^2J_3^2 + \ldots] \times \\ &\quad [\mathbf{I} - i\alpha'(t)J_3 - \tfrac{1}{2}(\alpha'(t))^2J_3^2 + \ldots] = \mathbf{I}, \end{aligned}$$

from which it follows that

$$R^+(t)R(t) = \mathbf{I} \tag{7,4.6}$$

and similarly that $R(t)R^+(t) = \mathbf{I}$. Thus $R^+(t)$ is the inverse $R^{-1}(t)$ of $R(t)$, and $R(t)$ is a unitary operator as defined in Appendix A. On comparing (7,4.5) with (7,4.1) we see that we may go from $R(t)$ to $R^{-1}(t)$ by making the replacements

$$\alpha' \to -\gamma', \quad \beta' \to -\beta', \quad \gamma' \to -\alpha'.$$

Since the matrix representative of R^+ is the Hermitian conjugate of the matrix representative of R,

$$D^j_{m'm}(-\gamma', -\beta', -\alpha') = D^{j*}_{mm'}(\alpha', \beta', \gamma'). \tag{7,4.7}$$

The second equation of (7,3.15) is an immediate consequence of (7,4.7), since the properties that we have established for $R(t)$ hold equally well for R defined by (7,3.9).

The rotation operator $R(t)$ brings the body frame at time zero to the body frame at time t. If we make the rotation $R(t)$ of the laboratory frame, the angles $\alpha(t)$, $\beta(t)$, $\gamma(t)$ become $\alpha(0)$, $\beta(0)$, $\gamma(0)$, respectively, because the orientation of the body frame at time t relative to the rotated laboratory frame is the same as the orientation of the body frame at time zero relative to the laboratory frame before it was rotated. We may therefore take over (7,3.12) with the replacements

$$R \to R(t), \quad \theta \to \beta(t), \quad \phi \to \alpha(t), \quad \theta' \to \beta(0), \quad \phi' \to \alpha(0),$$

so that

$$Y_{jm}(\beta(0), \alpha(0)) = R(t)\, Y_{jm}(\beta(t), \alpha(t).$$

Then from (7,4.6)

$$Y_{jm}(\beta(t), \alpha(t)) = R^{+}(t)\, Y_{jm}(\beta(0), \alpha(0)). \tag{7,4.8}$$

We return to Section 4.4 in order to establish some results that will facilitate the calculation of orientational correlation functions. According to (4,4.1) the correlation function of A and B for the steady state of a dynamical system,

$$\langle A^*(0)B(t)\rangle = \int du(0) f(0) A^*(0) B(t), \tag{7,4.9}$$

where $f(0)$ is the probability density function at time 0 of the system in the absence of an external field. This probability density will depend on both dynamical and orientational variables. In future applications to rotational motion of a Brownian particle we shall use Euler angles to describe the orientation of the particle, and for these the orientational probability density function g is given by

$$g(\alpha(0), \beta(0), \gamma(0)) = \frac{\sin\beta(0)\, d\alpha(0)\, d\beta(0)\, d\gamma(0)}{8\pi^2} \cdot \tag{7,4.10}$$

To establish this we see that in Fig. 7.1 the axis Oz' is specified by α and β, and that the element of solid angle in which Oz' is to be found is $\sin\beta d\alpha d\beta$. The probability of the particle being oriented in the element $d\gamma$ is proportional to $d\gamma$. Hence we have (7,4.10), the normalizing factor being chosen so that unity is obtained on integrating with respect to $\alpha(0)$ from 0 to 2π, to $\beta(0)$ from 0 to π and to $\gamma(0)$ from 0 to 2π.

The dynamical part of $f(0)$ will be the Maxwell–Boltzmann distribution and the energy that appears in this will be the kinetic energy of the Brownian particle, since it is subject to no external field. If we take for u the Euler angles and their conjugate momenta, the expression for the kinetic energy will involve both the angles and the components of angular momentum, and integrations with the

Maxwell–Boltzmann distribution will be difficult to perform (Van Vleck, 1932, pp. 32–36). This difficulty may be removed by taking for u the Euler angles and three components of the angular velocity of the body referred to orthogonal axes through its centre of mass. If the body is asymmetric, we take for these axes the principal axes of inertia and the Maxwell–Boltzmann distribution is then

$$h(\boldsymbol{\omega}(0)) = \frac{\exp\left[-\dfrac{I_1(\omega_1(0))^2 + I_2(\omega_2(0))^2 + I_3(\omega_3(0))^2}{2kT}\right]}{(I_1 I_2 I_3)^{1/2}(2\pi kT)^{3/2}}, \qquad (7,4.11)$$

the multiplying factor being chosen such that

$$\int_{-\infty}^{\infty} d\omega_1(0) \int_{-\infty}^{\infty} d\omega_2(0) \int_{-\infty}^{\infty} d\omega_3(0) h(\boldsymbol{\omega}(0)) = 1.$$

It will be seen in Sections 8.2, 9.1 and 10.1 that the kinetic energy for the disc, sphere and linear rotator at time 0 may be expressed as

$$\tfrac{1}{2}I(\omega_1(0))^2, \quad \tfrac{1}{2}I\{(\omega_1(0))^2 + (\omega_2(0))^2 + (\omega_3(0))^2\},$$
$$\tfrac{1}{2}I\{(\omega_1(0))^2 + (\omega_2(0))^2\},$$

respectively. Hence all cases that we shall study are comprised in (7,4.11), the normalizing factor being different if $\omega_2(0)$ or $\omega_3(0)$ is absent.

As a result of these considerations we conclude that we have chosen orientational and angular velocity probability density functions that are independent of one another. If then $C(t)$ is a steady-state random process depending on the angles, angular velocity components and the time, its ensemble average over angles and angular velocity variables in the absence of an external field,

$$\langle C(t)\rangle = \int\int\int g(\alpha(0), \beta(0), \gamma(0)) d\alpha(0) d\beta(0) d\gamma(0) \times$$
$$\int\int\int h(\boldsymbol{\omega}(0)) C(t) d\omega_1(0) d\omega_2(0) d\omega_3(0). \qquad (7,4.12)$$

Let us put

$$\iiint h(\boldsymbol{\omega}(0)) C(t) d\omega_1(0) d\omega_2(0) d\omega_3(0) = \langle C(t)\rangle_\omega, \qquad (7,4.13)$$

the ensemble average over the angular velocity space. Then (7,4.12) may, by (7,4.10), be expressed as

$$\langle C(t)\rangle = \frac{1}{8\pi^2}\int_0^{2\pi} d\gamma(0)\int_0^{2\pi} d\alpha(0)\int_0^{\pi} d\beta(0)\sin\beta(0)\langle C(t)\rangle_\omega . \quad (7,4.14)$$

If $\langle C(t)\rangle_\omega$ is independent of the angles $\alpha(0)$, $\beta(0)$, $\gamma(0)$,

$$\langle C(t)\rangle = \langle C(t)\rangle_\omega . \quad (7,4.15)$$

When $C(t)$ is identified with $A^*(0)B(t)$, we deduce from (7,4.14) that

$$\langle A^*(0)B(t)\rangle = \frac{1}{8\pi^2}\int_0^{2\pi} d\gamma(0)\int_0^{2\pi} d\alpha(0)\int_0^{\pi} d\beta(0)\sin\beta(0)\langle A^*(0)B(t)\rangle_\omega . \quad (7,4.16)$$

If $A(t)$ and $B(t)$ are independent of the Euler angle γ, (7,4.16) reduces to

$$\langle A^*(0)B(t)\rangle = \frac{1}{4\pi}\int_0^{2\pi} d\alpha(0)\int_0^{\pi} d\beta(0)\sin\beta(0)\langle A^*(0)B(t)\rangle_\omega . \quad (7,4.17)$$

When $\langle A^*(0)B(t)\rangle_\omega$ is independent of the angle variables, we have from (7,4.16)

$$\langle A^*(0)B(t)\rangle = \langle A^*(0)B(t)\rangle_\omega . \quad (7,4.18)$$

This situation arose, for example, in (5,3.8) which we express as

$$\langle \omega_i(0)\,\omega_j(t)\rangle = \frac{\delta_{ij}kT}{I_i}e^{-B_i|t|}. \quad (7,4.19)$$

In establishing this equation we were concerned only with the fluctuation of the angular velocity and not with configuration space coordinates, whether angle or linear, and there was no external field, so the ensemble average in (7,4.19) was one defined as in (7,4.13). Since, however, the right-hand side of (7,4.19) does not depend on the configuration space coordinates, we may employ (7,4.18) to justify leaving (7,4.19) as it stands.

We shall now explain how the stochastic rotation operator may be introduced into the calculation of orientational correlation functions. Suppose, for example, that we wish to calculate the correlation function of Y_{jm} and $Y_{jm'}$, where the arguments of the spherical harmonics are the spherical polar angles of the third body axis referred to the laboratory frame. Thus we must calculate the ensemble average $\langle Y^*_{jm}(\beta(0),\ \alpha(0))\,Y_{jm'}(\beta(t),\ \alpha(t))\rangle$. From (7,4.8) and (7,4.17)

$$\begin{aligned}
&\langle Y^*_{jm}(\beta(0), \alpha(0))\, Y_{jm'}(\beta(t), \alpha(t))\rangle \\
&= \frac{1}{4\pi}\int_0^{2\pi} d\alpha(0) \int_0^{\pi} d\beta(0)\, \sin\beta(0)\, \times \\
&\quad \langle Y^*_{jm}(\beta(0), \alpha(0)) R^+(t)\, Y_{jm'}(\beta(0), \alpha(0))\rangle_\omega \\
&= \frac{1}{4\pi}\int_0^{2\pi} d\alpha(0) \int_0^{\pi} d\beta(0)\, \sin\beta(0)\, Y^*_{jm}(\beta(0), \alpha(0))\, \times \\
&\quad \langle R^+(t)\rangle_\omega\, Y_{jm'}(\beta(0), \alpha(0)),
\end{aligned}$$

since Y^*_{jm} and $Y_{jm'}$ do not involve the angular velocities. This is $1/(4\pi)$ times the mm'-element of the matrix representatives of $\langle R^+(t)\rangle_\omega$ with reference to the basis Y_{js}. As in eqn (5,4.18), this element may be written $\langle (R^+(t))^j_{mm'}\rangle_\omega$ and, since $(R^+(t))^j_{mm'} = R^j_{m'm}(t)^*$,

$$\langle Y^*_{jm}(\beta(0), \alpha(0)) Y_{jm'}(\beta(t), \alpha(t))\rangle = \frac{1}{4\pi}\langle R^j_{m'm}(t)\rangle^*_\omega. \qquad (7,4.20)$$

If we put $t = 0$ so that, by (7,4.4), $R(t)$ becomes the identity operator, we obtain

$$\langle Y^*_{jm}(\beta(0), \alpha(0))\, Y_{jm'}(\beta(0), \alpha(0))\rangle = \frac{\delta_{mm'}}{4\pi},$$

since we saw in Appendix A that the identity operator is represented by the unit matrix. Hence the normalized autocorrelation function of Y_{jm};

$$\frac{\langle Y^*_{jm}(\beta(0), \alpha(0))\, Y_{jm}(\beta(t), \alpha(t))\rangle}{\langle Y^*_{jm}(\beta(0), \alpha(0))\, Y_{jm}(\beta(0), \alpha(0))\rangle} = \langle R^j_{mm}(t)\rangle^*_\omega. \qquad (7,4.21)$$

This is the complex conjugate of the ensemble average over the angular velocity space of the diagonal mm-element of the matrix representative of $R(t)$, defined by (7,4.1), with respect to the basis consisting of Y_{js}. It will be found, in all the cases that we shall investigate, that $\langle R^j_{mm}(t)\rangle_\omega$ is real.

We may express the above results in terms of $D^j_{m'm}$. From the discussion of (7,3.14) we deduce that

$$R^j_{m'm}(t) = D^j_{m'm}(\alpha'(t), \beta'(t), \gamma'(t)). \qquad (7,4.22)$$

On substitution of (7,4.22) into (7,4.20) and (7,4.21) we see that

$$\langle Y^*_{jm}(\beta(0), \alpha(0)) Y_{jm'}(\beta(t), \alpha(t))\rangle = \frac{1}{4\pi}\langle D^j_{m'm}(\alpha'(t), \beta'(t), \gamma'(t))\rangle^*_\omega, \qquad (7,4.23)$$

$$\frac{\langle Y^*_{jm}(\beta(0), \alpha(0)) Y_{jm}(\beta(t), \alpha(t))\rangle}{\langle Y^*_{jm}(\beta(0), \alpha(0)) Y_{jm}(\beta(0), \alpha(0))\rangle} = \langle D^j_{mm}(\alpha'(t), \beta'(t), \gamma'(t))\rangle^*_\omega . \tag{7,4.24}$$

The rotation operator may be employed to derive orientational correlation functions for quantities other than spherical harmonics, but the $R(t)$ will no longer be defined by (7,4.1). Examples of this will be found in Sections 8.3 and 12.5.

8

Introductory Calculations for Rotational Brownian Motion

8.1 Methods of Calculating Correlation Functions

We are now in a position to begin the part of the programme proposed in Section 3.1 concerned with the calculation of correlation functions for the rotational motion of Brownian particles, when the effects of their inertia are included. These correlation functions are both angular velocity and orientational ones. The orientational correlation functions in which we are interested are $\langle(\mathbf{n}(0)\cdot\mathbf{n}(t))\rangle$, where $\mathbf{n}$ is a unit vector in a direction fixed in the body frame of the Brownian particle, and, for the case of three-dimensional motion, the correlation functions of the spherical harmonics Y_{jm}.

As we saw in Chapter 7, there are two approaches to the solution of problems in rotational Brownian motion: we can start with Euler–Langevin stochastic differential equations or with a partial differential Fokker–Planck equation. In the first approach we solve (7,1.1) and from the solutions deduce angular velocity correlation functions. The stochastic rotation operator $R(t)$ for the problem in question satisfies the stochastic differential equation (7,3.8). To find the ensemble average of the stochastic rotation operator taken over the angular velocity variables $\langle R(t)\rangle_\omega$, which was discussed in Section 7.4, we employ the averaging method of Section 5.4 and eqn (5,3.8). Then the orientational correlation function may be deduced from $\langle R(t)\rangle_\omega$ by the method illustrated in Section 7.4 for spherical harmonics and applicable to other quantities when the appropriate expression is taken for the rotation operator.

Since solving a Fokker–Planck equation usually presents considerable mathematical difficulties and since the main application of the theory that we have in mind is to dielectric relaxation phenomena, we shall use the Fokker–Planck equation only to calculate polarizabilities for polar molecules in the linear approximation. We shall then connect these results with the correlation function $\langle\langle \mathbf{n}(0) \cdot \mathbf{n}(t)\rangle\rangle$ by the Kubo relation.

The stochastic differential equation method has several advantages over the Fokker–Planck equation method, as we present it. It is applicable to nonpolar as well as polar Brownian particles, it may be used to study Brownian particles of any shape, and it leads not only to the correlation function $\langle\langle \mathbf{n}(0) \cdot \mathbf{n}(t)\rangle\rangle$ but also to those for spherical harmonics. On the other hand the Fokker–Planck equation method allows polarizabilities to be calculated to a higher order of approximation than that at present attainable by the stochastic differential equation method.

As an introduction to other more elaborate calculations we first investigate the case of a rotating disc which, as explained in Section 7.1, is intended to designate a body which rotates about an axis that points in a fixed direction. When applied to a polar molecule, the dipole axis rotates in a plane perpendicular to the axis of rotation. This model of a polar molecule is of no great physical importance but it is useful in illustrating mathematical methods that will be applied later. Also it is of interest as being the only case in which we shall obtain a closed expression for an orientational correlation function. Following historical sequence we first study the disc by means of a Fokker–Planck equation.

8.2 Fokker–Planck Equation Study of the Rotating Disc

The problem of the rotating disc model of a polar molecule was studied in some detail by Sack (1953, 1957a) and by Gross (1955a, 1955b). The response of a dilute solution of polar molecules in a nonpolar compressed gas to an alternating field was examined by Gross (1955b). He assumed that the duration of a collision of a polar molecule with a gas molecule is negligible compared with the free rotation time τ_R defined in (1,2.10), the friction time τ_F defined in (1,2.9) and the period of the alternating field, and he argued from Gross (1955a) that these assumptions imply that the angular orientation of the molecule is unchanged by a collision but the angular velocity may change abruptly. He proposed several collision models, including the Brownian motion model. Since Sack's papers contain everything about Brownian motion that is of interest in Gross (1955b) and since he takes the calculations further, we shall not enter into the details of the paper of Gross.

According to (6,3.7) the Fokker–Planck equation for a disc, upon which there acts a couple due to an electric field F, is

$$\frac{\partial w(\theta, \omega, t)}{\partial t} + \omega \frac{\partial w}{\partial \theta} - \frac{\mu F \sin \theta}{I} \frac{\partial w}{\partial \omega} = \frac{\zeta}{I} \frac{\partial}{\partial \omega} \left(\omega w + \frac{kT}{I} \frac{\partial w}{\partial \omega} \right), \tag{8,2.1}$$

where we have replaced $\dot{\theta}$ in (6,3.7) by ω. Sack (1957a) makes a sequence of transformations of (8,2.1) that provides a differential equation, which enables relaxation effects in dielectrics to be calculated in the linear approximation. He first makes a Fourier transformation in ω-space

$$\Phi(\theta, u, t) = \int_{-\infty}^{\infty} w(\theta, \omega, t) e^{-iu\omega} \, d\omega, \tag{8,2.2}$$

which gives

$$\frac{\partial \Phi}{\partial t} + i \frac{\partial^2 \Phi}{\partial \theta \, \partial u} - \frac{i \mu F \sin \theta}{I} u \Phi = -\frac{\zeta}{I} \left(u \frac{\partial \Phi}{\partial u} + \frac{kT}{I} u^2 \Phi \right). \tag{8,2.3}$$

According to (8,2.2) and (3,2.16) the transform of the steady state distribution of (6,2.9), constant $\exp\{(-\frac{1}{2} I \omega^2 + \mu F \cos \theta)/(kT)\}$, is Φ_0 given by

$$\Phi_0 = \text{constant} \exp\left(\frac{\mu F \cos \theta}{kT} \right) \exp\left(-\frac{kTu^2}{2I} \right).$$

Then putting

$$\Psi(\theta, u, t) = \exp\left(\frac{kTu^2}{2I} \right) \Phi(\theta, u, t) \tag{8,2.4}$$

one deduces from (8,2.3)

$$\frac{\partial \Psi(\theta, u, t)}{\partial t} + i \frac{\partial^2 \Psi}{\partial \theta \, \partial u} - \frac{iu}{I} \left(kT \frac{\partial \Psi}{\partial \theta} + \mu F \sin \theta \Psi \right) = -\frac{\zeta u}{I} \frac{\partial \Psi}{\partial u}, \tag{8,2.5}$$

which is the above-mentioned differential equation.

We see from (1,3.1) that the polarization

$$P(t) = N\mu \int_0^\infty d\omega \int_0^{2\pi} w(\theta, \omega, t) \cos\theta d\theta \qquad (8,2.6)$$

$$= N\mu \lim_{u\to 0} \int_{-\infty}^{2\pi} \cos\theta d\theta \Phi(\theta, u, t),$$

by (8,2.2), and hence that

$$P(t) = N\mu \lim_{u\to 0} \int_0^{2\pi} \Psi(\theta, u, t) \cos\theta d\theta, \qquad (8,2.7)$$

by (8,2.4). At this stage we restrict our calculations to the linear approximation, in which terms of higher order than the first in $\mu F/(kT)$ are neglected in the expansion of $\exp[\mu F \cos\theta/(kT)]$. Then the steady state distribution for a time independent F,

$$w_0 = \left(\frac{I}{8\pi^3 kT}\right)^{1/2} \left(1 + \frac{\mu F \cos\theta}{kT}\right) \exp\left(-\frac{I\omega^2}{2kT}\right), \qquad (8,2.8)$$

the constant multiplier being chosen such that $\int_{-\infty}^{\infty} d\omega \int_0^{2\pi} d\theta w_0 = 1$. From (8,2.6) and (8,2.8) we deduce for a constant F the polarization

$$P(0) = \frac{N\mu^2 F}{2kT}, \qquad (8,2.9)$$

where we adopt the notation of Section 1.3 for polarization due to a constant field. Then from (8,2.2), (8,2.4), (8,2.8) and (3,2.16) the function Ψ corresponding to w_0,

$$\Psi_0 = \frac{1}{2\pi}\left(1 + \frac{\mu F \cos\theta}{kT}\right). \qquad (8,2.10)$$

On account of the periodicity in θ of w, and consequently of Φ and Ψ, it is allowable to make the expansion

$$\Psi(\theta, u, t) = \sum_{m=-\infty}^{\infty} e^{im\theta} \psi_m(u, t). \qquad (8,2.11)$$

Substituting into (8,2.7) we have

$$P(t) = N\mu \lim_{u\to 0} \int_0^{2\pi} \cos\theta d\theta \sum_{m=-\infty}^{\infty} e^{im\theta} \psi_m(u, t).$$

On expressing $e^{im\theta}$ as $\cos m\theta + i \sin m\theta$ it becomes clear that there will be a non-vanishing contribution only from $m = 1$ and $m = -1$, and that consequently

$$P(t) = N\mu \lim_{u \to 0} \int_0^{2\pi} [\psi_1(u, t) + \psi_{-1}(u, t)] \cos^2 \theta d\theta$$

$$= \pi N\mu[\psi_1(0, t) + \psi_{-1}(0, t)]. \tag{8,2.12}$$

Moreover for the steady state case we have from (8,2.10) and (8,2.11)

$$\sum_{m=-\infty}^{\infty} e^{im\theta} \psi_m(u, t) = \Psi_0 = \frac{1}{2\pi}\left[1 + \frac{\mu F(e^{i\theta} + e^{-i\theta})}{2kT}\right],$$

so that

$$\psi_1(u, t) = \psi_{-1}(u, t) = \frac{\mu F}{4\pi kT}. \tag{8,2.13}$$

We now calculate in linear approximation the polarization $P_a(t)$ for the after effect when a constant field F, which had been operating for a long time, is switched off at time zero, the system of polar molecules being then allowed to relax. On substitution from (8,2.11) into (8,2.5) with $F = 0$ we obtain for $m = 1$ and $m = -1$

$$\frac{\partial \psi_1(u, t)}{\partial t} - \frac{\partial \psi_1(u, t)}{\partial u} + \frac{kTu}{I} \psi_1(u, t) + \frac{\zeta u}{I} \frac{\partial \psi_1(u, t)}{\partial u} = 0, \tag{8,2.14}$$

$$\frac{\partial \psi_{-1}(u, t)}{\partial t} + \frac{\partial \psi_{-1}(u, t)}{\partial u} - \frac{kTu}{I} \psi_{-1}(u, t) + \frac{\zeta u}{I} \frac{\partial \psi_{-1}(u, t)}{\partial u} = 0. \tag{8,2.15}$$

These equations have to be solved for $\psi_1(u, t)$ and $\psi_{-1}(u, t)$ subject to the initial conditions

$$\psi_1(u, 0) = \psi_{-1}(u, 0) = \frac{\mu F}{4\pi kT} \tag{8,2.16}$$

obtained from (8,2.13), since at $t = 0$ we have a steady state due to F. When this has been done, we substitute from (8,2.12) to get $P_a(t)$ from

$$P_a(t) = \pi N\mu(\psi_1(0, t) + \psi_{-1}(0, t)). \tag{8,2.17}$$

Now on comparing (8,2.15) with (8,2.14) we see that $\psi_{-1}(u, t)$ and $\psi_1(-u, t)$ obey the same equation. Moreover we see from (8,2.16) that for $u = 0$, which is all that interests us, the functions $\psi_{-1}(u, t)$ and $\psi_1(-u, t)$ are equal for $t = 0$. They are therefore equal at all times for $u = 0$, and this allows us to change (8,2.17) to

$$P_a(t) = 2\pi N\mu\psi_1(0, t). \tag{8,2.18}$$

Thus we have to solve (8,2.14). Guided by (8,2.16) we put

$$\psi_1(u, t) = \frac{\mu F \exp\{Z(u, t)\}}{4\pi kT}, \tag{8,2.19}$$

where $Z(u, 0) = 0$. On substitution into (8,2.14) we deduce

$$\frac{\partial Z}{\partial t} + \frac{\partial Z}{\partial u}\left(\frac{\zeta u}{I} - 1\right) = -\frac{kT}{I}u.$$

Using standard methods of solving partial differential equations of the first order (Duff, 1956, Chap. 2) we find that

$$Z(u, t) = -\frac{kT}{\zeta}(u + t) + f\left\{\left(\frac{\zeta u}{I} - 1\right) \exp\left[-(\zeta/I)t\right]\right\},$$

where f is an arbitrary function. This is determined by the condition $Z(u, 0) = 0$, and the result is

$$Z(u, t) = -\frac{kT}{\zeta}(u + t) + \frac{kTI}{\zeta^2}\left\{1 + \left(\frac{\zeta u}{I} - 1\right) \exp\left[-(\zeta/I)t\right]\right\}.$$

On substitution into (8,2.19) and (8,2.18) we obtain

$$P_a(t) = \frac{N\mu^2 F}{2kT} \exp\left\{\frac{kTI}{\zeta^2}\left(1 - \frac{\zeta}{I}t - \exp\left[-(\zeta/I)t\right]\right)\right\}. \tag{8,2.20}$$

From (1,2.9) and (1,2.10)

$$\left(\frac{\tau_F}{\tau_R}\right)^2 = \frac{kTI}{\zeta^2}. \tag{8,2.21}$$

This dimensionless quantity will appear very frequently, and we write

$$\gamma = \left(\frac{\tau_F}{\tau_R}\right)^2. \tag{8,2.22}$$

Putting $\zeta = IB$, as in Section 5.2, we express (8,2.20) as

$$P_a(t) = \frac{N\mu^2 F}{2kT} \exp\{-\gamma(Bt - 1 + e^{-Bt})\}.$$

Hence from (8,2.9)

$$\frac{P_a(t)}{P(0)} = \exp\{-\gamma(Bt - 1 + e^{-Bt})\}. \tag{8,2.23}$$

Equations (2,1.14) and (2,1.15) then yield for the after-effect function

$$b(t) = \alpha_s \exp\{-\gamma(Bt - 1 + e^{-Bt})\}. \tag{8,2.24}$$

Since only orientational polarization is now being considered, we have from (2,2.24)

$$\langle(\mathbf{n}(0) \cdot \mathbf{n}(t))\rangle = \exp\{-\gamma(Bt - 1 + e^{-Bt})\}, \tag{8,2.25}$$

where $\mathbf{n}(t)$ is the unit vector in the direction of the dipole axis of the disc. As explained in Sections 4.4 and 7.4, the ensemble average is over both angle and angular velocity variables, and the normalized autocorrelation function of $\mathbf{n}(t)$ is the same as its correlation function $\langle(\mathbf{n}(0) \cdot \mathbf{n}(t))\rangle$. The result (8,2.25) was derived by Saito and Kato (1956) who used a Lagrangian formalism. As we see from (5,1.4), the expression $Bt - 1 + e^{-Bt}$ occurred in the Uhlenbeck and Ornstein theory.

We deduce from (2,1.11) and (8,2.24) that

$$\frac{\alpha(\omega)}{\alpha_s} = 1 - i\omega \int_0^\infty \exp\{-\gamma(Bt - 1 + e^{-Bt})\} e^{-i\omega t} dt. \tag{8,2.26}$$

It should be remembered that in this equation ω designates not the angular velocity $\dot{\theta}$ of the disc but 2π times the frequency of the periodic electric field that produces the response investigated in Chapter 2. If the exponent in (8,2.26) is expanded in the powers of e^{-Bt} and the integration is performed term-by-term, it will be found that

$$\frac{\alpha(\omega)}{\alpha_s} = 1 - \frac{i\omega\tau}{1 + i\omega\tau} \times$$

$$\left[1 + \frac{\gamma}{1 + \gamma + i\omega\tau_F} + \frac{\gamma^2}{(1 + \gamma + i\omega\tau_F)(2 + \gamma + i\omega\tau_F)} + \ldots\right], \tag{8,2.27}$$

where τ is the relaxation time $\zeta/(kT)$ which appears in the exponential of eqn (1,1.14) for the two-dimensional Debye theory. We obtain the Debye result (1,3.12), if we put γ equal to zero in (8,2.27). According to (8,2.21) and (8,2.22) the value of γ becomes very small when the frictional constant ζ becomes very large. We shall refer to this situation as the *overdamped case.*

8.3 Stochastic Differential Equation Study of the Rotating Disc

We take a unit vector $\mathbf{n}(t)$ fixed in the disc and perpendicular to the axis of rotation. If the disc is a model of a polar molecule, we take $\mathbf{n}(t)$ along the dipole axis. We also take the rectangular coordinate axes in the plane of $\mathbf{n}(t)$ and rotating with it. We consider these coordinate axes at time t and at time zero, and we regard the latter as the laboratory frame. Let $n_1(t), n_2(t)$ be the components of $\mathbf{n}(t)$ in the rotating frame. They are, of course, constant. Let $n_1(0)$, $n_2(0)$ be the components of the same vector in the laboratory frame. If the coordinate axes have rotated through an angle $\theta(t)$ between time zero and time t,

$$n_1(t) = n_1(0)\cos\theta(t) + n_2(0)\sin\theta(t),$$

$$n_2(t) = -n_1(0)\sin\theta(t) + n_2(0)\cos\theta(t).$$

We write these equations

$$\begin{bmatrix} n_1(t) \\ n_2(t) \end{bmatrix} = R(t)\begin{bmatrix} n_1(0) \\ n_2(0) \end{bmatrix}, \tag{8,3.1}$$

where

$$R(t) = \begin{bmatrix} \cos\theta(t) & \sin\theta(t) \\ -\sin\theta(t) & \cos\theta(t) \end{bmatrix}. \tag{8,3.2}$$

In the language of Section 7.3, the matrix $R(t)$ is the rotation operator for the rotation through angle $\theta(t)$ and the vector $\begin{bmatrix} n_1 \\ n_2 \end{bmatrix}$. According to Appendix A the adjoint operator is the Hermitian conjugate of the matrix in (8,3.2):

$$R^+(t) = \begin{bmatrix} \cos\theta(t) & -\sin\theta(t) \\ \sin\theta(t) & \cos\theta(t) \end{bmatrix}.$$

We see immediately that $R^{+}(t)R(t) = R(t)R^{+}(t) = \mathbf{I}$, and we deduce from (8,3.1) that

$$\begin{bmatrix} n_1(0) \\ n_2(0) \end{bmatrix} = R^{+}(t)\begin{bmatrix} n_1(t) \\ n_2(t) \end{bmatrix}. \tag{8,3.3}$$

Since $\dot{\theta}(t) = \omega(t)$, the angular velocity of the dipole,

$$\frac{dR(t)}{dt} = \begin{bmatrix} -\sin\theta(t) & \cos\theta(t) \\ -\cos\theta(t) & -\sin\theta(t) \end{bmatrix} \omega(t)$$

$$= \begin{bmatrix} 0 & 1 \\ -1 & 0 \end{bmatrix} R(t)\omega(t).$$

We write this

$$\frac{dR(t)}{dt} = -iJ\omega(t)R(t), \tag{8,3.4}$$

where

$$J = \begin{bmatrix} 0 & i \\ -i & 0 \end{bmatrix}. \tag{8,3.5}$$

We next consider the angular velocity $\omega(t)$. The Euler–Langevin equations (7,1.1) reduce for the present problem to the single equation

$$\frac{d\omega(t)}{dt} = -B\omega(t) + \frac{dW(t)}{dt}. \tag{8,3.6}$$

According to (5,2.20) the steady state solution of this equation is a centred Gaussian random variable satisfying

$$\langle\omega(t_1)\omega(t_2)\rangle = \frac{kT}{I} e^{-B|t_1 - t_2|}, \tag{8,3.7}$$

where the left-hand side is an ensemble average over the spaces of both θ and ω, as was explained in the discussion of (7,4.19). According to (8,2.21) and (8,2.22)

$$\gamma = \frac{kT}{IB^2}. \tag{8,3.8}$$

This is a dimensionless quantity whose experimental value does not seem to exceed more than a few per cent (Herzfeld, 1964; Leroy *et al.*, 1967). We shall often employ it as an expansion parameter.

Let us now apply the averaging method of Section 5.4 to deduce the value of $\langle R(t)\rangle$ from (8,3.4). This equation is obtainable from (5,4.1) by making the substitutions

$$x \to R(t), \quad \epsilon O(t) \to -i\omega(t)J. \tag{8,3.9}$$

In the second substitution

$$\epsilon = \left(\frac{kT}{IB^2}\right)^{1/2} \tag{8,3.10}$$

that is, $\gamma^{1/2}$. The parameter ϵ being a constant is independent of the stochastic operator $O(t)$, as was required in Section 5.4. Since $\omega(t)$ is centred and Gaussian and since J obviously commutes with itself, $O(t)$ has the properties (i) and (ii) of Section 5.4. We may therefore write (5,4.4) as

$$\frac{d\langle R(t_1)\rangle}{dt_1} = \epsilon^2 \Omega^{(2)}(t_1)\langle R(t_1)\rangle + \epsilon^4 \Omega^{(4)}(t_1)\langle R(t_1)\rangle + \ldots, \tag{8,3.11}$$

where $\epsilon^2 \Omega^{(2)}(t_1)$, $\epsilon^4 \Omega^{(4)}(t_1)$ are given by (5,4.15), (5,4.16) and (8,3.9). The solution of (8,3.11) will thus be expressible as a series in powers of γ. Equation (8,3.5) shows that J^2 is equal to the identity $\mathbf{I}$ and hence we deduce from (8,3.7) that

$$\begin{aligned}\epsilon^2 \Omega^{(2)}(t_1) &= -\int_0^{t_1} \langle \omega(t_1)\omega(t_2)\rangle dt_2 \mathbf{I} \\ &= -\frac{kT}{I}\int_0^{t_1} e^{-B(t_1-t_2)} dt_2 \mathbf{I},\end{aligned}$$

that is,

$$\epsilon^2 \Omega^{(2)}(t_1) = -B\gamma(1 - e^{-Bt_1})\mathbf{I}. \tag{8,3.12}$$

Then from (5,4.16)

$$\begin{aligned}\epsilon^4 \Omega^{(4)}(t_1) = \int_0^{t_1} dt_2 \int_0^{t_2} dt_3 \int_0^{t_3} dt_4 [&\langle \omega(t_1)\omega(t_2)\omega(t_3)\omega(t_4)\rangle \\ &- \langle \omega(t_1)\omega(t_2)\rangle\langle \omega(t_3)\omega(t_4)\rangle \\ &- \langle \omega(t_1)\omega(t_3)\rangle\langle \omega(t_2)\omega(t_4)\rangle \\ &- \langle \omega(t_1)\omega(t_4)\rangle\langle \omega(t_2)\omega(t_3)\rangle],\end{aligned}$$

which vanishes by (3,4.18). It is easily seen that $\epsilon^6 \Omega^{(6)}(t_1)$ also vanishes, and so, by (8,3.12), eqn (8,3.11) reduces to

$$\frac{d\langle R(t_1)\rangle}{dt_1} = -B\gamma(1 - e^{-Bt_1})\langle R(t_1)\rangle. \qquad (8,3.13)$$

This equation has to be solved with the initial condition

$$\langle R(0)\rangle = \langle \mathbf{I}\rangle = \mathbf{I},$$

the identity operator which in the present case is the two-dimensional unit matrix **I**. The solution is

$$\langle R(t)\rangle = \exp\left\{-B\gamma\int_0^t (1 - e^{-Bt_1})dt_1\right\}\mathbf{I},$$

that is,

$$\langle R(t)\rangle = \exp\{-\gamma[Bt - 1 + e^{-Bt}]\}\mathbf{I}. \qquad (8,3.14)$$

An alternative expression for $\langle R(t)\rangle$ will be derived in Section 12.2.

We employ (8,3.14) to calculate $\langle(\mathbf{n}(0)\cdot\mathbf{n}(t))\rangle$. If $\mathbf{n}(t)$ is the column vector with the constant elements $n_1(t)$, $n_2(t)$ and $n^{\mathrm{T}}(t)$ is the row vector with these elements, we have

$$(\mathbf{n}(0)\cdot\mathbf{n}(t)) = n^{\mathrm{T}}(0)n(t)$$

and, from (8,3.3),

$$n(0) = R^{+}(t)n(t).$$

Since $(R^{+}(t))^{\mathrm{T}} = R(t)^{*}$ and since the elements of both $R(t)$ and $n(t)$ are real, eqn (A.4) gives

$$\begin{aligned}\langle(\mathbf{n}(0)\cdot\mathbf{n}(t))\rangle &= \langle n^{\mathrm{T}}(t)R(t)n(t)\rangle \\ &= n^{\mathrm{T}}(t)\langle R(t)\rangle n(t).\end{aligned}$$

because $n(t)$ is non-stochastic. Hence from (8,3.14)

$$\langle(\mathbf{n}(0)\cdot\mathbf{n}(t))\rangle = \exp\{-\gamma[Bt - 1 + e^{-Bt}]\}n^{\mathrm{T}}(t)\mathbf{I}n(t),$$

so that

$$\langle(\mathbf{n}(0)\cdot\mathbf{n}(t))\rangle = \exp\{-\gamma[Bt - 1 + e^{-Bt}]\}, \qquad (8,3.15)$$

which agrees with (8,2.25). A more elementary but much lengthier derivation of this result from the Langevin equation was given by Lewis *et al.* (1976). Equation (8,3.15) and the Kubo relation (2,2.23) lead to (8,2.26) for the complex polarizability. Equation (8,3.15) may also be written

$$\langle \cos\theta(t)\rangle = \exp\{-\gamma[Bt - 1 + e^{-Bt}]\}. \qquad (8,3.16)$$

Since $\theta(0) = 0$, the autocorrelation function of $\cos\theta(t)$,

$$\langle \cos\theta(0)\cos\theta(t)\rangle = \exp\{-\gamma[Bt - 1 + e^{-Bt}]\}, \qquad (8,3.17)$$

and the normalized autocorrelation function of $\cos\theta(t)$ has the same value.

Since the axis of rotation of the disc is not free to move in three dimensions, spherical harmonics do not enter into the discussion. However, it may be useful to know the value of $\langle\cos m\theta(t)\rangle$. This may be deduced from the above calculation by considering a vector that rotates with angular velocity $m\omega(t)$, where $\omega(t)$ is still given by the Langevin equation (8,3.6). On replacing $\omega(t)$ by $m\omega(t)\lambda$ in (8,3.9), eqn (8,3.12) becomes

$$\epsilon^2\Omega^{(2)}(t_1) = -m^2 B\gamma(1 - e^{-Bt_1})\mathbf{I},$$

and (8,3.14) becomes

$$\langle R(t)\rangle = \exp\{-m^2\gamma[Bt - 1 + e^{-Bt}]\}\mathbf{I}.$$

From this it is deduced that

$$\langle \cos m\theta(t)\rangle = \exp\{-m^2\gamma[Bt - 1 + e^{-Bt}]\}. \qquad (8,3.18)$$

Since for m an integer $\cos m\theta(t)$ is expressible as a polynomial in $\cos\theta(t)$, we may use (8,3.18) to find the higher moments $\langle\cos^2\theta(t)\rangle$, $\langle\cos^3\theta(t)\rangle$, etc., of $\cos\theta(t)$. Putting $m = 2$ we obtain

$$\langle \cos 2\theta(t)\rangle = \exp\{-4\gamma[Bt - 1 + e^{-Bt}]\},$$

$$\langle \cos^2\theta(t)\rangle = \tfrac{1}{2} + \tfrac{1}{2}\exp\{-4\gamma[Bt - 1 + e^{-Bt}]\}.$$

The variance $V(\cos\theta(t))$ of $\cos\theta(t)$, defined in Section 3.2 as $\langle\cos^2\theta(t)\rangle - \langle\cos\theta(t)\rangle^2$, is given by

$$V(\cos\theta(t)) = \tfrac{1}{2}(1 - \exp\{-2\gamma[Bt - 1 + e^{-Bt}]\})^2,$$

which agrees with the result of an independent calculation of Lewis *et al.* (1976).

In the limit $t \gg \tau_F$, i.e. $Bt \gg 1$, (8,3.16) becomes

$$\langle \cos\theta(t)\rangle = \exp(-\gamma Bt) = \exp\left(-\frac{t}{2\tau_D}\right),$$

where τ_D is defined in (1,2.7). Now for the Debye theory of the disc (1,3.10) and (1,3.11) give

$$\frac{P_a(t)}{P(0)} = \exp\left(-\frac{t}{2\tau_D}\right),$$

and hence by (2,1.14), (2,1.15) and (2,2.24)

$$\langle(\mathbf{n}(0)\cdot\mathbf{n}(t))\rangle = \exp\left(-\frac{t}{2\tau_D}\right).$$

Thus in the limit $t \gg \tau_F$ eqn (8,3.16) gives the Debye result. This shows that inertial effects are negligible in the long time limit. In this limit we also have, from (8,3.18),

$$\langle\cos m\theta(t)\rangle = \exp\left(-\frac{m^2 t}{2\tau_D}\right).$$

In the short time limit $t \ll \tau_F$ we expand the exponents of (8,3.16) and (8,3.18) in powers of Bt and obtain using (1,2.10)

$$\langle\cos\theta(t)\rangle = \exp\left(-\frac{t^2}{2\tau_R^2}\right),$$

$$\langle\cos m\theta(t)\rangle = \exp\left(-\frac{m^2 t^2}{2\tau_R^2}\right), \tag{8,3.19}$$

which are independent of τ_F. To see whether these agree with the mean values for free motion, which would be attained when τ_F becomes infinite, we average $\cos m\theta(t)$ over values of ω from $-\infty$ to ∞, using the one-dimensional normalized Maxwell–Boltzmann density function $I^{1/2}/(2\pi kT)^{1/2}\exp(-I\omega^2/(2kT))$. On employing (3,2.16) we deduce that, since $\theta(t) = \omega t$ for free motion,

$$\begin{aligned}\langle\cos m\theta(t)\rangle &= \langle\cos m\omega t\rangle \\ &= \left(\frac{I}{2\pi kT}\right)^{1/2}\int_{-\infty}^{\infty}\cos m\omega t\exp\left(-\frac{I\omega^2}{2kT}\right)d\omega \\ &= \exp\left(-\frac{m^2 t^2 kT}{2I}\right),\end{aligned}$$

which agrees with (8,3.19) when we substitute from (1,2.10).

The calculation of $\langle \cos \theta(t) \rangle$ performed in the present section is more direct than that deduced from the Fokker–Planck equation. Moreover, since the latter has been solved only in the linear approximation, the method of the previous section does not provide an expression for $\langle \cos m\theta(t) \rangle$.

9 Rotational Brownian Motion of the Sphere

9.1 Fokker–Planck Equation Study of the Rotating Sphere

We examine the rotational Brownian motion of a homogeneous sphere of radius a and moment of inertia I about a diameter, which is subject to an external electric field of intensity F. The sphere has a dipole of moment μ at its centre, and the dipole axis is free to rotate in three-dimensional space. Let this axis make angle $\beta(t)$ with the direction of F. We take rotating orthogonal coordinate axes through the centre of the sphere, not fixed in the sphere but as shown in Fig. 9.1, the axes being numbered 1, 2, 3. We describe the orientation of the sphere by Euler angles $\alpha(t)$, $\beta(t)$, $\gamma(t)$, as was explained in Section 7.3. The sphere is subject to a frictional couple ζ, or IB, times the angular velocity and to a random driving couple in no preferential direction $I(dW(t)/dt)$, where $W(t)$ is a Wiener process. In terms of the Euler angles the components $(\omega_\beta, \omega_\alpha, \omega_\gamma)$ of the angular velocity of the sphere in the directions 1, 2, 3, respectively, are given by

$$\omega_\beta = \dot{\beta}, \omega_\alpha = \dot{\alpha} \sin \beta, \omega_\gamma = \dot{\gamma} + \dot{\alpha} \cos \beta, \qquad (9,1.1)$$

and the corresponding components of angular momentum are $(I\omega_\beta, I\omega_\alpha, I\omega_\gamma)$. The components of angular velocity of the axes 1, 2, 3, are, respectively, $(\dot{\beta}, \dot{\alpha} \sin \beta, \dot{\alpha} \cos \beta)$, since for their rotation the angle γ is unaltered. As in Section 7.1, we obtain the rate of change of angular momentum from $\dot{\mathbf{V}} + (\boldsymbol{\omega} \times \mathbf{V})$, where $\mathbf{V}$ denotes angular momentum and $\boldsymbol{\omega}$ the angular velocity of the axes. Equating the rate of change of angular momentum to the sum of the moments due to the field F, the white noise driving couple and the frictional drag, we deduce the equations of motion of the sphere

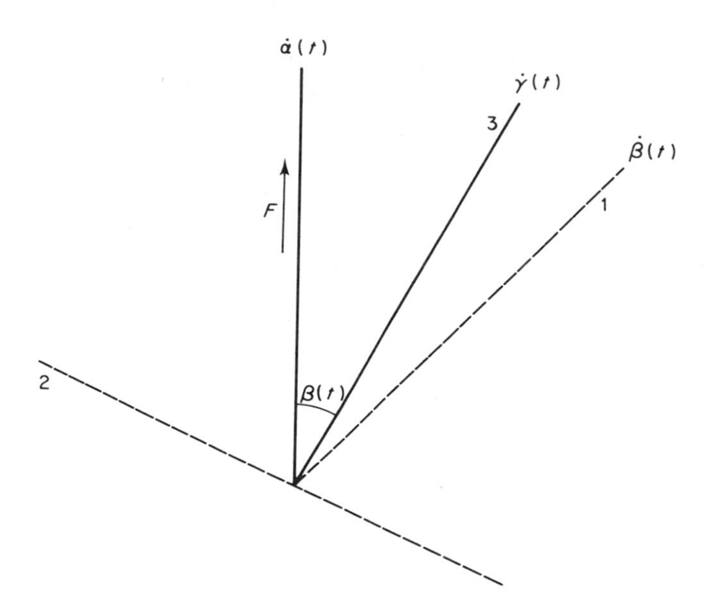

Fig. 9.1. The coordinate system for the rotating sphere.

$$I\ddot{\beta} + I\dot{\alpha}\dot{\gamma}\sin\beta = -\mu F\sin\beta - \zeta\dot{\beta} + I\,\frac{dW_1(t)}{dt},$$

$$I\,\frac{d(\dot{\alpha}\sin\beta)}{dt} - I\dot{\beta}\dot{\gamma} = -\zeta\dot{\alpha}\sin\beta + I\,\frac{dW_2(t)}{dt},$$

$$I\,\frac{d(\dot{\gamma} + \dot{\alpha}\cos\beta)}{dt} = -\zeta(\dot{\gamma} + \dot{\alpha}\cos\beta) + I\,\frac{dW_3(t)}{dt}.$$

We express these equations in terms of ω_β, ω_α, ω_γ by (9,1.1) as

$$d\omega_\beta = -\left[\omega_\alpha(\omega_\gamma - \omega_\alpha\cot\beta) + B\omega_\beta + \frac{\mu F\sin\beta}{I}\right]dt + dW_1(t),$$

$$d\omega_\alpha = [\omega_\beta(\omega_\gamma - \omega_\alpha\cot\beta) - B\omega_\alpha]\,dt + dW_2(t), \tag{9,1.2}$$

$$d\omega_\gamma = -B\omega_\gamma dt + dW_3(t).$$

From (9,1.1) we also have

$$d\beta = \omega_\beta dt,\ d\alpha = \frac{\omega_\alpha}{\sin\beta}\,dt,\ d\gamma = (\omega_\gamma - \omega_\alpha\cot\beta)\,dt. \tag{9,1.3}$$

Let us derive from (9,1.2) and (9,1.3) the Fokker–Planck equation for the probability density function w specified in terms of $\beta, \omega_\beta, \omega_\alpha, \omega_\gamma, t$, the function being independent of α and γ on account of the symmetry of the configuration. We write

$$\beta(t) = x_1, \alpha(t) = x_2, \gamma(t) = x_3, \omega_\beta = x_4, \omega_\alpha = x_5, \omega_\gamma = x_6,$$

and express (9,1.2) and (9,1.3) as

$$\begin{aligned} dx_1 &= \omega_\beta dt, dx_2 = \frac{\omega_\alpha}{\sin\beta} dt, dx_3 = (\omega_\gamma - \omega_\alpha \cot\beta)dt, \\ dx_4 &= -\left[\omega_\alpha(\omega_\gamma - \omega_\alpha \cot\beta) + B\omega_\beta + \frac{\mu F \sin\beta}{I}\right] dt + dW_1(t), \\ dx_5 &= [\omega_\beta(\omega_\gamma - \omega_\alpha \cot\beta) - B\omega_\alpha]\, dt + dW_2(t), \\ dx_6 &= -B\omega_\gamma dt + dW_3(t). \end{aligned} \tag{9,1.4}$$

On comparing (9,1.4) with (6,1.1) when $X(t_i)$ has been replaced by x_i we see that

$$\begin{aligned} f_1 &= \omega_\beta, \ f_2 = \frac{\omega_\alpha}{\sin\beta}, \ f_3 = \omega_\gamma - \omega_\alpha \cot\beta, \\ f_4 &= -\left[\omega_\alpha(\omega_\gamma - \omega_\alpha \cot\beta) + B\omega_\beta + \frac{\mu F \sin\beta}{I}\right], \\ f_5 &= \omega_\beta(\omega_\gamma - \omega_\alpha \cot\beta) - B\omega_\alpha, \ f_6 = -B\omega_\gamma, \\ G_{1l} &= G_{2l} = G_{3l} = 0, G_{4l} = \delta_{1l}, G_{5l} = \delta_{2l}, G_{6l} = \delta_{3l}, \end{aligned} \tag{9,1.5}$$

where δ_{il} is the Kronecker delta. On account of the spherical symmetry we use eqn (6,1.20). However the difficulty arises that the volume element is now $\sin\beta dx_1 dx_2 dx_3 dx_4 dx_5 dx_6$, whereas in the derivation of (6,1.20) it was assumed to be $dx_1 dx_2 \ldots dx_p$. This difficulty may be overcome by replacing the arbitrary function ϕ in the derivation of (6,1.20) by $\phi \sin\beta$. When we do this, we obtain in the present case

$$\sin\beta \frac{\partial w}{\partial t} + \sum_{i=1}^{6} \frac{\partial}{\partial x_i}(\sin\beta f_i w) - \tfrac{1}{2}\sigma^2 \sum_{i,j=4}^{6} \frac{\partial^2}{\partial x_i \partial x_j}[(GG^T)_{ij} \sin\beta w] = 0.$$

On substituting from (9,1.5) and remembering that w is independent of α and γ we get

$$\frac{\partial w}{\partial t} + \omega_\beta \frac{\partial w}{\partial \beta} + (\omega_\alpha \cot \beta - \omega_\gamma)\left(\omega_\alpha \frac{\partial w}{\partial \omega_\beta} - \omega_\beta \frac{\partial w}{\partial \omega_\alpha}\right)$$

$$- \frac{\mu F \sin \beta}{I} \frac{\partial w}{\partial \omega_\beta} - B\left(\omega_\beta \frac{\partial w}{\partial \omega_\beta} + \omega_\alpha \frac{\partial w}{\partial \omega_\alpha} + \omega_\gamma \frac{\partial w}{\partial \omega_\gamma}\right) \tag{9,1.6}$$

$$- 3Bw - \tfrac{1}{2}\sigma^2\left(\frac{\partial^2 w}{\partial \omega_\beta^2} + \frac{\partial^2 w}{\partial \omega_\alpha^2} + \frac{\partial^2 w}{\partial \omega_\gamma^2}\right) = 0.$$

As might be expected from (5,3.6) the value of σ^2 is $2BkT/I$. This value may also be found by substituting into (9,1.6) the Maxwell–Boltzmann probability density function for constant F:

$$w_0 = \text{constant} \exp \{[-\tfrac{1}{2} I(\omega_\beta^2 + \omega_\alpha^2 + \omega_\gamma^2) + \mu F \cos \beta] /(kT)\}. \tag{9,1.7}$$

On inserting the value of σ^2 into (9,1.6) we obtain the required Fokker–Planck equation for $w(\beta, \omega_\beta, \omega_\alpha, \omega_\gamma, t)$:

$$\frac{\partial w}{\partial t} + \omega_\beta \frac{\partial w}{\partial \beta} + (\omega_\alpha \cot \beta - \omega_\gamma)\left(\omega_\alpha \frac{\partial w}{\partial \omega_\beta} - \omega_\beta \frac{\partial w}{\partial \omega_\alpha}\right)$$

$$- \frac{\mu F \sin \beta}{I} \frac{\partial w}{\partial \omega_\beta} \tag{9,1.8}$$

$$= B \frac{\partial}{\partial \omega_\beta}\left[\omega_\beta w + \frac{kT}{I} \frac{\partial w}{\partial \omega_\beta}\right] + B \frac{\partial}{\partial \omega_\alpha}\left[\omega_\alpha w + \frac{kT}{I} \frac{\partial w}{\partial \omega_\alpha}\right]$$

$$+ B \frac{\partial}{\partial \omega_\gamma}\left[\omega_\gamma w + \frac{kT}{I} \frac{\partial w}{\partial \omega_\gamma}\right].$$

This is eqn (3.8) of Sack (1957b) combined with (2.3) of Sack (1957a).

We now proceed along the lines of the discussion of the rotating disk in Section 8.2. We first make the Fourier transformation in angular velocity space only

$$\Phi(\beta, u_\beta, u_\alpha, u_\gamma, t) \tag{9,1.9}$$
$$= \int_{-\infty}^{\infty}\int_{-\infty}^{\infty}\int_{-\infty}^{\infty} w(\beta, \omega_\beta, \omega_\alpha, \omega_\gamma, t) \exp\left(-i \sum_{l=1}^{3} u_l\omega_l\right) d\omega_1 d\omega_2 d\omega_3,$$

writing for convenience

$$\omega_\beta = \omega_1, \omega_\alpha = \omega_2, \omega_\gamma = \omega_3, u_\beta = u_1, u_\alpha = u_2, u_\gamma = u_3$$
$$\sum_{l=1}^{3} u_l^2 = u^2.$$

Equation (9,1.8) transforms to

$$\frac{\partial \Phi}{\partial t} + i \frac{\partial^2 \Phi}{\partial \beta \partial u_\beta} - i \left(\cot \beta \frac{\partial}{\partial u_\alpha} - \frac{\partial}{\partial u_\gamma}\right)\left(u_\beta \frac{\partial}{\partial u_\alpha} - u_\alpha \frac{\partial}{\partial u_\beta}\right)\Phi$$
$$- \frac{i\mu F}{I} \sin \beta u_\beta \Phi \tag{9,1.10}$$
$$= -B \sum_{l=1}^{3} u_l \frac{\partial \Phi}{\partial u_l} - \frac{BkTu^2}{I}\Phi.$$

The transform Φ_0 of w_0 in (9,1.7) is given by

$$\Phi_0(\beta, u_\beta, u_\alpha, u_\gamma) = \text{constant} \exp\left(\frac{\mu F \cos \beta}{kT}\right) \exp\left(-\frac{kTu^2}{2I}\right). \tag{9,1.11}$$

We multiply the transform Φ of w by $\exp(kTu^2/2I)$ writing

$$\Psi(\beta, u_\beta, u_\alpha, u_\gamma, t) = \exp\left(\frac{kTu^2}{2I}\right)\Phi(\beta, u_\beta, u_\alpha, u_\gamma, t), \tag{9,1.12}$$

and this leads to the equation

$$\frac{\partial \Psi}{\partial t} + i \frac{\partial^2 \Psi}{\partial \beta \partial u_\beta} - \frac{iu_\beta}{I}\left[kT \frac{\partial \Psi}{\partial \beta} + \mu F \sin \beta \Psi\right]$$
$$+ i\left[\cot \beta \frac{\partial}{\partial u_\alpha} - \frac{\partial}{\partial u_\gamma} - \frac{kT}{I}\left(\cot \beta u_\alpha - u_\gamma\right)\right]\left(u_\alpha \frac{\partial \Psi}{\partial u_\beta} - u_\beta \frac{\partial \Psi}{\partial u_\alpha}\right) \tag{9,1.13}$$

$$= -B \sum_{l=1}^{3} u_l \frac{\partial \Psi}{\partial u_l}.$$

This equation is more manageable than (9,1.8) for the examination of polarization effects in dielectrics.

Equation (1,3.1) applied to the space of the Euler angles and the angular velocities $\omega_\beta, \omega_\alpha, \omega_\gamma$ yields for the orientational polarization

$$P(t) = N\mu \int_0^{2\pi} d\gamma \int_0^{2\pi} d\alpha \times$$

$$\int_0^{\pi} \cos\beta \sin\beta d\beta \int_{-\infty}^{\infty} \int_{-\infty}^{\infty} \int_{-\infty}^{\infty} w(\beta, \omega_\beta, \omega_\alpha, \omega_\gamma, t) d\omega_\beta d\omega_\alpha d\omega_\gamma$$

$$= N\mu \lim_{u\to 0} \int \int \int \int \int \int w \cos\beta \sin\beta \times$$

$$\exp\left(-i \sum_{l=1}^{3} u_l \omega_l\right) d\omega_\beta d\omega_\alpha d\omega_\gamma d\beta d\alpha d\gamma$$

$$= N\mu \lim_{u\to 0} \int \int \int \Phi(\beta, u_\beta, u_\alpha, u_\gamma, t) \cos\beta \sin\beta d\beta d\alpha d\gamma,$$

by (9,1.9). Then by (9,1.12)

$$P(t) = N\mu \lim_{u\to 0} \int \int \int \Psi(\beta, u_\beta, u_\alpha, u_\gamma, t) \cos\beta \sin\beta d\beta d\alpha d\gamma. \tag{9,1.14}$$

When we know the solution of (9,1.13), we can find the polarization from (9,1.14).

If we choose the constant in (9,1.7) such that

$$\int_0^{2\pi} d\gamma \int_0^{2\pi} d\alpha \int_0^{\pi} \sin\beta d\beta \int_{-\infty}^{\infty} \int_{-\infty}^{\infty} \int_{-\infty}^{\infty} w_0 d\omega_\beta d\omega_\alpha d\omega_\gamma = 1,$$

we obtain

$$w_0 = \frac{1}{8\pi^2} \left(\frac{I}{2\pi kT}\right)^{3/2} \frac{\mu F \exp\left(-\dfrac{I\omega^2}{2kT}\right) \exp\left(\dfrac{\mu F \cos\beta}{kT}\right)}{kT \sinh\left(\dfrac{\mu F}{kT}\right)}.$$

From (9,1.9) and (9,1.12) we deduce that the corresponding value of Ψ, namely,

$$\Psi_0 = \frac{\mu F \exp\left(\dfrac{\mu F \cos\beta}{kT}\right)}{8\pi^2 kT \sinh\left(\dfrac{\mu F}{kT}\right)}.$$

At this stage we assume that $\mu F/(kT)$ is sufficiently small that we can take the linear approximation $1 + \mu F \cos\beta/(kT)$ for the exponential and $\mu F/(kT)$ for the sinh, and so write

$$8\pi^2 \Psi_0 = 1 + \frac{\mu F \cos\beta}{kT}. \tag{9,1.15}$$

To solve (9,1.13) in the linear approximation we try the expansion

$$\begin{aligned} 8\pi^2 \Psi = {} & 1 + \cos\beta \chi_1(u^2, t) + \sin\beta u_\beta \chi_2(u^2, t) \\ & + (\cos\beta u_\gamma^2 + \sin\beta u_\gamma u_\alpha)\chi_3(u^2, t) + \dots, \end{aligned} \tag{9,1.16}$$

where χ_1, χ_2, χ_3 are of order $\mu F/(kT)$ and terms of higher order are neglected. On substituting (9,1.16) into (9,1.13) and collecting terms proportional to

$$\cos\beta, \ \sin\beta u_\beta \quad \text{and} \quad \cos\beta u_\gamma^2 + \sin\beta u_\gamma u_\alpha$$

it is found that

$$\begin{aligned} & \frac{\partial \chi_1}{\partial t} + 2i\chi_2 + 2iz\frac{\partial \chi_2}{\partial z} - 2iz\chi_2 = -2Bz\frac{\partial \chi_1}{\partial z} \\ & \frac{\partial \chi_2}{\partial t} - \frac{ikT}{I}\frac{\partial \chi_1}{\partial z} + \frac{ikT}{I}\chi_1 - \frac{i\mu F}{I} + i\chi_3 \\ & = -2Bz\frac{\partial \chi_2}{\partial z} - B\chi_2 \\ & \frac{\partial \chi_3}{\partial t} - \frac{ikT}{I}\frac{\partial \chi_2}{\partial z} + \frac{ikT}{I}\chi_2 = -2Bz\frac{\partial \chi_3}{\partial z} - 2B\chi_3, \end{aligned} \tag{9,1.17}$$

where $z = \frac{1}{2}kTu^2/I$, a dimensionless variable. Details of these and of later calculations in this section were given by McConnell (1977). The solutions of (9,1.17) when substituted into (9,1.16) give the solution $\Psi(\beta, u_\beta, u_\alpha, u_\gamma, t)$ of (9,1.13) in the linear approximation.

For the steady state case of $F = F_0 e^{i\omega t}$ we put

$$\chi_1(u^2, t) = e^{i\omega t} \sum_{r=0}^{\infty} A_r z^r, \quad \chi_2(u^2, t) = e^{i\omega t} \sum_{r=0}^{\infty} B_r z^r, \tag{9,1.18}$$

$$\chi_3(u^2, t) = e^{i\omega t} \sum_{r=0}^{\infty} C_r z^r.$$

Substitution of (9,1.18) into (9,1.17) leads to recurrence relations for A_r, B_r, C_r. On eliminating A_r and C_r the relations give

$$\sum_l (L_{nl} + \gamma M_{nl}) B_l = \delta_{n0} \frac{i\mu F_0}{\zeta}$$

with

$$L_{nl} = (i\omega' + 2n + 1)\delta_{nl}, \tag{9,1.19}$$

$$M_{nl} = -\frac{2}{i\omega' + 2n}\delta_{n-1,l} + \left(\frac{2n+2}{i\omega' + 2n} + \frac{2n+3}{i\omega' + 2n + 2}\right)\delta_{nl}$$
$$- \frac{(n+1)(2n+5)}{i\omega' + 2n + 2}\delta_{n+1,l},$$

$$\omega' = \frac{\omega}{B}, \tag{9,1.20}$$

δ_{rs} the Kronecker delta and γ equal to $kT/(IB^2)$, as in (8,3.8). Regarding L_{nl} and M_{nl} as the nl-elements of matrices L and M, respectively, we deduce that

$$B_r = \frac{i\mu F_0}{\zeta}((L + \gamma M)^{-1})_{r0}, \tag{9,1.21}$$

$$C_r = \frac{\gamma\mu F_0}{(i\omega' + 2r + 2)I}\{-(r+1)((L+\gamma M)^{-1})_{r+1,0}$$
$$+ ((L + \gamma M)^{-1})_{r0}\}, \tag{9,1.22}$$

$$A_r = \frac{2\mu F_0 I}{(i\omega' + 2r)\zeta^2}\{(r+1)((L+\gamma M)^{-1})_{r0} - ((L+\gamma M)^{-1})_{r-1,0}\}, \tag{9,1.23}$$

and in particular

$$A_0 = \frac{2\mu F_0 I}{i\omega'\zeta^2}((L + \gamma M)^{-1})_{00}. \tag{9,1.24}$$

Equations (9,1.16), (9,1.18)–(9,1.24) provide in the linear approximation the value of $\Psi(\beta, u_\beta, u_\alpha, u_\gamma, t)$ in terms of the reciprocal of the matrix $L + \gamma M$, when $F = F_0 e^{i\omega t}$.

In studies of rotational Brownian motion, when correlation functions or other physical quantities cannot be expressed in closed form, they may often be expanded as power series in γ. We therefore try to expand A_r, B_r, C_r as such series up to a prescribed finite power s of γ. We may then truncate L and M to a finite order and so avoid convergence difficulties. In our approximation

$$(L + \gamma M)^{-1} = L^{-1} - \gamma L^{-1} M L^{-1} + \gamma^2 (L^{-1} M)^2 L^{-1} - \dots + (-)^s \gamma^s (L^{-1} M)^s L^{-1},$$

as we immediately verify on multiplying the right-hand side by $L + \gamma M$. Owing to the diagonal property of L^{-1}, the calculations are comparatively easy to perform.

According to (9,1.14) and (9,1.16) the polarization

$$P(t) = \tfrac{1}{3} N \mu \chi_1 (0, t). \tag{9,1.25}$$

If as in Section 1.3 we denote by $P(\omega) e^{i\omega t}$ the polarization due to $F_0 e^{i\omega t}$, we obtain from (9,1.18) and (9,1.25)

$$P(\omega) = \tfrac{1}{3} N \mu A_0.$$

On the other hand we deduce from (9,1.14) and (9,1.15) that the polarization $P(0)$ arising from the constant field F_0 is $\tfrac{1}{3} N \mu^2 F_0/(kT)$. Hence we have

$$\frac{\alpha(\omega)}{\alpha_s} = \frac{P(\omega)}{P(0)} = \frac{2\gamma B}{i\omega} ((L + \gamma M)^{-1})_{00}, \tag{9,1.26}$$

where we have used (9,1.24), (2,1.13), (2,1.14) and (8,3.8). The right-hand side of (9,1.26) may be expanded without much difficulty as a power series in γ as far as terms proportional to γ^6. To avoid a lengthy expression we give the result to the fourth power of γ:

$$\frac{\alpha(\omega)}{\alpha_s} = \frac{P(\omega)}{P(0)} = \frac{2\gamma}{i\omega'(i\omega' + 1)} - \frac{2\gamma^2}{i\omega'(i\omega' + 1)^2} \left\{ \frac{2}{i\omega'} + \frac{3}{i\omega' + 2} \right\}$$

$$+ \frac{2\gamma^3}{i\omega'(i\omega' + 1)^2} \left\{ \frac{4}{(i\omega')^2 (i\omega' + 1)} + \frac{12}{i\omega'(i\omega' + 1)(i\omega' + 2)} \right.$$

$$+ \frac{9}{(i\omega' + 1)(i\omega' + 2)^2} + \frac{10}{(i\omega' + 2)^2 (i\omega' + 3)}\Bigg\}$$

$$- \frac{2\gamma^4}{i\omega'(i\omega' + 1)^2} \Bigg\{ \frac{8}{(i\omega')^3 (i\omega' + 1)^2}$$

$$+ \frac{36}{(i\omega')^2 (i\omega' + 1)^2 (i\omega' + 2)} + \frac{54}{i\omega'(i\omega' + 1)^2 (i\omega' + 2)^2} \tag{9,1.27}$$

$$+ \frac{40}{i\omega'(i\omega' + 1)(i\omega' + 2)^2 (i\omega' + 3)} + \frac{27}{(i\omega' + 1)^2 (i\omega' + 2)^3}$$

$$+ \frac{60}{(i\omega' + 1)(i\omega' + 2)^3 (i\omega' + 3)} + \frac{40}{(i\omega' + 2)^3 (i\omega' + 3)^2}$$

$$+ \frac{50}{(i\omega' + 2)^2 (i\omega' + 3)^2 (i\omega' + 4)}\Bigg\} + \cdots .$$

In the overdamped case of very large ζ or B, both γ from (8,3.8) and ω' from (9,1.20) vanish but

$$\frac{\omega'}{\gamma} = \frac{IB\omega}{kT} = 2\omega\tau_D ,$$

where τ_D is the Debye relaxation time defined in (1,2.7). On writing out the matrix $L + \gamma M$ explicitly from (9,1.19) in this approximation we find that

$$((L + \gamma M)^{-1})_{00} = \frac{1}{1 + \dfrac{2\gamma}{i\omega'}} .$$

Then (9,1.26) gives $P(\omega)/P(0) = (1 + i\omega\tau_D)^{-1}$, which is the result (1,3.4) for the Debye theory of a spherical molecule when the dipole axis rotates in three-dimensional space.

Recurrence relations may be connected not only with matrices but also with continued fractions. Sack (1957b) takes for the expansion of Ψ not (9,1.16) but another equation, to which (9,1.16) is equivalent. He deduces equations analogous to (9,1.17), makes expansions analogous to (9,1.18) and obtains recurrence relations for the coefficients in the expansions. He solves these relations by a continued fraction method (Perron, 1913, pp. 290, 24), and the solution leads to

$$\frac{P(\omega)}{P(0)} = 1 - \frac{i\omega'|}{|i\omega'} + \frac{2\gamma|}{|1 + i\omega' + \frac{1}{2}\gamma(2 + i\omega')^{-1}} + \frac{(3-\frac{1}{2})\gamma|}{|2 + i\omega'} \tag{9,1.28}$$

$$+ \frac{4\gamma|}{|3 + i\omega' + \frac{1}{3}\gamma(2 + i\omega')^{-1}} + \frac{(5 - \frac{1}{3})\gamma|}{|4 + i\omega'} + \cdots.$$

The continued fraction may be expanded in powers of γ by taking successive convergents (Wall, 1948). When this is done as far as terms of order γ^4, there is exact agreement between (9,1.27) and (9,1.28).

9.2 Stochastic Differential Equation Study of the Rotating Sphere

We apply the Euler–Langevin equations (7,1.1) to the sphere putting $I_1 = I_2 = I_3 = I; B_1 = B_2 = B_3 = B = \zeta/I$, and obtaining

$$\frac{d\boldsymbol{\omega}(t)}{dt} = -B\boldsymbol{\omega}(t) + \frac{d\mathbf{W}(t)}{dt}. \tag{9,2.1}$$

This equation is referred to a body frame of coordinate axes labelled 1, 2, 3 through the centre of the sphere. According to (5,3.3) and (5,3.8) the steady state solution (9,2.1) is a centred Gaussian random variable, whose components obey the relation

$$\langle \omega_i(t_k)\omega_j(t_l)\rangle_\omega = \delta_{ij}\frac{kT}{I}e^{-B|t_k - t_l|}, \tag{9,2.2}$$

where, as we noted when discussing (7,4.19), the ensemble average was taken over the angular velocity space.

We saw in (7,4.20) that the correlation function for the spherical harmonics $Y_{jm}(\beta(t), \alpha(t))$, $Y_{jm'}(\beta(t), \alpha(t))$ is $1/(4\pi)\langle R^j_{m'm}(t)\rangle^*_\omega$, where $R(t)$ is the stochastic rotation operator that brings the body frame at time 0 to the body frame at time t, and $R^j_{m'm}(t)$ is the $m'm$-element of $R(t)$ in the representation with basis elements Y_{js}. The operator $R(t)$ for any rotation satisfies (7,3.8), namely

$$\frac{dR(t)}{dt} = -i(\mathbf{J}\cdot\boldsymbol{\omega}(t))R(t). \qquad (9,2.3)$$

To apply the method of Section 5.4 to solve this equation we make the substitutions

$$x(t)\to R(t),\ \epsilon O(t)\to -i(\mathbf{J}\cdot\boldsymbol{\omega}(t)) \qquad (9,2.4)$$

in (5,4.1). The stochastic properties of $O(t)$ reside in $\boldsymbol{\omega}(t)$. Since $\boldsymbol{\omega}(t)$ is a centred Gaussian random variable, $O(t)$ being a linear combination of $\omega_1(t)$, $\omega_2(t)$, $\omega_3(t)$ is a centred Gaussian stochastic operator. Then, since $O(t_1)O(t_2)\ldots O(t_{2n-1})$ is the sum of terms each of which contains an odd number of ω_i's, eqn (3,4.17) shows that $\langle O(t_1)O(t_2)\ldots O(t_{2n-1})\rangle$ vanishes. Hence $O(t)$ has the properties (i) and (ii) of Section 5.4, so that (5,4.4) becomes

$$\frac{d\langle R(t_1)\rangle}{dt_1} = (\epsilon^2\Omega^{(2)}(t_1)+\epsilon^4\Omega^{(4)}(t_1)+\epsilon^6\Omega^{(6)}(t_1)+\epsilon^8\Omega^{(8)}(t_1)$$

$$+\ldots)\langle R(t_1)\rangle, \qquad (9,2.5)$$

where the angular brackets denote ensemble average over both angle and angular velocity variables for the steady state of the system in the absence of an external field. As in the case of the disc, ϵ is given by (8,3.10) and the solution of (9,2.5) is expressible as a series in powers of γ. From (5,4.15), (5,4.16) and (9,2.4)

$$\epsilon^2\Omega^{(2)}(t_1) = -\int_0^{t_1}\langle(\mathbf{J}\cdot\boldsymbol{\omega}(t_1))(\mathbf{J}\cdot\boldsymbol{\omega}(t_2))\rangle dt_2,$$

$$\epsilon^4\Omega^{(4)}(t_1) =$$

$$\int_0^{t_1}dt_2\int_0^{t_2}dt_3\int_0^{t_3}dt_4\{\langle(\mathbf{J}\cdot\boldsymbol{\omega}(t_1))(\mathbf{J}\cdot\boldsymbol{\omega}(t_2))(\mathbf{J}\cdot\boldsymbol{\omega}(t_3))(\mathbf{J}\cdot\boldsymbol{\omega}(t_4))\rangle$$

$$-\langle(\mathbf{J}\cdot\boldsymbol{\omega}(t_1))(\mathbf{J}\cdot\boldsymbol{\omega}(t_2))\rangle\langle(\mathbf{J}\cdot\boldsymbol{\omega}(t_3))(\mathbf{J}\cdot\boldsymbol{\omega}(t_4))\rangle$$

$$-\langle(\mathbf{J}\cdot\boldsymbol{\omega}(t_1))(\mathbf{J}\cdot\boldsymbol{\omega}(t_3))\rangle\langle(\mathbf{J}\cdot\boldsymbol{\omega}(t_2))(\mathbf{J}\cdot\boldsymbol{\omega}(t_4))\rangle$$

$$-\langle(\mathbf{J}\cdot\boldsymbol{\omega}(t_1))(\mathbf{J}\cdot\boldsymbol{\omega}(t_4))\rangle\langle(\mathbf{J}\cdot\boldsymbol{\omega}(t_2))(\mathbf{J}\cdot\boldsymbol{\omega}(t_3))\rangle\}.$$

Since the only stochastic variables in these two equations are the ω_i's, the ensemble averages are equal to those taken over the angular velocity space alone, and we employ (9,2.2) to calculate them. With an obvious notation we write for brevity

$$\epsilon^2 \Omega^{(2)}(t_1) = -\int_0^{t_1} \langle 12 \rangle dt_2, \tag{9,2.6}$$

$$\epsilon^4 \Omega^{(4)}(t_1) = \int_0^{t_1} dt_2 \int_0^{t_2} dt_3 \int_0^{t_3} dt_4 \{\langle 1234 \rangle - \langle 12 \rangle\langle 34 \rangle - \langle 13 \rangle\langle 24 \rangle - \langle 14 \rangle\langle 23 \rangle\}. \tag{9,2.7}$$

The presence of the Kronecker delta and the exponential time factor in (9,2.2) allows further simplifications to be made. We see that, since $t_1 \geqslant t_2$,

$$\langle 12 \rangle = (J_1^2 + J_2^2 + J_3^2)\langle \omega_1(t_1)\omega_1(t_2)\rangle = \frac{kT}{I} J^2 e^{-B(t_1 - t_2)}, \tag{9,2.8}$$

where we have written J^2 for $J_1^2 + J_2^2 + J_3^2$, and that

$$\begin{aligned}
\langle 1234 \rangle &= \sum_{k,l,m,n} \langle J_k \omega_k(t_1) J_l \omega_l(t_2) J_m \omega_m(t_3) J_n \omega_n(t_4)\rangle \\
&= \sum_{k,l,m,n} J_k J_l J_m J_n \langle \omega_k(t_1)\omega_l(t_2)\omega_m(t_3)\omega_n(t_4)\rangle \\
&= \sum_{k,l,m,n} J_k J_l J_m J_n \{\langle \omega_k(t_1)\omega_l(t_2)\rangle\langle \omega_m(t_3)\omega_n(t_4)\rangle \\
&\quad + \langle \omega_k(t_1)\omega_m(t_3)\rangle\langle \omega_l(t_2)\omega_n(t_4)\rangle \\
&\quad + \langle \omega_k(t_1)\omega_n(t_4)\rangle\langle \omega_l(t_2)\omega_m(t_3)\rangle\},
\end{aligned}$$

by (3,4.18), the summations over k, l, m, n being from 1 to 3. We write the last result symbolically as

$$\langle 1234 \rangle = \langle \overline{12}\,\overline{34} \rangle + \langle \overline{1\underline{23}}\underline{4} \rangle + \langle \overline{1\underline{23}4} \rangle, \tag{9,2.9}$$

ensemble averages being taken for the linked pairs in each bracket. We saw in Appendix B that J^2 is $j(j+1)$ multiplied by the identity operator. Hence, from (9,2.8), $\langle 12 \rangle$ and similarly $\langle 13 \rangle$, $\langle 23 \rangle$, etc., are multiples of **I** and therefore commuting operators. Thus

$$\begin{aligned}
\langle \overline{12}\,\overline{34} \rangle &= \sum_{k,m} J_k^2 J_m^2 \langle \omega_k(t_1)\omega_k(t_2)\rangle\langle \omega_m(t_3)\omega_m(t_4)\rangle \\
&= \langle 12 \rangle \langle 34 \rangle = \langle 34 \rangle \langle 12 \rangle
\end{aligned}$$

$$\langle \overline{\underline{1}23\underline{4}} \rangle = \sum_{k,l} \langle J_k \omega_k(t_1) J_l^2 \omega_l(t_2) \omega_l(t_3) J_k \omega_k(t_4) \rangle$$
$$= \langle 1 \langle 23 \rangle 4 \rangle = \langle 14 \rangle \langle 23 \rangle,$$

and from (9,2.9)

$$\langle 1234 \rangle = \langle 12 \rangle \langle 34 \rangle + \langle 14 \rangle \langle 23 \rangle + \langle \overline{1\underline{23}4} \rangle.$$

Equation (9,2.7) then gives

$$\epsilon^4 \Omega^{(4)}(t_1) = \int_0^{t_1} dt_2 \int_0^{t_2} dt_3 \int_0^{t_3} dt_4 \{\langle \overline{1\underline{23}4} \rangle - \langle 13 \rangle \langle 24 \rangle\}. \tag{9,2.10}$$

In calculating $\langle \overline{1\underline{23}4} \rangle$ we must keep the J-operators in proper sequence, and so

$$\langle \overline{1\underline{23}4} \rangle - \langle 13 \rangle \langle 24 \rangle$$
$$= \left(\frac{kT}{I}\right)^2 e^{-B(t_1 + t_2 - t_3 - t_4)} \left\{ \sum_{k,l} J_k J_l J_k J_l - (J^2)^2 \right\}. \tag{9,2.11}$$

To perform summations over J_n's, which we call J-sums, we put down from (7,4.2)

$$J_k J_l - J_l J_k = -i \sum_n \epsilon_{kln} J_n. \tag{9,2.12}$$

Thence

$$\sum_k J_k J_l J_k - J_l J^2 = -i \sum_{k,n} \epsilon_{kln} J_n J_k$$
$$= -i \sum_{k,n} \epsilon_{lnk} J_n J_k, \quad \text{by (7,4.3)},$$
$$= -J_l,$$

by (9,2.12), and therefore

$$\sum_k J_k J_l J_k = (J^2 - \mathbf{I}) J_l. \tag{9,2.13}$$

This gives

$$\sum_{k,l} J_k J_l J_k J_l - (J^2)^2 = -J^2, \tag{9,2.14}$$

which is needed for (9,2.11). From (9,2.12) and (9,2.13)

$$\begin{aligned}\sum_k J_k J_l J_m J_k &= \sum_k J_l J_k J_m J_k - i \sum_{k,n} \epsilon_{kln} J_n J_m J_k \\ &= (J^2 - \mathbf{I}) J_l J_m - i \sum_{k,n} \epsilon_{kln} J_n (J_k J_m - i \sum_r \epsilon_{mkr} J_r) \\ &= (J^2 - \mathbf{I}) J_l J_m - J_l J_m - \sum_{k,n,r} \epsilon_{kln} \epsilon_{mkr} J_n J_r .\end{aligned}$$

Now

$$\sum_k \epsilon_{kln} \epsilon_{mkr} = -\sum_k \epsilon_{kln} \epsilon_{kmr} = -(\delta_{lm} \delta_{nr} - \delta_{lr} \delta_{nm}),$$

so

$$\sum_{k,n,r} \epsilon_{kln} \epsilon_{mkr} J_n J_r = -\delta_{lm} J^2 + J_m J_l,$$

$$\sum_k J_k J_l J_m J_k = (J^2 - 2\mathbf{I}) J_l J_m - J_m J_l + \delta_{lm} J^2 . \quad (9,2.15)$$

We deduce from (9,2.6) and (9,2.8) that

$$\epsilon^2 \Omega^{(2)}(t_1) = -\frac{kT}{I} J^2 \int_0^{t_1} e^{-B(t_1 - t_2)} dt_2 , \quad (9,2.16)$$

and from (9,2.10), (9,2.11) and (9,2.14) that

$$\epsilon^4 \Omega^{(4)}(t_1) = -\left(\frac{kT}{I}\right)^2 J^2 \int_0^{t_1} dt_2 \int_0^{t_2} dt_3 \int_0^{t_3} dt_4 e^{-B(t_1 + t_2 - t_3 - t_4)} . \quad (9,2.17)$$

We next take $\epsilon^6 \Omega^{(6)}(t_1)$ from (5,4.17). Employing (3,4.18) again we can clearly extend (9,2.9) to

$$\begin{aligned}\langle 123456 \rangle = {} & \langle \overline{12}\,\overline{34}56 \rangle + \langle \overline{12}\,\overline{345}6 \rangle + \langle \overline{12}\,\overline{3456} \rangle \\ & + \langle \overline{1\underline{23}}\underline{4}56 \rangle + \langle \overline{1\underline{23}}\underline{45}6 \rangle + \langle \overline{1\underline{23}}\underline{456} \rangle \\ & + \langle \overline{1\underline{23}4}56 \rangle + \langle \overline{1\underline{234}}\underline{5}6 \rangle + \langle \overline{1\underline{234}}\underline{56} \rangle \quad (9,2.18) \\ & + \langle \overline{1\underline{23}45}6 \rangle + \langle \overline{1\underline{234}5}6 \rangle + \langle \overline{1\underline{2345}}\underline{6} \rangle \\ & + \langle \overline{1\underline{23}456} \rangle + \langle \overline{1\underline{234}56} \rangle + \langle \overline{1\underline{2345}6} \rangle .\end{aligned}$$

For simplicity we have omitted one horizontal line in each bracket on the understanding that the only two surviving integers are linked. Using (9,2.18) and (9,2.9) we find cancellations in (5,4.17) that leave

$$
\begin{aligned}
\epsilon^6 \Omega^{(6)}(t_1) = -\int_0^{t_1} dt_2 \ldots \int_0^{t_5} dt_6 [&\langle \overline{1\underline{234}}\underline{5}6\rangle - \langle 14\rangle \langle \overline{2356}\rangle \\
&- \langle \overline{124}5\rangle \langle 36\rangle - \langle \overline{134}6\rangle \langle 25\rangle + 2\langle 14\rangle \langle 25\rangle \langle 36\rangle \\
&+ \langle \overline{1\underline{23}}\underline{45}6\rangle - \langle 13\rangle \langle \overline{245}6\rangle - \langle \overline{123}5\rangle \langle 46\rangle \\
&+ \langle 13\rangle \langle 25\rangle \langle 46\rangle + \langle \overline{1\underline{2345}}\underline{6}\rangle - \langle \overline{134}5\rangle \langle 26\rangle \\
&- \langle \overline{124}6\rangle \langle 35\rangle + \langle 14\rangle \langle 35\rangle \langle 26\rangle \qquad (9,2.19) \\
&+ \langle \overline{1\underline{234}5}6\rangle - \langle 15\rangle \langle \overline{234}6\rangle - \langle \overline{135}6\rangle \langle 24\rangle \\
&+ \langle 15\rangle \langle 24\rangle \langle 36\rangle + \langle \overline{1\underline{234}56}\rangle - \langle 16\rangle \langle \overline{234}5\rangle].
\end{aligned}
$$

In (9,2.19) the time dependence is the same when the same integers are linked; for example, it is the same for each of the first five terms of the integrand. The J-sums depend not on the integers but on the way in which they are linked in the angular brackets; for example, the J-sum is the same for $\langle \overline{134}6\rangle$ and $\langle \overline{245}6\rangle$, and we denote it just by $\langle \overline{\ldots}\ldots\rangle$. On employing (9,2.12)–(9,2.15) we find that the J-sums correspond to the different linkings in the following manner:

$$
\begin{aligned}
&\langle \ldots\rangle \sim J^2 ; \quad \langle \overline{\ldots}\ldots\rangle \sim J^2(J^2 - \mathbf{I}), \\
&\langle \overline{.\underline{\ldots}}\underline{.}.\rangle \sim J^2(J^2 - \mathbf{I})(J^2 - 2\mathbf{I}); \quad \langle \overline{.\underline{..}}\underline{..}.\rangle \sim J^2(J^2 - \mathbf{I})^2, \\
&\langle \overline{.\underline{\ldots}}\underline{..}\rangle \sim J^2(J^2 - \mathbf{I})^2 ; \quad \langle \overline{.\underline{\ldots}..}\rangle \sim J^2(J^2 - \mathbf{I})^2 .
\end{aligned}
$$

It is then deduced that

$$
\begin{aligned}
\epsilon^6 \Omega^{(6)}(t_1) = -\left(\frac{kT}{I}\right)^3 J^2 \int_0^{t_1} dt_2 \ldots \int_0^{t_5} dt_6 \times \\
\{4 \exp [-B(t_1 + t_2 + t_3 - t_4 - t_5 - t_6)] \\
+ \exp [-B(t_1 + t_2 - t_3 + t_4 - t_5 - t_6)]\}. \\
(9,2.20)
\end{aligned}
$$

As we proceed to find $\Omega^8(t_1)$ the calculations become very unwieldy. However they are greatly reduced by a graphical method developed by Ford *et al.* (1976) and independently by Pomeau and Weber (1976). This gives

$$\epsilon^8 \Omega^{(8)}(t_1) = -\left(\frac{kT}{I}\right)^4 J^2 \int_0^{t_1} dt_2 \ldots \int_0^{t_7} dt_8 \times$$

$$\{\exp[-B(t_1+t_2-t_3+t_4-t_5+t_6-t_7-t_8)]$$
$$+ 8\exp[-B(t_1+t_2+t_3-t_4-t_5+t_6-t_7-t_8)] \qquad (9,2.21)$$
$$+ [-4J^2+16\mathbf{I}]\exp[-B(t_1+t_2+t_3-t_4+t_5-t_6-t_7-t_8)]$$
$$+ [-10J^2+38\mathbf{I}]\exp[-B(t_1+t_2+t_3+t_4-t_5-t_6-t_7-t_8)]\}.$$

We see that the dependence of $\Omega^8(t_1)$ on J_1, J_2, J_3 is not simply J^2 but something more complicated. However, it is a polynomial in J^2 and so is a multiple of the identity.

We return to the problem of solving (9,2.5) with the initial condition $\langle R(0)\rangle = \mathbf{I}$. Since the Ω's given by (9,2.16), (9,2.17), (9,2.20) and (9,2.21) are all multiples of the identity and therefore commuting operators, the solution is

$$\langle R(t)\rangle = \exp\left\{\int_0^t dt_1 [\epsilon^2\Omega^{(2)}(t_1) + \epsilon^4\Omega^{(4)}(t_1) + \epsilon^6\Omega^{(6)}(t_1) + \epsilon^8\Omega^{(8)}(t_1) + \ldots]\right\}. \qquad (9,2.22)$$

From eqn (D.1) of Appendix D

$$\int_0^t \epsilon^2\Omega^{(2)}(t_1)dt_1 = -J^2 \frac{kT}{I} I^{(2)}(t), \qquad (9,2.23)$$

$$\int_0^t \epsilon^4\Omega^{(4)}(t_1)dt_1 = -J^2 \left(\frac{kT}{I}\right)^2 I_2^{(4)}(t), \qquad (9,2.24)$$

$$\int_0^t \epsilon^6\Omega^{(6)}(t_1)dt_1 = -J^2 \left(\frac{kT}{I}\right)^3 (I_3^{(6)}(t) + 4I_4^{(6)}(t)), \qquad (9,2.25)$$

$$\int_0^t \epsilon^8\Omega^{(8)}(t_1)dt_1 = -J^2 \left(\frac{kT}{I}\right)^4 (I_4^{(8)}(t) + 8I_5^{(8)}(t) + [-4J^2+16\mathbf{I}]I_7^{(8)}(t) + [-10J^2+38\mathbf{I}]I_8^{(8)}(t)), \qquad (9,2.26)$$

where the values of the $I_i^{(2r)}(t)$ which occur here and later are given in (D.2). We may note in particular that

$$I^{(2)}(t) = B^{-2}(Bt - 1 + e^{-Bt}). \qquad (9,2.27)$$

On substituting (9,2.23)–(9,2.26) into (9,2.22) we obtain

$$\begin{aligned}\langle R(t)\rangle = \exp\Bigg\{-J^2\Bigg[\frac{kT}{I}I^{(2)}(t) + \left(\frac{kT}{I}\right)^2 I_2^{(4)}(t) \\ + \left(\frac{kT}{I}\right)^3 (I_3^{(6)}(t) + 4I_4^{(6)}(t)) \qquad (9,2.28) \\ + \left(\frac{kT}{I}\right)^4 (I_4^{(8)}(t) + 8I_5^{(8)}(t) + [-4J^2 + 16\mathbf{I}]I_7^{(8)}(t) \\ + [-10J^2 + 38\mathbf{I}]I_8^{(8)}(t))\Bigg] + \ldots\Bigg\}.\end{aligned}$$

Equations (D.2) show that each $B^{2r}I_i^{(2r)}(t)$ is a power series in the dimensionless Bt. The exponent in (9,2.28) is therefore a power series in $kT/(IB^2)$, that is, γ. The higher dependence on J^2 of the γ^4-term was unexpected. It could give rise to convergence difficulties of the exponent for large values of j.

In order to discuss (9,2.28) we refer back to Section 7.4. The ensemble average $\langle R(t)\rangle$ is real and is a multiple of the identity operator. Hence from (7,4.22) and (9,2.28), and on replacing J^2 by $j(j+1)\mathbf{I}$

$$\begin{aligned}\langle D^j_{m'm}(\alpha'(t), \beta'(t), \gamma'(t))\rangle = \delta_{m'm}\exp\Bigg\{-j(j+1)\Bigg[\frac{kT}{I}I^{(2)}(t) \qquad (9,2.29) \\ + \left(\frac{kT}{I}\right)^2 I_2^{(4)}(t) + \left(\frac{kT}{I}\right)^3 (I_3^{(4)}(t) + 4I_4^{(4)}(t)) + \left(\frac{kT}{I}\right)^4 (I_4^{(8)}(t) + 8I_5^{(8)}(t) \\ + [-4j(j+1) + 16]I_7^{(8)}(t) + [-10j(j+1) + 38]I_8^{(8)}(t))\Bigg] + \ldots\Bigg\},\end{aligned}$$

where $\alpha'(t)$, $\beta'(t)$, $\gamma'(t)$ are the Euler angles of the body frame at time t referred to the body frame at time zero. Putting $m' = m = 0$, so that from (7,3.15) $D^j_{m'm}$ becomes P_j, we deduce from (9,2.29)

$$\langle\cos\beta'(t)\rangle = \exp\Bigg\{-2\Bigg[\frac{kT}{I}I^{(2)}(t) + \left(\frac{kT}{I}\right)^2 I_2^{(4)}(t)$$

$$+\left(\frac{kT}{I}\right)^3 (I_3^{(4)}(t) + 4I_4^{(4)}(t))$$

$$+\left(\frac{kT}{I}\right)^4 (I_4^{(8)}(t) + 8I_5^{(8)}(t)$$

$$+ 8I_7^{(8)}(t) + 18I_8^{(8)}(t))\Big] + \ldots\}, \qquad (9,2.30)$$

$$\langle P_2(\cos\beta'(t))\rangle = \exp\left\{-6\left[\frac{kT}{I} I^{(2)}(t) + \left(\frac{kT}{I}\right)^2 I_2^{(4)}(t)\right.\right.$$

$$+\left(\frac{kT}{I}\right)^3 (I_3^{(4)}(t) + 4I_4^{(4)}(t))$$

$$+\left(\frac{kT}{I}\right)^4 (I_4^{(8)}(t) + 8I_5^{(8)}(t)$$

$$\left.\left. - 8I_7^{(8)}(t) - 22I_8^{(8)}(t)\right] + \ldots\right\}. \qquad (9,2.31)$$

On expanding the exponential in (9,2.30) and employing (D.3) we find that

$$\langle\cos\beta'(t)\rangle = 1 - \frac{kT}{I} 2I^{(2)}(t) + \left(\frac{kT}{I}\right)^2 [4I_1^{(4)}(t) + 6I_2^{(4)}(t)]$$

$$-\left(\frac{kT}{I}\right)^3 [8I_1^{(6)}(t) + 24I_2^{(6)}(t) + 18I_3^{(6)}(t) + 20I_4^{(6)}(t)]$$

$$+\left(\frac{kT}{I}\right)^4 [16I_1^{(8)}(t) + 72I_2^{(8)}(t) + 108I_3^{(8)}(t) + 54I_4^{(8)}(t)$$

$$+ 120I_5^{(8)}(t) + 80I_6^{(8)}(t) + 80I_7^{(8)}(t) + 100I_8^{(8)}(t)]$$

$$- \ldots . \qquad (9,2.32)$$

This result, as far as terms proportional to $(kT/I)^3$, was derived independently by a time ordered expansion method (Lewis *et al.*, 1974, 1975, 1976).

We can use (9,2.32) to compare the results of this section with (9,1.27) and (9,1.28). Let us take the dipole in a polar spherical molecule to be along the third axis in the body frame. Then the scalar product $(\mathbf{n}(0)\cdot\mathbf{n}(t))$ in the Kubo relation (2,2.23) is just $\cos\beta'(t)$. Hence for orientational polarization

$$\frac{\alpha(\omega)}{\alpha_s} = 1 - i\omega \int_0^\infty \langle\cos\beta'(t)\rangle e^{-i\omega t}dt. \qquad (9,2.33)$$

This integral is the value for $s = i\omega$ of the Laplace transform $\int_0^\infty e^{-st}\langle\cos\beta'(t)\rangle dt$ with $\langle\cos\beta'(t)\rangle$ given by (9,2.32). The transforms of the $I_i^{(2r)}$ are given in (D.2). On substitution into (9,2.33) we get (9,1.27) and therefore agreement also with Sack's result (9,1.28).

We may also derive the correlation functions of $Y_{jm}(\beta(t), \alpha(t))$ and $Y_{jm'}(\beta(t), \alpha(t))$, where $\beta(t)$, $\alpha(t)$ are the spherical polar angles at time t of the third body axis referred to the laboratory frame. Since ensemble averages are equivalent to averages over angular velocity space in the present problem, we see from (7,4.20) and (9,2.28) that

$$\langle Y_{jm}^*(\beta(0), \alpha(0))Y_{jm'}(\beta(t), \alpha(t))\rangle$$

$$= \frac{\delta_{mm'}}{4\pi}\exp\left\{-j(j+1)\left[\frac{kT}{I}I^{(2)}(t) + \left(\frac{kT}{I}\right)^2 I_2^{(4)}(t) + \left(\frac{kT}{I}\right)^3(I_3^{(6)}(t)\right.\right.$$

$$+ 4I_4^{(6)}(t)) + \left(\frac{kT}{I}\right)^4(I_4^{(8)}(t) + 8I_5^{(8)}(t) + [-4j(j+1)+16]I_7^{(8)}(t)$$

$$\left.\left. + [-10j(j+1)+38]I_8^{(8)}(t))\right] + \dots\right\}. \qquad (9,2.34)$$

We then deduce from (9,2.34) that the normalized autocorrelation function of $Y_{jm}(\beta(t), \alpha(t))$,

$$\frac{\langle Y_{jm}^*(\beta(0), \alpha(0))Y_{jm}(\beta(t), \alpha(t))\rangle}{\langle Y_{jm}^*(\beta(0), \alpha(0))Y_{jm}(\beta(0), \alpha(0))\rangle}$$

$$= \exp\left\{-j(j+1)\left[\frac{kT}{I}I^{(2)}(t) + \left(\frac{kT}{I}\right)^2 I_2^{(4)}(t) + \left(\frac{kT}{I}\right)^3(I_3^{(6)}(t)\right.\right.$$

$$+ 4I_4^{(6)}(t)) + \left(\frac{kT}{I}\right)^4(I_4^{(8)}(t) + 8I_5^{(8)}(t) + [-4j(j+1)+16]I_7^{(8)}(t)$$

$$\left.\left. + [-10j(j+1)+38]I_8^{(8)}(t)\right] + \dots\right\}, \qquad (9,2.35)$$

which is independent of m.

9.3 Comparison with Other Studies of the Rotating Sphere

In the long time limit $t \gg \tau_F$, or $Bt \gg 1$, the $I_i^{(2n)}(t)$'s that appear in (9,2.28) are seen from (D.2) to have asymptotic values $aB^{-2n}Bt$, where $0 < a \leqslant 1$, the value of a tending to decrease with increasing n. If then we take the limits $Bt \gg 1, \gamma \ll 1$, we deduce form (9,2.27) and (9,2.28)

$$\langle R(t) \rangle = \exp\{-J^2 \gamma B t\}, \tag{9,3.1}$$

so that (9,2.29) becomes

$$\langle D^j_{m'm}(\alpha'(t), \beta'(t), \gamma'(t)) \rangle = \delta_{m'm} \exp\{-j(j+1)\gamma B t\}. \tag{9,3.2}$$

If we put $m = m' = 0$, this yields

$$\langle P_j(\cos \beta'(t)) \rangle = \exp\{-j(j+1)\gamma B t\},$$

$$\langle \cos \beta'(t) \rangle = e^{-t/\tau_D}, \tag{9,3.3}$$

by (1,2.7). In the Debye theory of relaxation of a polar spherical molecule (1,3.3) and (1,3.8) give

$$\frac{P_a(t)}{P(0)} = e^{-t/\tau_D}$$

and (2,2.25) then gives (9,3.3). Hence in the long time limit we get the Debye result from our theory. We may look on (9,3.1) as expressing a result for the Debye theory with the value of j not necessarily restricted to unity, and we may define $IB/[j(j+1)kT]$ as the Debye time for this more general situation.

In the limit $t \ll \tau_F$ we have from (D.2) for all values of i the approximations

$$I^{(2)}(t) \doteqdot \frac{t^2}{2!}, \; I_i^{(4)}(t) \doteqdot \frac{t^4}{4!}, \; I_i^{(6)}(t) \doteqdot \frac{t^6}{6!}, \; I_i^{(8)}(t) \doteqdot \frac{t^8}{8!}.$$

We then obtain from (9,2.28)

$$\langle R(t) \rangle = \exp\{-J^2 \left[\frac{1}{2!}\left(\frac{t}{\tau_R}\right)^2 + \frac{1}{4!}\left(\frac{t}{\tau_R}\right)^4\right.$$

$$+\frac{5}{6!}\left(\frac{t}{\tau_R}\right)^6+\frac{63I-14J^2}{8!}\left(\frac{t}{\tau_R}\right)^8\bigg]+\ldots\bigg\},$$

(9,3.4)

where τ_R is the free rotation time defined in (1,2.10). From (9,2.28), (9,2.29) and (9,3.4) we deduce that

$$\langle D^j_{m'm}(\alpha'(t),\beta'(t),\gamma'(t))\rangle$$

$$=\delta_{m'm}\exp\left\{-j(j+1)\left[\frac{1}{2!}\left(\frac{t}{\tau_R}\right)^2+\frac{1}{4!}\left(\frac{t}{\tau_R}\right)^4+\frac{5}{6!}\left(\frac{t}{\tau_R}\right)^6\right.\right. \quad (9,3.5)$$

$$\left.\left.+\frac{63-14j(j+1)}{8!}\left(\frac{t}{\tau_R}\right)^8\right]+\ldots\right\},$$

which is independent of τ_F. When τ_F becomes very large, the friction is negligible. Then both the white noise driving couple and the frictional drag disappear, and we have free motion. Consequently (9,3.5) should apply to the free motion of the sphere for all values of t. By a direct calculation for the free motion, Hubbard (1972, eqn (6.18)) gives a result which in our notation reads

$$\langle D^j_{m'm}(\alpha'(t),\beta'(t),\gamma'(t))\rangle$$

$$=\frac{\delta_{m'm}}{2j+1}\sum_{s=-j}^{j}\left[1-\left(\frac{st}{\tau_R}\right)^2\right]\exp\left[-\frac{1}{2}\left(\frac{st}{\tau_R}\right)^2\right]. \quad (9,3.6)$$

On expanding the exponential of (9,3.5) in powers of t/τ_R as far as $(t/\tau_R)^8$ it is verified for $j=1$ and $j=2$, which are the cases of chief physical interest, that (9,3.5) agrees with (9,3.6).

While (9,3.1) and (9,3.4) give the asymptotic values of $\langle R(t)\rangle$ for $t\gg\tau_F$ and $t\ll\tau_F$, we have so far not derived an expression for $\langle R(t)\rangle$ valid for all times that will allow a direct comparison with the Debye result. To do this, further mathematical methods have to be developed. This will be done in Section 11.5. The methods will then be applied to the rotating sphere in Section 12.3, where a new expression for $\langle R(t)\rangle$ will be derived.

Steele (1963b) obtained expressions for the normalized auto-correlation functions of the spherical harmonics with $j=1$ and $j=2$. He did this by using a probability density function W, which in a previous paper (Steele, 1963a) is said to satisfy

$$\frac{\partial W}{\partial t} = \sum_{i=x,y,z} R_{ii}(t)\, \frac{\partial^2 W}{\partial \psi_i^2}, \tag{9,3.7}$$

where ψ_x, ψ_y, ψ_z are angles and

$$R_{ii}(t) = \int_{-\infty}^{\infty} \langle \omega_i(0)\omega_i(t') \rangle \, dt' = \frac{kT}{\zeta}\left[1 - \exp\left(-\frac{\zeta t}{I}\right)\right]. \tag{9,3.8}$$

Equations (2.34) and (2.35) of Steele 1963b are, by (7,4.24), expressible for the sphere as

$$\langle D^1_{00}(\alpha'(t), \beta'(t), \gamma'(t)) \rangle = \exp\{-2\gamma(Bt - 1 + e^{-Bt})\}, \tag{9,3.9}$$

$$\langle D^2_{mm}(\alpha'(t), \beta'(t), \gamma'(t)) \rangle = \exp\{-6\gamma(Bt - 1 + e^{-Bt})\}. \tag{9,3.10}$$

We see that these come from (9,2.29) with $j = 1$ and $j = 2$, if we take only the first term in the exponent; they are first order corrections to the Debye theory. Steele made an independent calculation of $\langle D^1_{00}(\alpha'(t), \beta'(t), \gamma'(t)) \rangle$ and $\langle D^2_{mm}(\alpha'(t), \beta'(t), \gamma'(t)) \rangle$ for free motion, obtaining results which agreed with (9,3.6) with $j = 1$ and $j = 2$ and, not surprisingly, finding disagreement with (9,3.9) and (9,3.10) in the limit of small B. The justification of the Smoluchowski equation (9,3.7) with a time dependent diffusion coefficient $R_{ii}(t)$ given in (9,3.8) is not evident from Steele's papers. Since the derivation of the inaccurate results (9,3.9) and (9,3.10) depends on (9,3.7), it seems that this equation may not be generally true, though for $t = \infty$ the diffusion coefficient has the value which appeared in Sections 1.1 and 1.2.

Fixman and Rider (1969) examined the value of $\langle P_j(\cos \beta'(t)) \rangle$ for $j = 1, 2$. Denoting by $\mathbf{u}$ the third component of a unit vector in the laboratory frame with respect to a body frame and by $\mathbf{L}$ the angular momentum, they derived a differential equation

$$\frac{\partial f}{\partial t} - iLSf = \mathscr{L} f \equiv \frac{\partial f}{\partial t} + \nabla_u \cdot (\dot{\mathbf{u}} f)$$

for the probability density $f(\mathbf{u}, \mathbf{L}, t)$ in terms of which

$$\langle P_j(\cos \beta'(t)) \rangle = \int P_j(\cos \theta_u) f(\mathbf{u}, \mathbf{L}, t)\, d\mathbf{u} d\mathbf{L}.$$

In these equations

$$S = -iL^{-1}(\mathbf{L} \cdot (\mathbf{u} \times \nabla_u))$$

and θ_u is the polar angle of $\mathbf{u}$. On writing

$$I_j(\omega) = \frac{1}{2\pi}\int_{-\infty}^{\infty}\langle P_j(\cos\beta'(t))\rangle e^{-i\omega t}dt \qquad (9,3.11)$$

they found that

$$I_j(\omega) = \frac{1}{\pi}\,\mathrm{Re}\int Y_{j0}(\mathbf{u})[i(\omega - LS) - \mathscr{L}]^{-1}\,Y_{j0}(\mathbf{u})f_{\mathrm{eq}}(L)d\mathbf{u}d\mathbf{L},$$

where $f_{\mathrm{eq}}(L)$ is the Maxwell-Boltzmann equilibrium density function. Then the value of $\langle P_j(\cos\beta'(t))\rangle$ was deduced by inverting (9,3.11). By the use of a computer for the calculations, graphs of $\langle\cos\beta'(t)\rangle$ and $\langle P_2(\cos\beta'(t))\rangle$ as functions of t were obtained for both the sphere and the linear rotator.

Hubbard (1972) applied (7,2.4) and its Fourier transformed equation to the sphere. He derived expressions for the normalized autocorrelation function of spherical harmonics in three cases:

(i) free motion, from which we took (9,3.6),
(ii) weak interactions, which in our notation means $\gamma \gg 1$,
(iii) strong interactions, namely $\gamma \ll 1$, which is the case on which we have focused attention.

For strong interactions he obtained (9,2.35) up to terms proportional to $(kT/I)^3$ in the exponent.

To relate the results of the previous section to earlier investigations we confine our attention to the case of $j = 1$ relevant to dielectric relaxation phenomena. We have already seen that the results of the last section are then in agreement with those of Sack as expressed in (9,1.28). If in this equation we take the second convergent, we have

$$\frac{\alpha(\omega)}{\alpha_s} = \frac{P(\omega)}{P(0)} = 1 - \frac{i\omega'}{i\omega' + 2\gamma(1 + i\omega')^{-1}}$$

$$= \frac{1}{1 + \dfrac{i\omega' - \omega'^2}{2\gamma}}$$

$$= \frac{1}{1 + i\omega\tau_D - \omega^2\tau_D\tau_F}.$$

This is the result (2,5.9) of Powles when $\lambda = 1$. Taking the second convergent is approximately equivalent to truncating the series in (9,1.27) so that the equation reads

$$\frac{\alpha(\omega)}{\alpha_s} = \frac{2\gamma}{i\omega'(i\omega'+1)} - \frac{4\gamma^2}{(i\omega')^2(i\omega'+1)^2}. \tag{9,3.12}$$

9.4 Further Consideration of the Stochastic Rotation Operator

It is obvious from Section 9.2 that the stochastic rotation operator $R(t)$ is of central importance in the study of rotational Brownian motion. It may be helpful at this stage to make some remarks about $R(t)$ in the light of our experience of employing it. The operator satisfies (9,2.3), which is an equation that does not involve the angle variables explicitly. We did not solve (9,2.3) for $R(t)$ but we applied the averaging method of Section 5.4 to it in order to find $\langle R(t)\rangle$. The angular velocity $\boldsymbol{\omega}(t)$ for a steady state is a centred Gaussian random variable, and its components satisfy (9,2.2). From these properties we deduced (9,2.5) and the values of $\epsilon^2\Omega^{(2)}(t_1)$, $\epsilon_4\Omega^{(4)}(t_1)$, etc. Strictly speaking, (9,2.5) is a differential equation for $\langle R(t)\rangle_\omega$. However, since the solution is based entirely on (9,2.2), angle variables will be nowhere introduced and so, by (7,4.15), $\langle R(t)\rangle_\omega$ and $\langle R(t)\rangle$ are equal. By a similar argument this is true not only for the sphere but also for the disc, linear rotator and asymmetric rotator.

Hence what we need for the calculation of orientational correlation functions according to the method explained in Section 7.4 is $\langle R(t)\rangle$. Equations (7,4.20), (7,4.21), (7,4.23) and (7,4.24) may thus be replaced by

$$\begin{aligned}&\langle Y^*_{jm}(\beta(0),\alpha(0))\,Y_{jm'}(\beta(t),\alpha(t))\rangle\\ &= \frac{1}{4\pi}\,\langle R^j_{m'm}(t)\rangle^* = \frac{1}{4\pi}\,\langle D^j_{m'm}(\alpha'(t),\beta'(t),\gamma'(t))\rangle^*,\end{aligned} \tag{9,4.1}$$

$$\begin{aligned}&\frac{\langle Y^*_{jm}(\beta(0),\alpha(0))Y_{jm}(\beta(t),\alpha(t))\rangle}{\langle Y^*_{jm}(\beta(0),\alpha(0))\,Y_{jm}(\beta(0),\alpha(0))\rangle}\\ &= \langle R^j_{mm}(t)\rangle^* = \langle D^j_{mm}(\alpha'(t),\beta'(t),\gamma'(t))\rangle^*.\end{aligned} \tag{9,4.2}$$

Since $D^j_{m'm}$ is angle dependent, it might seem strange that $\langle D^j_{m'm} \rangle = \langle D^j_{m'm} \rangle_\omega$. However, it must be remembered that $\alpha'(t)$, $\beta'(t)$, $\gamma'(t)$ are independent of $\alpha(0)$, $\beta(0)$, $\gamma(0)$, as we may see by referring to Fig. 7.2 for $\beta'(t)$ and $\beta(0)$: the angles $\alpha'(t)$, $\beta'(t)$, $\gamma'(t)$ relate only to the rotation of the body between time 0 and time t and have nothing to do with the initial orientation of the body. Since $D^j_{m'm}(\alpha'(t), \beta'(t), \gamma'(t))$ is independent of $\alpha(0)$, $\beta(0)$, $\gamma(0)$, so also is $\langle D^j_{m'm}(\alpha'(t), \beta'(t), \gamma'(t)) \rangle_\omega$ and, by (7,4.15),

$$\langle D^j_{m'm}(\alpha'(t), \beta'(t), \gamma'(t)) \rangle = \langle D^j_{m'm}(\alpha'(t), \beta'(t), \gamma'(t)) \rangle_\omega .$$

10

Brownian Motion of the Linear Rotator

10.1 Fokker–Planck Equation Study of the Linear Rotator

We examine the rotational Brownian motion of a polar molecule whose shape is approximated by a thin rod, or needle, and which is subject to an external electric field of intensity F. Let us again take rotating axes as in Fig. 9.1, the rotator lying along axis 3 with its mid point at the origin. We assume that the angular velocity of the rotator about its line of symmetry is zero, and we denote by I the moment of inertia of the rotator about any line through the origin perpendicular to the line of symmetry. We define the Euler angles α, β as in Sections 7.3 and 9.1, the angle γ no longer appearing. The components of angular velocity of the axes are $(\dot{\beta}, \dot{\alpha}\sin\beta, \dot{\alpha}\cos\beta)$ and the components of angular momentum of the body are $(I\dot{\beta}, I\dot{\alpha}\sin\beta, 0)$. Then employing the formula $\dot{\mathbf{V}} + (\boldsymbol{\omega} \times \mathbf{V})$ for the rate of change of angular momentum $\mathbf{V}$ and recalling from Section 1.1 that the electric field produces a couple of moment $-\mu F \sin\beta$ about axis 1 we have the equations of motion

$$I\ddot{\beta} - I\dot{\alpha}^2 \sin\beta\cos\beta = -\mu F\sin\beta - \zeta\dot{\beta} + I\frac{dW_1(t)}{dt},$$

$$I\frac{d(\dot{\alpha}\sin\beta)}{dt} + I\dot{\alpha}\dot{\beta}\cos\beta = -\zeta\dot{\alpha}\sin\beta + I\frac{dW_2(t)}{dt}. \qquad (10,1.1)$$

We write

$$\dot{\beta} = \omega_\beta, \quad \dot{\alpha} = \frac{\omega_\alpha}{\sin\beta}, \qquad (10,1.2)$$

and

$$\beta = x_1, \quad \alpha = x_2, \quad \omega_\beta = x_3, \quad \omega_\alpha = x_4,$$

which allows us to express (10,1.1) and (10,1.2) as

$$dx_1 = \omega_\beta dt, \quad dx_2 = \frac{\omega_\alpha}{\sin\beta} dt,$$

$$dx_3 = -\left[B\omega_\beta + \frac{\mu F \sin\beta}{I} - \omega_\alpha^2 \cot\beta\right] dt + dW_1(t),$$

$$dx_4 = -[B\omega_\alpha + \omega_\alpha\omega_\beta \cot\beta]\, dt + dW_2(t).$$

The volume element in the space of these x_i's is $\sin\beta\, dx_1 dx_2 dx_3 dx_4$. The kinetic energy of the rotator,

$$\tfrac{1}{2} I(\dot{\beta}^2 + \dot{\alpha}^2 \sin^2\beta) = \tfrac{1}{2} I(\omega_\beta^2 + \omega_\alpha^2),$$

and so the Maxwell–Boltzmann probability density function for constant F,

$$w_0 = \text{constant} \exp\{[-\tfrac{1}{2} I(\omega_\beta^2 + \omega_\alpha^2) + \mu F \cos\beta]/(kT)\}.$$

Proceeding as in Section 9.1 we deduce the Fokker–Planck equation for $w(\beta, \omega_\beta, \omega_\alpha, t)$:

$$\frac{\partial w}{\partial t} + \omega_\beta \frac{\partial w}{\partial \beta} + \omega_\alpha \cot\beta \left(\omega_\alpha \frac{\partial w}{\partial \omega_\beta} - \omega_\beta \frac{\partial w}{\partial \omega_\alpha}\right) - \frac{\mu F \sin\beta}{I} \frac{\partial w}{\partial \omega_\beta}$$

$$= B \frac{\partial}{\partial \omega_\beta}\left(\omega_\beta w + \frac{kT}{I}\frac{\partial w}{\partial \omega_\beta}\right) + B \frac{\partial}{\partial \omega_\alpha}\left(\omega_\alpha w + \frac{kT}{I}\frac{\partial w}{\partial \omega_\alpha}\right).$$

(10,1.3)

This agrees with eqn (2.9) of Sack (1957b) combined with (2.3) of Sack (1957a).

We next make the transformation

$$\Phi(\beta, u_\beta, u_\alpha, t)$$

$$= \int_{-\infty}^{\infty}\int_{-\infty}^{\infty} w(\beta, \omega_\beta, \omega_\alpha, t) \exp[-i(u_\beta\omega_\beta + u_\alpha\omega_\alpha)]\, d\omega_\beta d\omega_\alpha,$$

which when substituted into (10,1.3) yields

$$\frac{\partial \Phi}{\partial t} + i \frac{\partial^2 \Phi}{\partial \beta \partial u_\beta} - i \cot\beta \frac{\partial}{\partial u_\alpha}\left(u_\beta \frac{\partial \Phi}{\partial u_\alpha} - u_\alpha \frac{\partial \Phi}{\partial u_\beta}\right) - \frac{i\mu F}{I} \sin\beta\, u_\beta \Phi$$

$$= -B\left(\omega_\beta \frac{\partial \Phi}{\partial \omega_\beta} + \omega_\alpha \frac{\partial \Phi}{\partial \omega_\alpha}\right) - \frac{BkTu^2}{I}\Phi,$$

where

$$u^2 = u_\beta^2 + u_\alpha^2.$$

Then the transformation

$$\Psi(\beta, u_\beta, u_\alpha, t) = \exp\left(\frac{kTu^2}{2I}\right) \Phi(\beta, u_\beta, u_\alpha, t)$$

leads to

$$\frac{\partial \Psi}{\partial t} + i\frac{\partial^2 \Psi}{\partial \beta \partial u_\beta} - \frac{iu_\beta}{I}\left(kT\frac{\partial \Psi}{\partial \beta} + \mu F \sin \beta \Psi\right)$$

$$+ i \cot \beta \left(\frac{\partial}{\partial u_\alpha} - \frac{kT}{I} u_\alpha\right) \left(u_\alpha \frac{\partial \Psi}{\partial u_\beta} - u_\beta \frac{\partial \Psi}{\partial u_\alpha}\right)$$

$$= -B\left(u_\beta \frac{\partial \Psi}{\partial u_\beta} + u_\alpha \frac{\partial \Psi}{\partial u_\alpha}\right). \qquad (10,1.4)$$

The orientational polarization $P(t)$ may then be found from

$$P(t) = N\mu \lim_{u \to 0} \int_0^{2\pi} d\alpha \int_0^{\pi} \Psi(\beta, u_\beta, u_\alpha, t) \cos \beta \sin \beta d\beta.$$

Working in the linear approximation we expand the solution of (10,1.4):

$$4\pi \Psi = 1 + \cos \beta \chi_1(u^2, t) + \sin \beta u_\beta \chi_2(u^2, t) + \ldots, \qquad (10,1.5)$$

where χ_1, χ_2 are of order $\mu F/(kT)$. On substitution of (10,1.5) into (10,1.4) it is found that

$$\frac{\partial \chi_1}{\partial t} + 2i\chi_2 + 2iz\frac{\partial \chi_2}{\partial z} - 2iz\chi_2 = -2Bz\frac{\partial \chi_1}{\partial z},$$

$$\frac{\partial \chi_2}{\partial t} - \frac{ikT}{I}\frac{\partial \chi_1}{\partial z} + \frac{ikT}{I}\chi_1 - \frac{i\mu F}{I} = -2Bz\frac{\partial \chi_2}{\partial z} - B\chi_2, \qquad (10,1.6)$$

where $z = \frac{1}{2}kTu^2/I$. For the steady state case of $F = F_0 e^{i\omega t}$ we put

$$\chi_1(u^2, t) = e^{i\omega t} \sum_{r=0}^{\infty} A_r z^r, \quad \chi_2(u^2, t) = e^{i\omega t} \sum_{r=0}^{\infty} B_r z^r. \qquad (10,1.7)$$

Substituting this into (10,1.6) we find that

$$\sum (L_{nl} + \gamma M_{nl})B_l = \delta_{n0}\frac{i\mu F_0}{\zeta} \qquad (\gamma = kT/(IB^2)),$$

where

$$L_{nl} = (i\omega' + 2n + 1)\delta_{nl} \qquad (\omega' = \omega/B), \qquad (10,1.8)$$

$$M_{nl} = -\frac{2}{i\omega' + 2n}\delta_{n-1,l} + \left(\frac{2n+2}{i\omega' + 2n} + \frac{2}{i\omega' + 2n + 2}\right)\delta_{nl} - \frac{(2n+4)}{i\omega' + 2n + 2}\delta_{n+1,l},$$

and that

$$(i\omega' + 2n)A_n = \frac{2i}{B}[B_{n-1} - (n+1)B_n] \qquad (n \neq 0),$$

$$\omega' A_0 = -\frac{2B_0}{B}.$$

Equations (9,1.25) and (9,1.26) are still true. Using them and proceeding as in Section 9.1 we deduce that

$$\begin{aligned}
\frac{\alpha(\omega)}{\alpha_s} = \frac{P(\omega)}{P(0)} &= \frac{2\gamma}{i\omega'(i\omega'+1)} - \frac{2\gamma^2}{i\omega'(i\omega'+1)^2}\left\{\frac{2}{i\omega'} + \frac{2}{i\omega'+2}\right\} \\
&+ \frac{2\gamma^3}{i\omega'(i\omega'+1)^2}\left\{\frac{4}{(i\omega')^2(i\omega'+1)} + \frac{8}{i\omega'(i\omega'+1)(i\omega'+2)}\right. \\
&\left. + \frac{4}{(i\omega'+1)(i\omega'+2)^2} + \frac{8}{(i\omega'+2)^2(i\omega'+3)}\right\} \\
&- \frac{2\gamma^4}{i\omega'(i\omega'+1)^2}\left\{\frac{8}{(i\omega')^3(i\omega'+1)^2} + \frac{24}{(i\omega')^2(i\omega'+1)^2(i\omega'+2)}\right. \\
&+ \frac{24}{i\omega'(i\omega'+1)^2(i\omega'+2)^2} + \frac{32}{i\omega'(i\omega'+1)(i\omega'+2)^2(i\omega'+3)} \\
&+ \frac{8}{(i\omega'+1)^2(i\omega'+2)^3} + \frac{32}{(i\omega'+1)(i\omega'+2)^3(i\omega'+3)} \\
&\left. + \frac{32}{(i\omega'+2)^3(i\omega'+3)^2} + \frac{16}{(i\omega'+2)^2(i\omega'+3)^2(i\omega'+4)}\right\} + \cdots.
\end{aligned} \qquad (10,1.9)$$

In the overdamped case it is found that $P(\omega)/P(0) = (1 + i\omega\tau_D)^{-1}$, as was found in Section 9.1 for the rotating sphere.

In his investigation of the linear rotator Sack (1957b) took in place of (10,1.5)

$$4\pi\Psi = 1 + \cos\beta\,\psi_1(u,t) + \sin\beta\cos\phi_u\,\psi_2(u,t) + \ldots,$$

where ϕ_u is defined by

$$u_\beta = u\cos\phi_u, \quad u_\alpha = u\sin\phi_u.$$

He expanded ψ_1 and ψ_2 for $F = F_0 e^{i\omega t}$ as $e^{i\omega t}$ multiplied by power series in u, and deduced three-term recurrence relations for the coefficients in the power series. On solving by the continued fraction method he found that

$$\frac{P(\omega)}{P(0)} = 1 - \frac{i\omega'|}{|i\omega'} + \frac{2\gamma\ |}{|1+i\omega'} + \frac{2\gamma\ |}{|2+i\omega'} + \frac{4\gamma\ |}{|3+i\omega'}$$

$$+ \frac{4\gamma\ |}{|4+i\omega'} + \frac{6\gamma\ |}{|5+i\omega'} + \ldots. \tag{10,1.10}$$

When the fraction is expanded in powers of γ, agreement with (10,1.9) is obtained.

10.2 Stochastic Differential Equation Study of the Linear Rotator

To establish stochastic differential equations for the Brownian motion of a linear rotator subject to no external forces or couples we take Cartesian axes fixed in the body with origin at its centre, the third axis being along the rotator. Since the third component of angular velocity of the body is zero and since the body is symmetrical about the third axis, the Euler–Langevin equations (7,1.1) reduce to

$$\begin{aligned} \frac{d\omega_1(t)}{dt} &= -B\omega_1(t) + \frac{dW_1(t)}{dt}, \\ \frac{d\omega_2(t)}{dt} &= -B\omega_2(t) + \frac{dW_2(t)}{dt}. \end{aligned} \tag{10,2.1}$$

We see from (5,3.3) and (5,3.8) that the steady state solutions of (10,2.1) are centred Gaussian random variables obeying

$$\langle \omega_i(t_k)\omega_j(t_l)\rangle_\omega = \delta_{ij}\frac{kT}{I}e^{-B|t_k-t_l|} \qquad (i,j=1,2), \tag{10,2.2}$$

which is (9,2.2) restricted to the values 1 and 2 of i and j. This restriction gives rise to difficulties in the calculations that were not present in those of Section 9.2 for the rotating sphere. In contrast, the Fokker–Planck equation study was simpler for the linear rotator than for the sphere.

We retain as far as possible the notation of Section 9.2. Equation (9,2.3) becomes

$$\frac{dR(t)}{dt} = -i(J_1\omega_1(t)+J_2\omega_2(t))R(t).$$

Then $\langle R(t)\rangle$ obeys (9,2.5), namely,

$$\frac{d\langle R(t_1)\rangle}{dt_1} = (\epsilon^2\Omega^{(2)}(t_1)+\epsilon^4\Omega^{(4)}(t_1)+\epsilon^6\Omega^{(6)}(t_1)+\ldots)\langle R(t_1)\rangle. \tag{10,2.3}$$

In the subsequent equations (9,2.6) and (9,2.7) for $\epsilon^2\Omega^{(2)}(t_1)$ and $\epsilon^4\Omega^{(4)}(t_1)$,

$$\epsilon^2\Omega^{(2)}(t_1) = -\int_0^{t_1}\langle 12\rangle dt_2, \tag{10,2.4}$$

$$\epsilon^4\Omega^{(4)}(t_1) = \int_0^{t_1}dt_2\int_0^{t_2}dt_3\int_0^{t_3}dt_4 \times$$
$$\{\langle 1234\rangle-\langle 12\rangle\langle 34\rangle-\langle 13\rangle\langle 24\rangle-\langle 14\rangle\langle 23\rangle\}, \tag{10,2.5}$$

an integer l in the angular brackets is an abbreviation for $J_1\omega_1(t_l)+J_2\omega_2(t_l)$, so that

$$\begin{aligned}\langle 12\rangle &= \langle (J_1\omega_1(t_1)+J_2\omega_2(t_1))(J_1\omega_1(t_2)+J_2\omega_2(t_2))\rangle \\ &= \frac{kT}{I}(J_1^2+J_2^2)e^{-B(t_1-t_2)},\end{aligned} \tag{10,2.6}$$

by (10,2.2), since as in the case of the sphere ensemble averages may be taken over the angular velocity space only. In the present section we adopt the convention that a repeated index is summed over 1 and 2. Then writing J^2 for $J_1^2+J_2^2+J_3^2$ we have

$$J_k^2 = J_1^2+J_2^2 = J^2-J_3^2,$$

and from (10,2.4) and (10,2.6)

$$\epsilon^2 \Omega^{(2)}(t_1) = -\frac{kT}{I}(J^2 - J_3^2) \int_0^{t_1} e^{-B(t_1 - t_2)} dt_2 . \quad (10,2.7)$$

It is shown in Appendix B that J^2 is a multiple of the identity operator. It therefore commutes with J_1, J_2, J_3 and for this reason it is now singled out in these calculations.

We next evaluate $\Omega^{(4)}(t_1)$ from (10,2.5). We have

$$\begin{aligned}\langle 1234 \rangle &= \langle J_i \omega_i(t_1) J_k \omega_k(t_2) J_l \omega_l(t_3) J_n \omega_n(t_4) \rangle \\ &= \langle \omega_i(t_1) \omega_k(t_2) \omega_l(t_3) \omega_n(t_4) \rangle J_i J_k J_l J_n .\end{aligned}$$

On expressing, by (3,4.18), the mean value as the sum of products of mean values taken in pairs and employing (10,2.6) we deduce that

$$\begin{aligned}\langle 1234 \rangle = \left(\frac{kT}{I}\right)^2 &\{\exp[-B(t_1 - t_2 + t_3 - t_4)] J_i^2 J_k^2 \\ &+ \exp[-B(t_1 + t_2 - t_3 - t_4)](J_i J_k J_i J_k + J_i J_k^2 J_i)\}.\end{aligned}$$

From (10,2.6) we find that

$$\begin{aligned}&\langle 12 \rangle \langle 34 \rangle + \langle 13 \rangle \langle 24 \rangle + \langle 14 \rangle \langle 23 \rangle \\ &= \left(\frac{kT}{I}\right)^2 \{\exp[-B(t_1 - t_2 + t_3 - t_4)] \\ &\quad + 2\exp[-B(t_1 + t_2 - t_3 - t_4)]\} J_i^2 J_k^2 ,\end{aligned}$$

and hence

$$\begin{aligned}&\langle 1234 \rangle - \langle 12 \rangle \langle 34 \rangle - \langle 13 \rangle \langle 24 \rangle - \langle 14 \rangle \langle 23 \rangle \\ &= \left(\frac{kT}{I}\right)^2 \exp[-B(t_1 + t_2 - t_3 - t_4)] \times \\ &\quad (J_i J_k J_i J_k + J_i J_k^2 J_i - 2 J_i^2 J_k^2).\end{aligned}$$

The J-sums may be calculated using (9,2.12)

$$J_k J_l - J_l J_k = -i \sum_{n=1}^{3} \epsilon_{kln} J_n ,$$

and derived relations to be found in McConnell (1978), where it is shown that

$$\begin{aligned}\epsilon^4 \Omega^{(4)}(t_1) = -\left(\frac{kT}{I}\right)^2 (2J^2 - 5J_3^2) \int_0^{t_1} dt_2 \int_0^{t_2} dt_3 \int_0^{t_3} dt_4 \times \\ \exp[-B(t_1 + t_2 - t_3 - t_4)]. \quad (10,2.8)\end{aligned}$$

The evaluation of $\epsilon^6 \Omega^{(6)}(t_1)$ is quite tedious. We have to work from (5,4.17), which no longer reduces to a simpler form like (9,2.19) for the sphere. The calculations are further complicated by the fact that J_k^2 does not commute with J_1 and J_2. Finally it is found that

$$\epsilon^6 \Omega^{(6)}(t_1) =$$

$$-\left(\frac{kT}{I}\right)^3 \{[9(J^2 - J_3^2)J_3^2 + 4J^2 - 16J_3^2]$$

$$\times \int_0^{t_1} dt_2 \ldots \int_0^{t_5} dt_6 \exp[-B(t_1 + t_2 - t_3 + t_4 - t_5 - t_6)]$$

$$+ [21(J^2 - J_3^2)J_3^2 + 20J^2 - 68J_3^2]$$

$$\times \int_0^{t_1} dt_2 \ldots \int_0^{t_5} dt_6 \exp[-B(t_1 + t_2 + t_3 - t_4 - t_5 - t_6)]\}. \tag{10,2.9}$$

It would be very difficult to extend the calculations to $\epsilon^8 \Omega^{(8)}(t_1)$, especially since the graphical method employed for the sphere is not applicable to the linear rotator.

We return to the calculation of $\langle R(t) \rangle$ from (10,2.3). Since the only operators that appear in $\epsilon^2 \Omega^{(2)}(t_1)$, $\epsilon^4 \Omega^{(4)}(t_1)$, $\epsilon^6 \Omega^{(6)}(t_1)$ are J^2 and J_3^2, which commute with one another, we can integrate (10,2.3) immediately to obtain

$$\langle R(t) \rangle = \exp\left\{\int_0^t dt_1 [\epsilon^2 \Omega^{(2)}(t_1) + \epsilon^4 \Omega^{(4)}(t_1) + \epsilon^6 \Omega^{(6)}(t_1) + \ldots]\right\}. \tag{10,2.10}$$

Now from (10,2.7)–(10,2.9) and (D.1)

$$\int_0^t \epsilon^2 \Omega^{(2)}(t_1) dt_1 = -\frac{kT}{I}(J^2 - J_3^2) I^{(2)}(t), \tag{10,2.11}$$

$$\int_0^t \epsilon^4 \Omega^{(4)}(t_1) dt_1 = -\left(\frac{kT}{I}\right)^2 (2J^2 - 5J_3^2) I_2^{(4)}(t), \tag{10,2.12}$$

$$\int_0^t \epsilon^6 \Omega^{(6)}(t_1) dt_1 = -\left(\frac{kT}{I}\right)^3 \{[9(J^2 - J_3^2)J_3^2 + 4J^2 - 16J_3^2] I_3^{(6)}(t)$$

$$+ [21(J^2 - J_3^2)J_3^2 + 20J^2 - 68J_3^2] I_4^{(6)}(t)\}, \tag{10,2.13}$$

and so

$$\langle R(t)\rangle = \exp\left\{-\left[\frac{kT}{I}(J^2 - J_3^2)I^{(2)}(t) + \left(\frac{kT}{I}\right)^2 (2J^2 - 5J_3^2)I_2^{(4)}(t)\right.\right.$$
$$+ \left(\frac{kT}{I}\right)^3 ([9(J^2 - J_3^2)J_3^2 + 4J^2 - 16J_3^2]I_3^{(6)}(t)$$
$$\left.\left. + [21(J^2 - J_3^2)J_3^2 + 20J^2 - 68J_3^2]I_4^{(6)}(t)) + \ldots\right]\right\}. \tag{10,2.14}$$

If we interpret $R(t)$ as the rotation operator that brings the body frame at time 0 to the body frame at time t, we have from (7,3.14) and (7,4.1)

$$\langle D^j_{m'm}(\alpha'(t), \beta'(t), \gamma'(t))\rangle$$
$$= \int_0^{2\pi} d\alpha(0) \int_0^{\pi} \sin\beta(0) Y^*_{jm'}(\beta(0), \alpha(0))\langle R(t)\rangle Y_{jm}(\beta(0), \alpha(0)) d\beta(0), \tag{10,2.15}$$

with the value of $\langle R(t)\rangle$ taken from (10,2.14). This is the $m'm$-element of $\langle R(t)\rangle$ in the $(2j+1)$-dimensional representation with basis vectors Y_{js}. For an operator J^2 in (10,2.14) this is $j(j+1)\delta_{m'm}$. For $\hbar J_3$ it is the $m'm$-element of the third component of angular momentum of the rotator, and that is just zero. In writing down $\langle D^j_{m'm}(\alpha'(t), \beta'(t), \gamma'(t))\rangle$ from (10,2.15) we therefore replace $(J^2)_{m'm}$ by $j(j+1)\delta_{m'm}$ and J_3^2 by 0 in $\langle R(t)\rangle$ and obtain

$$\langle D^j_{m'm}(\alpha'(t), \beta'(t), \gamma'(t))\rangle$$
$$= \delta_{m'm} \exp\left\{-j(j+1)\left[\frac{kT}{I}I^{(2)}(t) + \left(\frac{kT}{I}\right)^2 2I_2^{(4)}(t)\right.\right.$$
$$\left.\left. + \left(\frac{kT}{I}\right)^3 (4I_3^{(6)}(t) + 20I_4^{(6)}(t))\right] + \ldots\right\}. \tag{10,2.16}$$

In analogy with (9,2.30), (9,2.31), (9,2.32), (9,2.34) and (9,2.35) for the spherical rotator, eqn (10,2.16) gives

$$\langle\cos\beta'(t)\rangle = \exp\left\{-2\left[\frac{kT}{I}I^{(2)}(t) + 2\left(\frac{kT}{I}\right)^2 I_2^{(4)}(t)\right.\right.$$
$$\left.\left. + \left(\frac{kT}{I}\right)^3 (4I_3^{(6)}(t) + 20I_4^{(6)}(t))\right] + \ldots\right\}, \tag{10,2.17}$$

$$\langle P_2(\cos\beta'(t))\rangle = \exp\left\{-6\left[\frac{kT}{I}I^{(2)}(t) + 2\left(\frac{kT}{I}\right)^2 I_2^{(4)}(t) + \left(\frac{kT}{I}\right)^3 (4I_3^{(6)}(t) + 20I_4^{(6)}(t))\right] + \ldots\right\},$$

$$\langle\cos\beta'(t)\rangle = 1 - 2\frac{kT}{I}I^{(2)}(t) + 4\left(\frac{kT}{I}\right)^2 (I_1^{(4)}(t) + I_2^{(4)}(t)) - 8\left(\frac{kT}{I}\right)^3 (I_1^{(6)}(t) + 2I_2^{(6)}(t) + I_3^{(6)}(t) + 2I_4^{(6)}(t)) + \ldots, \quad (10,2.18)$$

$$\langle Y_{jm}^*(\beta(0),\alpha(0))Y_{jm'}(\beta(t),\alpha(t))\rangle = \frac{\delta_{mm'}}{4\pi}\exp\left\{-j(j+1)\left[\frac{kT}{I}I^{(2)}(t) + 2\left(\frac{kT}{I}\right)^2 I_2^{(4)}(t) + \left(\frac{kT}{I}\right)^3 (4I_3^{(6)}(t) + 20I_4^{(6)}(t))\right] + \ldots\right\},$$

$$\frac{\langle Y_{jm}^*(\beta(0),\alpha(0))Y_{jm}(\beta(t),\alpha(t))\rangle}{\langle Y_{jm}^*(\beta(0),\alpha(0))Y_{jm}(\beta(0),\alpha(0))\rangle} = \exp\left\{-j(j+1)\left[\frac{kT}{I}I^{(2)}(t) + 2\left(\frac{kT}{I}\right)^2 I_2^{(4)}(t) + \left(\frac{kT}{I}\right)^3 (4I_3^{(6)}(t) + 20I_4^{(6)}(t))\right] + \ldots\right\}. \quad (10,2.19)$$

On inserting into the Kubo relation as expressed by (9,2.33),

$$\frac{\alpha(\omega)}{\alpha_s} = 1 - i\omega\int_0^\infty \langle\cos\beta'(t)\rangle e^{-i\omega t}dt,$$

the value of $\langle\cos\beta'(t)\rangle$ given by (10,2.18) we obtain agreement with (10,1.9) to terms proportional to γ^3, which is as far as we have taken the calculations in the present section. We likewise have agreement with (10,1.10) derived by Sack (1957b). If we truncate the series on the right-hand side of (10,1.9) after the first term proportional to γ^2, we obtain (9,3.12). Hence in this approximation the expression for complex polarizability is formally the same for the sphere and the linear rotator. This implies that the same is true, if the calculations are restricted to terms proportional to γ. Indeed it is clear from (9,2.27), (9,2.35) and (10,2.19) that in this approximation the normalized autocorrelation function of Y_{jm} for both the sphere and the linear rotator is $\exp[-j(j+1)\gamma(Bt-1+e^{-Bt})]$. In particular, the autocorrelation function of $\cos\beta'(t)$ is $\exp[-2\gamma(Bt-1+e^{-Bt})]$. This may be compared with the corresponding exact correlation function $\exp[-\gamma(Bt-1+e^{-Bt})]$ found in (8,3.16) for the rotating disc.

In Section 12.4 a new expression for $\langle R(t)\rangle$ will be derived.

Kluk and Powles (1975) investigated the Brownian motion of the linear rotator employing the equation

$$\frac{\partial w(\theta,\phi,t)}{\partial t} = D(t)\Delta_s w(\theta,\phi,t), \tag{10,2.20}$$

where Δ_s is the Laplacian operator on the surface of a sphere and $D(t)$ is the $R_{ii}(t)$ of (9,3.8). If D were a constant, (10,2.20) would be a known Smoluchowski equation that neglects inertial effects (Abragam, 1961, p. 298). They deduce from (10,2.20) that

$$\frac{\langle Y^*_{jm}(\beta(0),\alpha(0))Y_{jm}(\beta(t),\alpha(t))\rangle}{\langle Y^*_{jm}(\beta(0),\alpha(0))Y_{jm}(\beta(0),\alpha(0))\rangle} = \exp[-j(j+1)(Bt-1+e^{-Bt})]. \tag{10,2.21}$$

We see from (9,4.2) that (10,2.21) for $j=1$ and $j=2$ gives Steele's results (9,3.9) and (9,3.10) for the sphere.

11

Brownian Motion of the Asymmetric Rotator

11.1 Angular Velocity Components for the Asymmetric Rotator

We study the rotational Brownian motion of a body that may have no axis of symmetry (Ford *et al.*, 1977, 1978b, 1979). The results will, of course, be applicable also to a symmetric rotator, that is, a body having an axis of symmetry through its centre of mass. We employ stochastic differential equations rather than a Fokker–Planck equation, since the former have been found more useful for the calculation of correlation functions in the previous three chapters.

Let us return to Section 7.1. We assume that the problem is a hydrodynamical one and that the frictional couple about a principal axis of inertia is proportional to the component of angular velocity about that axis. The equations of motion are then the Euler–Langevin equations (7, 1.1), a typical one of which is

$$\dot{\omega}_i(t) - \frac{I_j - I_k}{I_i}\,\omega_j(t)\omega_k(t) = -B_i\omega_i(t) + \frac{dW_i(t)}{dt}, \tag{11,1.1}$$

where i, j, k is a cyclic permutation of 1, 2, 3. If the left-hand side were just $\dot{\omega}_i(t)$, then from (5,3.3) the steady state solution of (11,1.1) would be

$$\omega_i(t) = \int_{-\infty}^{t} e^{-B_i(t-y)}\,\frac{dW_i(y)}{dy}\,dy, \tag{11,1.2}$$

a centred Gaussian random variable. Moreover from (5,3.2), (5,3.4) and (5,3.6)

$$\left\langle \frac{dW_i(t)}{dt} \frac{dW_l(s)}{ds} \right\rangle_\omega = \delta_{il} \frac{2B_i kT}{I_i} \delta(t-s),$$

the ensemble average being taken over angular velocity space only. To put the dependence of $\omega_i(t)$ on the physical constants more in evidence we introduce $W_i'(t)$ by

$$W_i(t) = \left(\frac{2B_i kT}{I_i}\right)^{1/2} W_i'(t),$$

so that

$$\left\langle \frac{dW_i'(t)}{dt} \frac{dW_l'(s)}{ds} \right\rangle_\omega = \delta_{il}\delta(t-s) \qquad (11,1.3)$$

and (11,1.1) becomes

$$\dot{\omega}_i(t) - \frac{I_j - I_k}{I_i} \omega_j(t)\omega_k(t) = -B_i\omega_i(t) + \left(\frac{2B_i kT}{I_i}\right)^{1/2} \frac{dW_i'(t)}{dt}. \qquad (11,1.4)$$

If the second term on the left-hand side vanishes, the steady state solution is, from (11,1.2),

$$\omega_i(t) = 2^{1/2} B_i^{3/2} \left(\frac{kT}{I_i B_i^2}\right)^{1/2} \int_{-\infty}^{t} e^{-B_i(t-y)} \frac{dW_i'(y)}{dy} dy. \qquad (11,1.5)$$

In the investigation of the disc, sphere and linear rotator we have assumed that $kT/(I_i B_i^2) \ll 1$. We retain this assumption for $i = 1, 2, 3$. Introducing the dimensionless quantity ϵ by the obvious generalization of (8,3.10):

$$\epsilon = \frac{(kT)^{1/2}}{(I_1 I_2 I_3 B_1^2 B_2^2 B_3^2)^{1/6}} \qquad (11,1.6)$$

we then see that $\epsilon^2 \ll 1$. To simplify (11,1.4) further we put

$$\lambda_i = \frac{I_j - I_k}{I_i}, \quad \mu_i = \left(\frac{I_j I_k B_j^2 B_k^2}{I_i^2 B_i^4}\right)^{1/6} \qquad (11,1.7)$$

and employing (11,1.6) obtain

$$\dot{\omega}_i(t) - \lambda_i \omega_j(t)\omega_k(t) = -B_i\omega_i(t) + 2^{1/2}B_i^{3/2}\mu_i\epsilon\frac{dW_i'(t)}{dt}. \tag{11,1.8}$$

When $\lambda_i \neq 0$, these equations are nonlinear.

We seek a steady state solution of (11,1.8) in the form of a power series in ϵ:

$$\omega_i(t) = \omega_i^{(0)}(t) + \epsilon\omega_i^{(1)}(t) + \epsilon^2\omega_i^{(2)}(t) + \epsilon^3\omega_i^{(3)}(t) + \dots. \tag{11,1.9}$$

On substituting into (11,1.8) and equating the coefficients of ϵ^0 we deduce that

$$\dot{\omega}_i^{(0)}(t) - \lambda_i\omega_j^{(0)}(t)\omega_k^{(0)}(t) = -B_i\dot{\omega}_i^{(0)}(t).$$

These equations describe the motion of the rotator under the influence of the frictional couple only, the driving couple being absent. In the steady state there is then no motion. Hence $\omega_i^{(0)}(t) = 0$ and (11,1.9) reduces to

$$\omega_i(t) = \epsilon\omega_i^{(1)}(t) + \epsilon^2\omega_i^{(2)}(t) + \epsilon^3\omega_i^{(3)}(t) + \dots. \tag{11,1.10}$$

On substituting into (11,1.8) and equating the terms proportional to ϵ we have

$$\epsilon\dot{\omega}_i(t) = -B\epsilon\omega_i^{(1)}(t) + 2^{1/2}B_i^{3/2}\mu_i\epsilon\frac{dW_i'(t)}{dt}.$$

Thus $\epsilon\omega_i^{(1)}(t)$ satisfies (11,1.8) with $\lambda_i = 0$, and we deduce from (11,1.5) that the steady state solution is

$$\epsilon\omega_i^{(1)}(t) = 2^{1/2}B_i^{3/2}\left(\frac{kT}{I_iB_i^2}\right)^{1/2}\int_{-\infty}^{t} e^{-B_i(t-t_1)}\frac{dW_i'(t)}{dt_1}dt_1. \tag{11,1.11}$$

Hence, by (11,1.3), we have for a steady state

$$\epsilon^2\langle\omega_l^{(1)}(t)\omega_m^{(1)}(t)\rangle_{\underline{\omega}} = 0 \qquad (l \neq m),$$

so that the cross-correlation function of any two of $\omega_1^{(1)}(t)$, $\omega_2^{(1)}(t)$, $\omega_3^{(1)}(t)$ vanishes. Since they are centred, their covariance vanishes by (3,3.15) and they are therefore uncorrelated, by definition. Since they are also Gaussian, they are independent by a theorem of

Section 3.4. Besides, since $\epsilon\omega_i^{(1)}(t)$ satisfies the linear Langevin equation, eqn (5,3.8) gives

$$\epsilon^2\langle\omega_l^{(1)}(t)\omega_m^{(1)}(s)\rangle_\omega = \delta_{lm}(kT/I_l)e^{-B_l(t-s)}. \qquad (11,1.12)$$

To obtain $\omega_i^{(2)}(t)$, $\omega_i^{(3)}(t)$, etc., we substitute (11,1.10) into (11,1.8) and equate coefficients of ϵ^2, ϵ^3, etc. The white noise term of (11, 1.8) does not appear in the equations, which are derived from

$$\begin{aligned}
&\epsilon^2\dot{\omega}_i^{(2)}(t) + \epsilon^3\dot{\omega}_i^{(3)}(t) + \epsilon^4\dot{\omega}_i^{(4)}(t) + \ldots \\
&= -B_i(\epsilon^2\omega_i^{(2)}(t) + \epsilon^3\omega_i^{(3)}(t) + \epsilon^4\omega_i^{(4)}(t) + \ldots) \\
&\quad + \lambda_i(\epsilon\omega_j^{(1)}(t) + \epsilon^2\omega_j^{(2)}(t) + \epsilon^3\omega_j^{(3)}(t) + \ldots) \times \\
&\qquad (\epsilon\omega_k^{(1)}(t) + \epsilon^2\omega_k^{(2)}(t) + \epsilon^3\omega_k^{(3)}(t) + \ldots).
\end{aligned}$$

From this we deduce that

$$\dot{\omega}_i^{(2)}(t) = -B_i\omega_i^{(2)}(t) + \lambda_i\omega_j^{(1)}(t)\omega_k^{(1)}(t), \qquad (11,1.13)$$

$$\dot{\omega}_i^{(3)}(t) = -B_i\omega_i^{(3)}(t) + \lambda_i(\omega_j^{(1)}(t)\omega_k^{(2)}(t) + \omega_j^{(2)}(t)\omega_k^{(1)}(t)) \qquad (11,1.14)$$

$$\begin{aligned}
\dot{\omega}_i^{(4)}(t) &= -B_i\omega_i^{(4)}(t) + \lambda_i(\omega_j^{(1)}(t)\omega_k^{(3)}(t) + \omega_j^{(2)}(t)\omega_k^{(1)}(t)), \\
&\quad + \omega_j^{(3)}(t)\omega_k^{(1)}(t)),
\end{aligned} \qquad (11,1.15)$$

etc. In order to find the steady state solution of (11,1.13) we proceed as in Section 5.2, expressing the equation as

$$\int_{t_2=-\infty}^{t_1} e^{B_it_2}d\omega_i^{(2)}(t_2) = -B_i\int_{-\infty}^{t_1}\omega_i^{(2)}(t_2)e^{B_it_2}dt_2$$

$$+ \lambda_i\int_{-\infty}^{t_1}\omega_j^{(1)}(t_2)\omega_k^{(1)}(t_2)e^{B_it_2}dt_2.$$

On integrating by parts the first term on the right-hand side we deduce that

$$\omega_i^{(2)}(t_1) = \lambda_i\int_{-\infty}^{t_1} e^{-B_i(t_1-t_2)}\omega_j^{(1)}(t_2)\omega_k^{(1)}(t_2)dt_2 \qquad (11,1.16)$$

with $\omega_j^{(1)}(t_2)$, $\omega_k^{(1)}(t_2)$ given by (11,1.11). Using this result we write down immediately from (11,1.14) and (11,1.15)

$$\omega_i^{(3)}(t_1) = \lambda_i \int_{-\infty}^{t_1} e^{-B_i(t_1 - t_2)} \times$$

$$(\omega_j^{(1)}(t_2)\omega_k^{(2)}(t_2) + \omega_j^{(2)}(t_2)\omega_k^{(1)}(t_2))dt_2, \tag{11,1.17}$$

$$\omega_i^{(4)}(t_1) = \lambda_i \int_{-\infty}^{t_1} e^{-B_i(t_1 - t_2)} \times$$

$$(\omega_j^{(1)}(t_2)\omega_k^{(3)}(t_2) + \omega_j^{(2)}(t_2)\omega_k^{(2)}(t_2) + \omega_j^{(3)}(t_2)\omega_k^{(1)}(t_2))dt_2. \tag{11,1.18}$$

Expressions for $\omega_i^{(5)}(t_1)$, $\omega_i^{(6)}(t_1)$, etc., may likewise be written down.

Since the product of two or more Gaussian random variables may not be Gaussian, we see from (11,1.16) that $\omega_i^{(2)}(t)$ need not be Gaussian. Similarly, we see from (11,1.17) and (11,1.18) that $\omega_i^{(3)}(t)$, $\omega_i^{(4)}(t)$ may not be Gaussian, and hence from (11,1.10) the angular velocity $\omega(t)$ may not be Gaussian. It is, however, clear that the integrand for $\omega_i^{(n)}(t)$ will contain the sum of products of n variables taken from $\omega_1^{(1)}$, $\omega_2^{(1)}$, $\omega_3^{(1)}$. When n is odd, each such product must contain an odd number of the variables and so, by (3,4.17), $\langle\omega_i^{(n)}(t)\rangle_\omega$ vanishes for n odd. When $n = 2$, we see from (11,1.16) that, since $\omega_j^{(1)}$ and $\omega_k^{(1)}$ in this equation are independent,

$$\langle\omega_i^{(2)}(t_1)\rangle_\omega = \lambda_i \int e^{-B_i(t_1 - t_2)}\langle\omega_j^{(1)}(t_2)\rangle\langle\omega_k^{(1)}(t_2)\rangle dt_2 = 0.$$

Then eqns (11,1.16)–(11,1.18) show that in the multiple integral expression for $\omega_i^{(4)}(t_1)$ in terms of the $\omega_i^{(1)}$'s each term of the integrand contains an odd number of $\omega_1^{(1)}$ or $\omega_2^{(1)}$ or $\omega_3^{(1)}$, and so $\langle\omega_i^{(4)}(t_1)\rangle_\omega$ vanishes. This is all we need for our subsequent investigations but it is not difficult to show quite generally that, for n odd or even,

$$\langle\omega_i^{(n)}(t)\rangle_\omega = 0. \tag{11,1.19}$$

We conclude from (11,1.10) that $\omega_i(t)$ is a centred and, in general, non-Gaussian random variable.

11.2 Averaging Method Study of the Rotation Operator

As in the investigation of the disc, sphere and linear rotator, we examine the mean value of the stochastic rotation operator $R(t)$ defined in Sections 7.3 and 4. No matter what is the shape of the body, (7,3.8) gives

$$\frac{dR(t)}{dt} = -i(\mathbf{J}\cdot\boldsymbol{\omega}(t))R(t), \tag{11,2.1}$$

where J_1, J_2, J_3 satisfy (7,4.2),

$$J_k J_l - J_l J_k = -i\sum_{n=1}^{3} \epsilon_{kln} J_n .$$

In the present case the operator $-i(\mathbf{J}\cdot\boldsymbol{\omega}(t))$ may no longer be supposed Gaussian. Consequently, if we identify this operator with $\epsilon O(t)$ in Section 5.4, we cannot accept eqns (5,4.15)–(5,4.17). Equations (11,1.10) and (11,2.1) yield

$$\frac{dR(t)}{dt} = -i\sum_{n=1}^{\infty} \epsilon^n(\mathbf{J}\cdot\boldsymbol{\omega}^{(n)}(t))R(t). \tag{11,2.2}$$

Guided by this equation we shall repeat some of the calculations of Section 5.4 and express results in a form suitable for application to the asymmetric rotator.

Let us put

$$K^{(n)}(t) = -i(\mathbf{J}\cdot\boldsymbol{\omega}^{(n)}(t)), \tag{11,2.3}$$

and express (11,2.2) as

$$\frac{dR(t)}{dt} = (\epsilon K^{(1)}(t) + \epsilon^2 K^{(2)}(t) + \epsilon^3 K^{(3)}(t) + \ldots)R(t). \tag{11, 2.4}$$

As in the previous four chapters we employ angular brackets without the suffix ω to denote an ensemble average over both angle and angular velocity variables for the steady state of the system in the absence of an external field. Since (11,2.1)–(11,2.4) involve only angular velocity variables and since (11,1.12), which we shall employ to calculate averages over angular velocity variables, does not involve angles, the ensemble averages that will occur in the course of calculating $\langle R(t)\rangle$ from (11,2.1) are equal to averages over angular velocity space alone, and we may use (11,1.12) and (11,1.19) to calculate them. Thus we can omit the suffix ω from angular brackets, as we claimed in Section 9.4 could be done for $\langle R(t)\rangle$ in the case of the asymmetric top. We then deduce from (11,2.3) that

$$\langle K^{(n)}(t)\rangle = 0. \tag{11,2.5}$$

We make an expansion similar to (5,4.2)

$$R(t) = (\mathbf{I} + \epsilon F^{(1)}(t) + \epsilon^2 F^{(2)}(t) + \ldots)\langle R(t)\rangle, \tag{11,2.6}$$

where

$$\langle F^{(i)}(t)\rangle = 0, \quad F^{(i)}(0) = 0, \tag{11,2.7}$$

the latter equation resulting from $R(0)$ being the identity operator **I**. As in (5,4.4) we suppose that $\langle R(t)\rangle$ satisfies an equation

$$\frac{d\langle R(t)\rangle}{dt} = (\epsilon\Omega^{(1)}(t) + \epsilon^2\Omega^{(2)}(t) + \epsilon^3\Omega^{(3)}(t) + \ldots)\langle R(t)\rangle, \tag{11,2.8}$$

where $\Omega^{(1)}$, $\Omega^{(2)}$, etc., are non-stochastic operators. On substituting (11,2.6) into (11,2.4), employing (11,2.8) and equating terms with the same power of ϵ we deduce that

$$\Omega^{(1)} + \dot{F}^{(1)} = K^{(1)}, \tag{11,2.9}$$

$$\Omega^{(2)} + \dot{F}^{(2)} = K^{(2)} + K^{(1)}F^{(1)} - F^{(1)}\Omega^{(1)}, \tag{11,2.10}$$

$$\Omega^{(3)} + \dot{F}^{(3)} = K^{(3)} + K^{(2)}F^{(1)} + K^{(1)}F^{(2)} - F^{(1)}\Omega^{(2)} - F^{(2)}\Omega^{(1)}, \tag{11,2.11}$$

$$\Omega^{(4)} + \dot{F}^{(4)} = K^{(4)} + K^{(3)}F^{(1)} + K^{(2)}F^{(2)} + K^{(1)}F^{(3)} - F^{(1)}\Omega^{(3)} - F^{(2)}\Omega^{(2)} - F^{(3)}\Omega^{(1)}, \tag{11,2.12}$$

etc.

We see from (11,2.5), (11,2.7) and (11,2.9) that, since $\Omega^{(1)}$ is non-stochastic,

$$\Omega^{(1)}(t) = 0. \tag{11,2.13}$$

Hence from (11,2.9)

$$F^{(1)}(t) = \int_0^t dt_1\, K^{(1)}(t_1),$$

where we have again used (11,2.7). Then from (11,2.10) and (11,2.13)

$$\Omega^{(2)}(t_1) = \int_0^{t_1} dt_2 \, \langle K^{(1)}(t_1) K^{(1)}(t_2) \rangle, \tag{11,2.14}$$

$$\dot{F}^{(2)}(t_1) = K^{(2)}(t_1) - \Omega^{(2)}(t_1) + \int_0^{t_1} dt_2 K^{(1)}(t_1) K^{(1)}(t_2)$$

$$F^{(2)}(t) = \int_0^t dt_1 K^{(2)}(t_1)$$

$$+ \int_0^t dt_1 \int_0^{t_1} dt_2 \, [K^{(1)}(t_1) K^{(1)}(t_2) - \langle K^{(1)}(t_1) K^{(1)}(t_2) \rangle].$$

We deduce from (11,2.11) that

$$\begin{aligned} \Omega^{(3)}(t) &= \langle \Omega^{(3)}(t) \rangle = \langle K^{(2)}(t) F^{(1)}(t) \rangle + \langle K^{(1)}(t) F^{(2)}(t) \rangle \\ &= 0, \end{aligned}$$

since $K^{(2)}(t)F^{(1)}(t)$ and $K^{(1)}(t)F^{(2)}(t)$ are expressible as sums of terms which contain three $\omega_i^{(1)}$'s. By extending (11,2.9)–(11,2.12) further it may easily be shown that $\Omega^{(2k+1)}(t)$ will vanish. There is no difficulty in deducing from (11,2.11) and (11,2.12) that

$$\begin{aligned} F^{(3)}(t) = & \int_0^t dt_1 K^{(3)}(t_1) + \int_0^t dt_1 \int_0^{t_1} dt_2 \times \\ & [K^{(2)}(t_1) K^{(1)}(t_2) + K^{(1)}(t_1) K^{(2)}(t_2)] \\ & + \int_0^t dt_1 \int_0^{t_1} dt_2 \int_0^{t_2} dt_3 \times \\ & [K^{(1)}(t_1) K^{(1)}(t_2) K^{(1)}(t_3) - K^{(1)}(t_1) \langle K^{(1)}(t_2) K^{(1)}(t_3) \rangle \\ & - K^{(1)}(t_2) \langle K^{(1)}(t_1) K^{(1)}(t_3) \rangle \\ & - K^{(1)}(t_3) \langle K^{(1)}(t_1) K^{(1)}(t_2) \rangle], \end{aligned}$$

$$\Omega^{(4)}(t_1) = \int_0^{t_1} dt_2 \times$$

$$\langle K^{(3)}(t_1)K^{(1)}(t_2) + K^{(2)}(t_1)K^{(2)}(t_2) + K^{(1)}(t_1)K^{(3)}(t_2)\rangle$$

$$+ \int_0^{t_1} dt_2 \int_0^{t_2} dt_3 \, \langle K^{(2)}(t_1)K^{(1)}(t_2)K^{(1)}(t_3) + K^{(1)}(t_1)K^{(2)}(t_2)K^{(1)}(t_3)$$

$$+ K^{(1)}(t_1)K^{(1)}(t_2)K^{(2)}(t_3)\rangle$$

$$+ \int_0^{t_1} dt_2 \int_0^{t_2} dt_3 \int_0^{t_3} dt_4 \, [\langle K^{(1)}(t_1)K^{(1)}(t_2)K^{(1)}(t_3)K^{(1)}(t_4)\rangle \qquad (11,2.15)$$

$$- \langle K^{(1)}(t_1)K^{(1)}(t_2)\rangle\langle K^{(1)}(t_3)K^{(1)}(t_4)\rangle$$
$$- \langle K^{(1)}(t_1)K^{(1)}(t_3)\rangle\langle K^{(1)}(t_2)K^{(1)}(t_4)\rangle$$
$$- \langle K^{(1)}(t_1)K^{(1)}(t_4)\rangle\langle K^{(1)}(t_2)K^{(1)}(t_3)\rangle] .$$

This is as far as we shall take these calculations, though it is evident that the method may be extended to higher orders of approximation. Since $\Omega^{(2k+1)}$ vanishes, (11,2.8) reduces to

$$\frac{d\langle R(t)\rangle}{dt} = (\epsilon^2 \Omega^{(2)}(t) + \epsilon^4 \Omega^{(4)}(t) + \epsilon^6 \Omega^{(6)}(t) + \ldots)\langle R(t)\rangle. \qquad (11,2.16)$$

11.3 Calculation of $\Omega^{(2)}(t)$ and $\Omega^{(4)}(t)$

We substitute into (11,2.14) and (11,2.15) the explicit expressions for $K^{(1)}$, $K^{(2)}$, $K^{(3)}$ given by (11,2.3). Thus

$$\Omega^{(2)}(t_1) = -\int_0^{t_1} \langle (\mathbf{J} \cdot \boldsymbol{\omega}^{(1)}(t_1))(\mathbf{J} \cdot \boldsymbol{\omega}^{(1)}(t_2))\rangle$$

and (11,1.12) gives

$$\epsilon^2 \Omega^{(2)}(t_1) = \sum_{r=1}^{3} \frac{kT}{I_r} J_r^2 \int_0^{t_1} e^{-B_r(t_1 - t_2)} dt_2 ,$$

that is

$$\epsilon^2 \Omega^{(2)}(t_1) = -\frac{kT}{I_1 B_1}(1 - e^{-B_1 t_1})J_1^2 - \frac{kT}{I_2 B_2}(1 - e^{-B_2 t_1})J_2^2$$

$$-\frac{kT}{I_3 B_3}(1-e^{-B_3 t_1})J_3^2. \tag{11,3.1}$$

We express $\Omega^{(4)}(t)$ by

$$\Omega^{(4)}(t_1) = \Omega_2^{(4)}(t_1) + \Omega_3^{(4)}(t_1) + \Omega_4^{(4)}(t_1), \tag{11,3.2}$$

where $\Omega_2^{(4)}(t_1)$ is the part of the right-hand side of (11,2.15) whose integrand contains the product of two K's, $\Omega_3^{(4)}(t_1)$ is the part where the integrand contains the product of three K's and $\Omega_4^{(4)}(t_1)$ is that where the integrand contains the product of four K's. Explicitly we have

$$\begin{aligned}\Omega_2^{(4)}(t_1) = -\int_0^{t_1} dt_2\, \{ &\langle \omega_1^{(1)}(t_1)\omega_1^{(3)}(t_2) + \omega_1^{(2)}(t_1)\omega_1^{(2)}(t_2) \\ &+ \omega_1^{(3)}(t_1)\omega_1^{(1)}(t_2)\rangle J_1^2 \\ &+ \langle \omega_2^{(1)}(t_1)\omega_2^{(3)}(t_2) + \omega_2^{(2)}(t_1)\omega_2^{(2)}(t_2) \\ &+ \omega_2^{(3)}(t_1)\omega_2^{(1)}(t_2)\rangle J_2^2 \\ &+ \langle \omega_3^{(1)}(t_1)\omega_3^{(3)}(t_2) + \omega_3^{(2)}(t_1)\omega_3^{(2)}(t_2) \\ &+ \omega_3^{(3)}(t_1)\omega_3^{(1)}(t_2)\rangle J_3^2 \} ,\end{aligned} \tag{11,3.3}$$

$$\begin{aligned}\Omega_3^{(4)}(t_1) = i\int_0^{t_1} dt_2 \int_0^{t_2} dt_3 \times \\ \{ &\langle \omega_1^{(2)}(t_1)\omega_2^{(1)}(t_2)\omega_3^{(1)}(t_3) + \omega_1^{(1)}(t_1)\omega_2^{(2)}(t_2)\omega_3^{(1)}(t_3) \\ &+ \omega_1^{(1)}(t_1)\omega_2^{(1)}(t_2)\omega_3^{(2)}(t_3)\rangle J_1 J_2 J_3 \\ &+ \langle \omega_1^{(2)}(t_1)\omega_3^{(1)}(t_2)\omega_2^{(1)}(t_3) + \omega_1^{(1)}(t_1)\omega_3^{(2)}(t_2)\omega_2^{(1)}(t_3) \\ &+ \omega_1^{(1)}(t_1)\omega_3^{(1)}(t_2)\omega_2^{(2)}(t_3)\rangle J_1 J_3 J_2 \\ &+ \langle \omega_2^{(2)}(t_1)\omega_3^{(1)}(t_2)\omega_1^{(1)}(t_3) + \omega_2^{(1)}(t_1)\omega_3^{(2)}(t_2)\omega_1^{(1)}(t_3) \\ &+ \omega_2^{(1)}(t_1)\omega_3^{(1)}(t_2)\omega_1^{(2)}(t_3)\rangle J_2 J_3 J_1 \\ &+ \langle \omega_2^{(2)}(t_1)\omega_1^{(1)}(t_2)\omega_3^{(1)}(t_3) + \omega_2^{(1)}(t_1)\omega_1^{(2)}(t_2)\omega_3^{(1)}(t_3) \\ &+ \omega_2^{(1)}(t_1)\omega_1^{(1)}(t_2)\omega_3^{(2)}(t_3)\rangle J_2 J_1 J_3 \\ &+ \langle \omega_3^{(2)}(t_1)\omega_1^{(1)}(t_2)\omega_2^{(1)}(t_3) + \omega_3^{(1)}(t_1)\omega_1^{(2)}(t_2)\omega_2^{(1)}(t_3) \\ &+ \omega_3^{(1)}(t_1)\omega_1^{(1)}(t_2)\omega_2^{(2)}(t_3)\rangle J_3 J_1 J_2\end{aligned}$$

$$+\langle\omega_3^{(2)}(t_1)\omega_2^{(1)}(t_2)\omega_1^{(1)}(t_3)+\omega_3^{(1)}(t_1)\omega_2^{(2)}(t_2)\omega_1^{(1)}(t_3)$$
$$+\omega_3^{(1)}(t_1)\omega_2^{(1)}(t_2)\omega_1^{(2)}(t_3)\rangle J_3J_2J_1\},\qquad(11,3.4)$$

$$\Omega_4^{(4)}(t_1)=\int_0^{t_1}dt_2\int_0^{t_2}dt_3\int_0^{t_3}dt_4\times$$
$$\left\{\sum_{r,s,u,v=1}^{3}\langle\omega_r^{(1)}(t_1)\omega_s^{(1)}(t_2)\omega_u^{(1)}(t_3)\omega_v^{(1)}(t_4)\rangle J_rJ_sJ_uJ_v\right.$$
$$(11,3.5)$$
$$-\sum_{r,u=1}^{3}(\langle\omega_r^{(1)}(t_1)\omega_r^{(1)}(t_2)\rangle\langle\omega_u^{(1)}(t_3)\omega_u^{(1)}(t_4)\rangle$$
$$+\langle\omega_r^{(1)}(t_1)\omega_r^{(1)}(t_3)\rangle\langle\omega_u^{(1)}(t_2)\omega_u^{(1)}(t_4)\rangle$$
$$\left.+\langle\omega_r^{(1)}(t_1)\omega_r^{(1)}(t_4)\rangle\langle\omega_u^{(1)}(t_2)\omega_u^{(1)}(t_3)\rangle)J_r^2J_u^2\right\}.$$

The integrations involved in the last three equations are elementary but tedious. We multiply by ϵ^4, take $\epsilon\omega_i^{(1)}(t)$ from (11,1.11), $\epsilon^2\omega_i^{(2)}(t)$ from (11,1.16) combined with (11,1.11), $\epsilon^3\omega_i^{(3)}(t)$ from (11,1.17) combined with (11,1.11) and (11,1.16). Since $\omega_i^{(1)}(t)$ is a centred Gaussian random variable, we employ (3,4.17), (3,4.18) and (11,1.12) to calculate the ensemble average of a continued product of $\omega_i^{(1)}$'s. The evaluation of $\epsilon^4\Omega_2^{(4)}(t_1)$, $\epsilon^4\Omega_3^{(4)}(t_1)$, $\epsilon^4\Omega_4^{(4)}(t_1)$ is outlined in Appendix E. The resulting expression for $\epsilon^4\Omega^{(4)}(t_1)$ obtained from (11,3.2) is extremely lengthy. Fortunately for our subsequent applications we need only the asymptotic value of $\epsilon^4\Omega^{(4)}(t_1)$ as t_1 tends to infinity. This is given by

$$\epsilon^4\Omega^{(4)}(\infty)=\frac{(kT)^2}{I_1I_2I_3}\left\{-\frac{i}{3}P\left[I_1\frac{B_2-B_3}{B_2^2B_3^2}+I_2\frac{B_3-B_1}{B_3^2B_1^2}+I_3\frac{B_1-B_2}{B_1^2B_2^2}\right]\right.$$
$$-\sum J_1^2\left[I_1\frac{2B_2B_3(B_2+B_3)-B_1(B_2^2+B_2B_3+B_3^2)}{B_1B_2^2B_3^2(B_2+B_3)}\right.$$
$$(11,3.6)$$
$$+I_2\frac{B_2(B_2+B_3)-2B_3^2}{B_1B_2B_3^2(B_2+B_3)}+I_3\frac{B_3(B_2+B_3)-2B_2^2}{B_1B_2^2B_3(B_2+B_3)}$$
$$\left.\left.-\frac{(I_2-I_3)^2}{I_1B_1^2(B_2+B_3)}\right]\right\},$$

where Σ denotes the sum over cyclic permutations of 1, 2, 3 and P is the symmetric operator defined by

$$P = J_1J_2J_3 + J_1J_3J_2 + J_2J_3J_1 + J_2J_1J_3 + J_3J_1J_2 + J_3J_2J_1. \tag{11,3.7}$$

11.4 Correlation Functions for Angular Velocity Components

Before proceeding with the solution of the differential equation (11,2.16) we employ results obtained in the course of the calculations of the previous section to derive correlation functions for $\omega_1(t)$, $\omega_2(t)$, $\omega_3(t)$. For $i, l = 1, 2, 3$ we have from (11,1.10)

$$\begin{aligned}\langle\omega_i(t)\omega_l(s)\rangle &= \epsilon^2\langle\omega_i^{(1)}(t)\omega_l^{(1)}(s)\rangle \\ &+ \epsilon^3\langle\omega_i^{(1)}(t)\omega_l^{(2)}(s) + \omega_i^{(2)}(t)\omega_l^{(1)}(s)\rangle \\ &+ \epsilon^4\langle\omega_i^{(1)}(t)\omega_l^{(3)}(s) + \omega_i^{(2)}(t)\omega_l^{(2)}(s) \\ &+ \omega_i^{(3)}(t)\omega_l^{(1)}(s)\rangle + \dots .\end{aligned} \tag{11,4.1}$$

If l, m, n is a cyclic permutation of 1, 2, 3, eqn (11,1.16) gives

$$\langle\omega_i^{(1)}(t)\omega_l^{(2)}(s)\rangle = \lambda_i\int_{-\infty}^{s} \exp\left[-B_l(s-u)\right] \times \langle\omega_i^{(1)}(t)\omega_m^{(1)}(u)\omega_n^{(1)}(u)\rangle du = 0,$$

by (3,4.17). Similarly $\langle\omega_i^{(2)}(t)\omega_l^{(1)}(s)\rangle$ vanishes, so that (11,4.1) reduces to

$$\begin{aligned}\langle\omega_i(t)\omega_l(s)\rangle &= \epsilon^2\langle\omega_i^{(1)}(t)\omega_l^{(1)}(s)\rangle \\ &+ \epsilon^4\langle\omega_i^{(1)}(t)\omega_l^{(3)}(s) + \omega_i^{(2)}(t)\omega_l^{(2)}(s) + \omega_i^{(3)}(t)\omega_l^{(1)}(s)\rangle + \dots .\end{aligned} \tag{11,4.2}$$

Let us next consider $\langle\omega_1(t)\omega_2(s)\rangle$. On account of the independence of $\omega_1^{(1)}(t)$ and $\omega_2^{(1)}(s)$, $\langle\omega_1^{(1)}(t)\omega_2^{(1)}(s)\rangle$ vanishes. Moreover from (11,1.17) and (11,1.16)

$$\langle\omega_1^{(1)}(t)\omega_2^{(3)}(s)\rangle = \lambda_2\int_{-\infty}^{s} \exp\left[-B_2(s-u)\right] \times \langle\omega_1^{(1)}(t)(\omega_3^{(1)}(u)\omega_1^{(2)}(u) + \omega_3^{(2)}(u)\omega_1^{(1)}(u))\rangle du$$

$$= \lambda_1 \lambda_2 \int_{-\infty}^{s} du \int_{-\infty}^{u} dv \exp\left[-B_2(s-u) - B_1(u-v)\right] \times$$

$$\langle \omega_1^{(1)}(t)\omega_3^{(1)}(u)\omega_2^{(1)}(v)\omega_3^{(1)}(v)\rangle$$

$$+ \lambda_2 \lambda_3 \int_{-\infty}^{s} du \int_{-\infty}^{u} dv \exp\left[-B_2(s-u) - B_3(u-v)\right] \times$$

$$\langle \omega_1^{(1)}(t)\omega_1^{(1)}(v)\omega_2^{(1)}(v)\omega_1^{(1)}(u)\rangle = 0,$$

and similarly $\langle \omega_1^{(2)}(t)\omega_2^{(2)}(s)\rangle$, $\langle \omega_1^{(3)}(t)\omega_2^{(1)}(s)\rangle$ vanish. Hence

$$\langle \omega_i(t)\omega_l(s)\rangle = 0 \qquad (i \neq l), \tag{11,4.3}$$

and we need concern ourselves only with the case of $i = l$.

We therefore calculate $\langle \omega_1(t)\omega_1(s)\rangle$, where for the present we take $t \geqslant s$. Then from (11,1.12) and (11,4.2)

$$\langle \omega_1(t)\omega_1(s)\rangle = (kT/I_1)e^{-B_1(t-s)} + \epsilon^4 \langle \omega_1^{(1)}(t)\omega_1^{(3)}(s) + \omega_1^{(2)}(t)\omega_1^{(2)}(s) + \omega_1^{(3)}(t)\omega_1^{(1)}(s)\rangle + \ldots .$$

Referring to Appendix E we deduce from eqns (E.2), (E.4) and (E.6) that

$$\epsilon^3 \langle \omega_1^{(1)}(t)\omega_1^{(3)}(s) + \omega_1^{(2)}(t)\omega_1^{(2)}(s) + \omega_1^{(3)}(t)\omega_1^{(1)}(s)\rangle$$

$$= \frac{(\lambda_1 kT)^2}{I_2 I_3} \left\{ \frac{\exp\left[-(B_2+B_3)(t-s)\right]}{(B_1-B_2-B_3)(B_1+B_2+B_3)} - \frac{(B_2+B_3)\exp\left[-B_1(t-s)\right]}{B_1(B_1-B_2-B_3)(B_1+B_2+B_3)} - \frac{\exp\left[-B_1(t-s)\right]}{2B_1(B_1+B_2+B_3)} + \frac{(3B_1-B_2-B_3)\exp\left[-B_1(t-s)\right]}{2B_1(B_1-B_2-B_3)^2} + \frac{(t-s)\exp\left[-B_1(t-s)\right]}{B_1-B_2-B_3} - \frac{2B_1\exp\left[-(B_2+B_3)(t-s)\right]}{(B_1+B_2+B_3)(B_1-B_2-B_3)^2} \right\}.$$

On collecting terms, dropping the restrictions $i = l$, $t \geqslant s$ and noting (11,4.3) we find that

$$\langle \omega_i(t)\omega_l(s)\rangle = \delta_{il}\left\{(kT/I_i)\exp\left[-B_i|t-s|\right] + \left(\frac{I_j - I_k}{I_i}\right)^2 \frac{(kT)^2}{I_j I_k} \times\right.$$

$$\left.\frac{\exp[-B_i|t-s|]\{1-(B_j+B_k-B_i)|t-s|-\exp[-(B_j+B_k-B_i)|t-s|]\}}{(B_j+B_k-B_i)^2}\right\}$$

$$+\ldots. \qquad (11,4.4)$$

This agrees with a result obtained by Hubbard (1977, eqn (6.2)) from a Fokker–Planck equation.

We see from (11,4.4) that

$$\langle\omega_i(0)\omega_i(0)\rangle = \frac{kT}{I_i}.$$

Hence the normalized autocorrelation function of $\omega_i(t)$,

$$\frac{\langle\omega_i(0)\omega_i(t)\rangle}{\langle\omega_i(0)\omega_i(0)\rangle} = \exp[-B_i|t|]$$

$$+\frac{(I_j-I_k)^2 kT\exp[-B_i|t|]\{1-(B_j+B_k-B_i)|t|-\exp[-(B_j+B_k-B_i)|t|]\}}{I_1 I_2 I_3(B_j+B_k-B_i)^2}$$

$$+\ldots. \qquad (11,4.5)$$

11.5 Solution of the Differential Equation for $\langle R(t)\rangle$

Equation (11,2.16) for $\langle R(t)\rangle$ has the same form as (9,2.5) for the sphere and the linear rotator. However $\Omega^{(2)}(t)$, $\Omega^{(4)}(t)$, etc., for the sphere and linear rotator involve only the operators J^2 and J_3, so they commute with one another. On the other hand we see from (11,3.1) and (11,3.6) that $\Omega^{(2)}(t)$ and $\Omega^{(4)}(t)$ do not commute in the case of the asymmetric rotator. To deal with this altered situation we employ a method of solution due to Krylov and Bologliubov (1947), and later presented by Bogoliubov and Mitropolsky (1961) and by Case (1966). We follow more closely the presentation of Case.

The method is essentially the same as that used in Sections 5.4 and 11.2 for the solution of stochastic differential equations, though we are now concerned only with non-stochastic operators. Indeed the technique used in the earlier sections is just an extension of the Krylov–Bogoliubov method to stochastic variables. Let us write

$$\epsilon^2\Omega^{(2)}(t)+\epsilon^4\Omega^{(4)}(t)+\ldots = \Omega(t), \qquad (11,5.1)$$

so that we have to find a solution of

$$\frac{d}{dt}\langle R(t)\rangle = \Omega(t)\langle R(t)\rangle, R(0) = \mathbf{I}. \qquad (11,5.2)$$

We express the solution as

$$\langle R(t)\rangle = V(t)U(t), \tag{11,5.3}$$

where $U(t)$ satisfies equations

$$\frac{dU(t)}{dt} = GU(t), \qquad U(0) = \mathbf{I}, \tag{11,5.4}$$

G being a time independent operator. $V(t)$ gives a small correction to the value $U(t)$ of $\langle R(t)\rangle$. On substituting (11,5.3) into (11,5.2) and using (11,5.4) we find that $V(t)$ satisfies

$$\frac{dV(t)}{dt} = \Omega(t)V(t) - V(t)G. \tag{11,5.5}$$

Since $\Omega(t)$ is expanded in (11,5.1) as a power series in ϵ^2, we now put

$$G = \epsilon^2 G^{(2)} + \epsilon^4 G^{(4)} + \ldots \tag{11,5.6}$$

$$V(t) = \mathbf{I} + \epsilon^2 V^{(2)}(t) + \epsilon^4 V^{(4)}(t) + \epsilon^6 V^{(6)}(t) + \ldots . \tag{11,5.7}$$

Then $V^{(2)}(t)$, $V^{(4)}(t)$, etc., must be bounded for all values of t and, from (11,5.3) and (11,5.4), $V^{(2)}(0)$, $V^{(4)}(0)$, etc, must vanish.

On substituting (11,5.6) and (11,5.7) into (11,5.5) and equating coefficients of ϵ^2, ϵ^4, etc., we obtain the following equations to determine $V^{(2)}(t)$, $V^{(4)}(t)$, etc.:

$$\frac{dV^{(2)}(t)}{dt} = \Omega^{(2)}(t) - G^{(2)}, \tag{11,5.8}$$

$$\frac{dV^{(4)}(t)}{dt} = \Omega^{(4)}(t) - G^{(4)} + \Omega^{(2)}(t)V^{(2)}(t) - V^{(2)}(t)G^{(2)}, \tag{11,5.9}$$

$$\begin{aligned}\frac{dV^{(6)}(t)}{dt} = {} & \Omega^{(6)}(t) - G^{(6)} + \Omega^{(4)}(t)V^{(2)}(t) - V^{(2)}(t)G^{(4)}, \\ & + \Omega^{(2)}(t)V^{(4)}(t) - V^{(4)}(t)G^{(2)},\end{aligned} \tag{11,5.10}$$

etc. Since $V^{(2)}(t)$ is bounded for t large, we must choose $G^{(2)}$ to satisfy

$$G^{(2)} = \Omega^{(2)}(\infty). \tag{11,5.11}$$

Thus, since $V^{(2)}(0)$ vanishes,

$$V^{(2)}(t) = \int_0^t (\Omega^{(2)}(t_1) - \Omega^{(2)}(\infty))dt_1, \qquad (11,5.12)$$

and (11,3.1) gives

$$\epsilon^2 V^{(2)}(t) = \frac{kT}{I_1 B_1^2}(1 - e^{-B_1 t})J_1^2 + \frac{kT}{I_2 B_2^2}(1 - e^{-B_2 t})J_2^2 + \frac{kT}{I_3 B_3^2}(1 - e^{-B_3 t})J_3^2 . \qquad (11,5.13)$$

Equations (11,5.9), (11,5.11) and (11,5.12) show that

$$\frac{dV^{(4)}(t)}{dt} = \Omega^{(4)}(t) - G^{(4)} + \Omega^{(2)}(t)\int_0^t (\Omega^{(2)}(t_1) - \Omega^{(2)}(\infty))dt_1 - \int (\Omega^{(2)}(t_1) - \Omega^{(2)}(\infty))dt_1 \Omega^{(2)}(\infty).$$

Since $V^{(4)}(t)$ is bounded for large t, the right-hand side of the last equation must vanish for $t = \infty$, and so

$$G^{(4)} = \Omega^{(4)}(\infty) + \int_0^\infty [\Omega^{(2)}(\infty), \Omega^{(2)}(t_1)]\, dt_1, \qquad (11,5.14)$$

the integrand being a commutator. This is as far as we take the general calculations of $V(t)$ and G.

Since G is time independent, the solution of (11,5.4) is

$$U(t) = e^{Gt}. \qquad (11,5.15)$$

Hence the solution of (11,5.2) is

$$\langle R(t)\rangle = (I + \epsilon^2 V^{(2)}(t) + \ldots) \exp\{[\epsilon^2 G^{(2)} + \epsilon^4 G^{(4)} + \ldots]\, t\} \qquad (11,5.16)$$

with $\epsilon^2 V^{(2)}(t)$ given by (11,5.13). From (11,3.1) and (11,5.11)

$$\epsilon^2 G^{(2)} = -D_1^{(1)} J_1^2 - D_2^{(1)} J_2^2 - D_3^{(1)} J_3^2 , \qquad (11,5.17)$$

where

$$D_1^{(1)} = \frac{kT}{I_1 B_1}, \quad D_2^{(1)} = \frac{kT}{I_2 B_2}, \quad D_3^{(1)} = \frac{kT}{I_3 B_3}. \qquad (11,5.18)$$

The $D_1^{(1)}$, $D_2^{(1)}$, $D_3^{(1)}$ are *first order diffusion coefficients*, generalizations of the diffusion coefficients $kT/(IB)$, that is, kT/ζ which first appeared in eqn (1,1.8) for the Debye theory of the rotational Brownian motion of the sphere.

We see from (11,3.1) that

$$\epsilon^4 [\Omega^{(2)}(\infty), \Omega^{(2)}(t_1)] = -\left[\sum_{i=1}^{3} \frac{kT}{I_i B_i} J_i^2, \sum_{j=1}^{3} \frac{kT}{I_j B_j} e^{-B_j t_1} J_j^2\right]. \tag{11,5.19}$$

Now from eqn (E.19) of Appendix E

$$J_2^2 J_3^2 - J_3^2 J_2^2 = J_3^2 J_1^2 - J_1^2 J_3^2 = J_1^2 J_2^2 - J_2^2 J_1^2 = -\frac{2i}{3} P \tag{11,5.20}$$

with P defined by (11,3.7). On substitution of (11,5.20) into (11,5.19) we deduce that

$$\begin{aligned}&\epsilon^4 [\Omega^{(2)}(\infty), \Omega^{(2)}(t_1)]\\ &= \frac{2i}{3} P \frac{(kT)^2}{I_1 I_2 I_3 B_1 B_2 B_3} \{I_1 B_1 (e^{-B_3 t_1} - e^{-B_2 t_1})\\ &+ I_2 B_2 (e^{-B_1 t_1} - e^{-B_3 t_1}) + I_3 B_3 (e^{-B_2 t_1} - e^{-B_1 t_1})\},\end{aligned}$$

and therefore that

$$\begin{aligned}&\epsilon^4 \int_0^\infty [\Omega^{(2)}(\infty), \Omega^{(2)}(t_1)]\, dt_1\\ &= \frac{2i}{3} P \frac{(kT)^2}{I_1 I_2 I_3} \left\{\frac{I_1 (B_2 - B_3)}{B_2^2 B_3^2} + \frac{I_2 (B_3 - B_1)}{B_3^2 B_1^2}\right.\\ &\left.+ \frac{I_3 (B_1 - B_2)}{B_1^2 B_2^2}\right\}.\end{aligned} \tag{11,5.21}$$

On calculating $\epsilon^4 G^{(4)}$ from (11,5.14) we see that the presence of the commutator changes the sign of the term proportional to P in (11,3.6) but leaves the rest unaltered. Just as in (11,5.17) we denoted the coefficient of $-J_i^2$ by $D_i^{(1)}$, so now we denote the coefficient of $-J_i^2$ in $\epsilon^4 G^{(4)}$ by $D_i^{(2)}$ and call $D_1^{(2)}$, $D_2^{(2)}$, $D_3^{(2)}$ the *second order diffusion coefficients.* Hence from (11,3.6), (11,5.14) and (11,5.21)

$$\epsilon^4 G^{(4)} = -D_1^{(2)} J_1^2 - D_2^{(2)} J_2^2 - D_3^{(2)} J_3^2 + \frac{i}{3} P \frac{(kT)^2}{I_1 I_2 I_3} \left\{ I_1 \frac{B_2 - B_3}{B_2^2 B_3^2} + I_2 \frac{B_3 - B_1}{B_3^2 B_1^2} + I_3 \frac{B_1 - B_2}{B_1^2 B_2^2} \right\}, \tag{11,5.22}$$

where

$$D_1^{(2)} = \frac{(kT)^2}{I_1 I_2 I_3} \left[I_1 \frac{2B_2 B_3 (B_2 + B_3) - B_1 (B_2^2 + B_2 B_3 + B_3^2)}{B_1 B_2^2 B_3^2 (B_2 + B_3)} + I_2 \frac{B_2 (B_2 + B_3) - 2B_3^2}{B_1 B_2 B_3^2 (B_2 + B_3)} + I_3 \frac{B_3 (B_2 + B_3) - 2B_2^2}{B_1 B_2^2 B_3 (B_2 + B_3)} - \frac{(I_2 - I_3)^2}{I_1 B_1^2 (B_2 + B_3)} \right], \tag{11,5.23}$$

etc. Collecting our results we deduce from (11,5.13), (11,5.16), (11,5.17) and (11,5.22) that

$$\langle R(t) \rangle = \left\{ \mathbf{I} + \sum_{i=1}^{3} \frac{kT}{I_i B_i^2} (1 - e^{-B_i t}) J_i^2 + \dots \right\} \times \exp\left\{ \left(-\sum_{i=1}^{3} (D_i^{(1)} + D_i^{(2)}) J_i^2 + \frac{i}{3} P \frac{(kT)^2}{I_1 I_2 I_3} \sum_{i=1}^{3} \frac{B_j - B_k}{B_j^2 B_k^2} + \dots \right) t \right\}, \tag{11,5.24}$$

i, j, k being a cyclic permutation of 1, 2, 3.

Let us now consider some implications of (11,5.24). If we neglect everything except $\mathbf{I}$ in the first set of braces and the terms proportional to $D_i^{(1)}$ in the second set, we obtain

$$\langle R(t) \rangle = \exp\left[-\sum_{i=1}^{3} \frac{kT}{I_i B_i} J_i^2 t \right]. \tag{11,5.25}$$

This is a generalization of the result

$$\langle R(t)\rangle = \exp\left[-\frac{kT}{IB}J^2 t\right]$$

found in (9,3.1) for the sphere in the Debye theory. We therefore say that (11,5.25) gives the value of $\langle R(t)\rangle$ in the *Debye approximation* for the asymmetric rotator.

For convenience we shall in this paragraph omit the subscripts of I's and the B's, and shall denote $kT/(IB^2)$ by γ. In approximating (11,5.24) by (11,5.25) we have omitted in the first set of braces terms of relative order of magnitude γ. In the exponent within the second set of braces $D_i^{(1)}$ is of order $kT/(IB)$, $D_i^{(2)}$ is of order $\gamma kT/(IB)$ and so is the coefficient of P. Hence (11,5.24) is a correction to (11,5.25) of first order in γ. To this order we were justified in calculating the series of (11,5.7) only as far as $\epsilon^2 V^{(2)}(t)$ and in calculating the series of (11,5.6) only as far as $\epsilon^4 G^{(4)}$. The time variation in the first factor of the right-hand side of (11,5.24) depends on e^{-Bt} and in the second factor approximately on $\exp(-\gamma Bt)$. Since $\gamma \ll 1$, the second exponential will decay much less rapidly than the first. When $B_i t \gg 1$ for $i = 1, 2, 3$, eqn (11,5.24) gives

$$\langle R(t)\rangle = \left\{\mathbf{I} + \sum \frac{kT}{I_i B_i^2} J_i^2\right\} e^{Gt} \quad (B_i t \gg 1). \tag{11,5.26}$$

This is the long time correction to the Debye approximation. It has a time exponent which differs from that in (11,5.25) by a quantity of relative order γ, and a constant multiplying factor which differs from $\mathbf{I}$ by a quantity of the same order. For short times we have in the multiplying factor an additional time dependent correction of this order.

The calculation of $\langle R(t)\rangle$ can be taken to any order of approximation by the methods employed in the present chapter, but it would clearly be very tedious to take it even to one higher order. Having calculated $\langle R(t)\rangle$ we may use it to evaluate orientational correlation functions as was explained in Section 7.4. We deduce from (19,4.1), (5,4.18) and (11,5.26) that

$$\lim_{t\to\infty} \langle Y_{jm}^*(\beta(0), \alpha(0)) Y_{jm'}(\beta(t), \alpha(t))\rangle$$

$$= \frac{1}{4\pi}\left\{\lim_{t\to\infty}(e^{Gt})^{j*}_{m'm} + \lim_{t\to\infty}\sum_{i=1}^{3}\frac{kT}{I_i B_i^2}(J_i^2 e^{Gt})^{j*}_{m'm}\right\}. \tag{11,5.27}$$

When t becomes very great, the two spherical harmonics become uncorrelated and the left-hand side vanishes. The same is therefore true for the right-hand side of (11,5.27). Since its second term is of order γ relative to the first, we must have

$$\lim_{t \to \infty} (e^{Gt})^{j*}_{m'm} = 0.$$

Hence every element of the representation of $\lim_{t \to \infty} e^{Gt}$ with respect to the basis Y_{js} vanishes. It then follows from a theorem at the end of Appendix A that

$$\lim_{t \to \infty} e^{Gt} = 0. \qquad (11,5.28)$$

11.6 Other Studies of the Asymmetric Rotator

Perrin (1934) extended the Debye theory of translational and rotational Brownian motion of the sphere to that of a body, not necessarily homogeneous, whose surface is symmetric with respect to three mutually perpendicular planes. The results of his investigations are often applied to a body with an ellipsoidal surface. Neglecting inertial effects he derived a Smoluchowski equation for rotational Brownian motion, and he solved it for a molecule with permanent dipole of moment μ in the presence of a periodic electric field which we write $F_0 e^{i\omega t}$. For orientational polarization this leads to (Perrin, 1934, eqn (89))

$$P(\omega) = \frac{NF_0}{3kT}\left\{\frac{\mu_1^2}{1+i\omega\tau_{D,1}} + \frac{\mu_2^2}{1+i\omega\tau_{D,2}} + \frac{\mu_3^2}{1+i\omega\tau_{D,3}}\right\}, \qquad (11,6.1)$$

where $P(\omega)$ is defined in Section 1.3, μ_1, μ_2, μ_3 are the components of electric moment along the coordinate axes which are chosen as the intersections of the planes of symmetry, and

$$\tau_{D,1} = \left(\frac{kT}{I_2 B_2} + \frac{kT}{I_3 B_3}\right)^{-1}, \text{ etc.} \qquad (11,6.2)$$

On comparing (11,6.1) with (1,3.2) we see that we have now three relaxation times $\tau_{D,1}, \tau_{D,2}, \tau_{D,3}$, which by (11,6.2) are independent of the direction of the dipole moment. We also have from (11,6.1) that $P(0) = N\mu^2 F_0/(3kT)$, so that by (2,1.13) and (2,1.14)

$$\alpha(\omega) = \alpha_s \left\{ \frac{n_1^2}{1 + i\omega\tau_{D,1}} + \frac{n_2^2}{1 + i\omega\tau_{D,2}} + \frac{n_3^2}{1 + i\omega\tau_{D,3}} \right\}, \quad (11,6.3)$$

where n_1, n_2, n_3 are the direction cosines of the dipole axis. This is the Perrin generalization of (2,1.17).

Budó *et al.* (1939) applied the Perrin theory to rigid ellipsoidal polar molecules. They tabulated the ratios of $\tau_{D,1}$, $\tau_{D,2}$, $\tau_{D,3}$ to the Debye relaxation time for a sphere with the volume of the ellipsoid, for different values of the ratios of the lengths of the major axes of the ellipsoid.

Furry (1957) discussed rotational Brownian motion of an asymmetric body without including inertial effects. He employed quaternion notation and established a Smoluchowski equation in terms of orthogonal coordinates introduced to specify the vector part of the quaternion that gives the rotation of the body. He obtained a spherically symmetric solution of the equation.

Favro (1960) developed the ideas of Furry but he worked with angular momentum operators. In place of diffusion coefficients he employed the rotational diffusion tensor D_{jk} defined by

$$D_{jk} = \frac{1}{2\Delta t} \int p(\boldsymbol{\epsilon}, \Delta t)\epsilon_j \epsilon_k \, d^3\epsilon,$$

where $p(\boldsymbol{\epsilon}, \Delta t)d^3\epsilon$ is the probability of the body undergoing a rotation $\boldsymbol{\epsilon}$ in time Δt and ϵ_j, ϵ_k are the components of $\boldsymbol{\epsilon}$ with respect to a body fixed frame. When D_{jk} is known, it is possible to deduce how the axis of a polar molecule relaxes as a result of the rotational Brownian motion of the molecule.

Ivanov (1963) investigated the rotational Brownian motion of molecules of arbitrary shape as a random walk problem. He employed the rotational diffusion tensor, and deduced an expression for the orientational probability density in the form of a series.

Morita (1978) studied the rotational Brownian motion of an asymmetric polar molecule when inertial effects are included. Denoting by $m_z'(t)$ the projection at time t of the dipole moment on an axis fixed in space he reported an approximate expression for the Laplace transform of the correlation function of $m_z'(t)$.

12 Complex Polarizability for Molecular Models

12.1 Further Investigation of Complex Polarizability

In Chapter 2 we introduced the concept of complex polarizability $\alpha(\omega)$ and related it to complex permittivity $\epsilon(\omega)$. We then explained how the latter could provide information about the absorption of electromagnetic rays in a dielectric medium. We calculated $\alpha(\omega)$ in the Debye theory from a solution (1,2.5) of the diffusion equation (1,2.3).

Later in Chapters 8, 9 and 10 we derived expressions for complex polarizability in the case of dipolar polarization when the disc, sphere and linear rotators are chosen as models of a polar molecule. Equation (8,2.27) for the disc was a consequence of the solution of the Fokker–Planck equation (8,2.1). Equations (9,1.27) and (9,1.28) for the spherical model were consequences of the solution of the Fokker–Planck equation (9,1.8). It was pointed out that the expansion (9,2.32) of the exponential expression in (9,2.30) for $\langle \cos \beta'(t) \rangle$, when substituted into the Kubo relation (2,2.23), leads to (9,1.27) and (9,1.28). However, since the expansion of an exponential with a negative exponent is a series of alternative positive and negative terms, the term-by-term integration of the series for $\langle \cos \beta'(t) \rangle e^{-i\omega t}$ may not yield a good approximation of the integral. Indeed, it may yield a divergent series (McConnell, 1977, Section 4). On the other hand there is no obvious way to calculate the integral, if the value of $\langle \cos \beta'(t) \rangle$ is taken from (9,2.30). A similar situation is present in the derivation of eqn (10,1.9) for $\alpha(\omega)$ in the case of the linear rotator. It was obtained from the Fokker–Planck equation (10,1.3), it may be deduced from the Kubo relation when $\langle \cos \beta'(t) \rangle$ is expanded as in (10,2.18), and it would not be feasible to calculate the integral of $\langle \cos \beta'(t) \rangle e^{-i\omega t}$ from (10,2.17).

In view of these difficulties it seems worthwhile to examine whether the Krylov–Bogoliubov method described in Section 11.5 might be employed to deduce in a more satisfactory way the complex polarizability from the autocorrelation function $\langle(\mathbf{n}(0)\cdot\mathbf{n}(t))\rangle$. The operators that occur in the course of deriving expressions for $\langle R(t)\rangle$ in the cases of the sphere and linear rotator commute with one another. In the case of the disc only one operator apart from $\mathbf{I}$, the J of eqn (8,3.5), is present in the calculations. It is therefore permissible to write (11,5.8)–(11,5.10) for these three models as

$$\frac{dV^{(2)}(t)}{dt} = \Omega^{(2)}(t) - G^{(2)}, \tag{12,1.1}$$

$$\frac{dV^{(4)}(t)}{dt} = \Omega^{(4)}(t) - G^{(4)} + (\Omega^{(2)}(t) - G^{(2)})V^{(2)}(t), \tag{12,1.2}$$

$$\frac{dV^{(6)}(t)}{dt} = \Omega^{(6)}(t) - G^{(6)} + (\Omega^{(4)}(t) - G^{(4)})V^{(2)}(t) + (\Omega^{(2)}(t) - G^{(2)})V^{(4)}(t). \tag{12,1.3}$$

Since $V^{(2)}(t)$ is bounded for large t and since $V^{(2)}(0)$ vanishes, we have as in (11,5.11) and (11,5.12)

$$\Omega^{(2)}(\infty) = G^{(2)}, \quad V^{(2)}(t) = \int_0^t (\Omega^{(2)}(t_1) - \Omega^{(2)}(\infty))dt_1. \tag{12,1.4}$$

Then, since $V^{(4)}(t)$ is bounded for large t, we deduce from (12,1.2) that $\Omega^{(4)}(\infty) = G^{(4)}$. Since $V^{(6)}(t)$ is bounded for large t, we likewise deduce from (12,1.3) that $\Omega^{(6)}(\infty) = G^{(6)}$. We conclude that for the disc, sphere and linear rotator

$$G^{(2)} = \Omega^{(2)}(\infty), \quad G^{(4)} = \Omega^{(4)}(\infty), \quad G^{(6)} = \Omega^{(6)}(\infty), \text{ etc.} \tag{12,1.5}$$

We shall first derive expressions for complex polarizability for each of these models and shall later examine the case of the asymmetric rotator, for which eqn (12,1.5) are no longer true.

12.2 Complex Polarizability for the Rotating Disc Model

Equation (8,3.12) for the disc gives

$$\epsilon^2 \Omega^{(2)}(t) = -B\gamma(1 - e^{-Bt})\mathbf{I}, \tag{12,2.1}$$

and we also saw that $\Omega^{(4)}(\infty)$, $\Omega^{(6)}(\infty)$ vanish. Hence from (12,1.5)

$$\epsilon^2 G^{(2)} = -B\gamma \mathbf{I}, \quad \epsilon^4 G^{(4)} = 0, \quad \epsilon^6 G^{(6)} = 0,$$

and therefore, from (11,5.6),

$$G = -B\gamma \mathbf{I}. \tag{12,2.2}$$

From (12,1.4) and (12,2.1)

$$\epsilon^2 V^{(2)}(t) = \int_0^t B\gamma e^{-Bt_1}\, dt_1 \mathbf{I} = \gamma(1 - e^{-Bt})\mathbf{I}. \tag{12,2.3}$$

Equation (12,1.2) reduces to

$$\frac{dV^{(4)}(t)}{dt} = (\Omega^{(2)}(t) - G^{(2)})V^{(2)}(t)$$

and, since $V^{(4)}(0)$ vanishes,

$$\epsilon^4 V^{(4)}(t) = \gamma^2 B \int_0^t e^{-Bt_1}(1 - e^{-Bt_1})dt_1 \mathbf{I}, \tag{12,2.4}$$

that is

$$\epsilon^4 V^{(4)}(t) = \gamma(\tfrac{1}{2} - e^{-Bt} + \tfrac{1}{2}e^{-2Bt})\mathbf{I}.$$

Similarly (12,1.3) reduces to

$$\frac{dV^{(6)}(t)}{dt} = (\Omega^{(2)}(t) - G^{(2)})V^{(4)}(t),$$

from which it is deduced that

$$\epsilon^6 V^{(6)}(t) = \gamma^3(\tfrac{1}{6} - \tfrac{1}{2}e^{-Bt} + \tfrac{1}{2}e^{-2Bt} - \tfrac{1}{6}e^{-3Bt})\mathbf{I}. \tag{12,2.5}$$

Now from (11,5.4) and (12,2.2)

$$\frac{dU(t)}{dt} = GU(t) = -B\gamma U(t)$$

and therefore, since $U(0) = \mathbf{I}$,

$$U(t) = \exp[-\gamma Bt]\mathbf{I}$$

$$= \exp\left[-\frac{kT}{IB}t\right]\mathbf{I}. \tag{12,2.6}$$

From (11,5.3) and (11,5.7)

$$\langle R(t)\rangle = (\mathbf{I} + \epsilon^2 V^{(2)}(t) + \epsilon^4 V^{(4)}(t) + \epsilon^6 V^{(6)}(t) + \ldots)U(t),$$

which by employing (12,2.3)–(12,2.6) now becomes

$$\begin{aligned}\langle R(t)\rangle = & \{1+\gamma(1-e^{-Bt})+\gamma^2(\tfrac{1}{2}-e^{-Bt}+\tfrac{1}{2}e^{-2Bt}) \\ & +\gamma^3(\tfrac{1}{6}-\tfrac{1}{2}e^{-Bt}+\tfrac{1}{2}e^{-2Bt}-\tfrac{1}{6}e^{-3Bt})+\dots\}\times \\ & \exp\left[-\frac{kT}{IB}t\right]\mathbf{I}. \end{aligned} \tag{12,2.7}$$

This could have been deduced from (8,3.14), namely,

$$\langle R(t)\rangle = \exp\{-\gamma[Bt-1+e^{-Bt}]\}\mathbf{I}$$

by writing it

$$\langle R(t)\rangle = \exp[-\gamma Bt]\exp[\gamma(1-e^{-Bt})]\mathbf{I}$$

and expanding the last exponential in powers of γ.

We use eqn (12,2.7) to calculate the polarizability for orientational polarization from the Kubo formula (2,2.23),

$$\frac{\alpha(\omega)}{\alpha_s} = 1-i\omega\int_0^\infty\langle(\mathbf{n}(0)\cdot\mathbf{n}(t))\rangle e^{-i\omega t}dt.$$

In the notation of Section 8.3, $(\mathbf{n}(0)\cdot\mathbf{n}(t)) = \cos\theta(t)$ and from (8,3.2) $\cos\theta(t) = R_{11}(t)$, so that

$$\frac{\alpha(\omega)}{\alpha_s} = 1-i\omega\int_0^\infty\langle R_{11}(t)\rangle e^{-i\omega t}dt. \tag{12,2.8}$$

As in (5,4.18), $\langle R_{11}(t)\rangle = \langle R(t)\rangle_{11}$, so $\langle R_{11}(t)\rangle$ is just the multiplier of the identity operator in (12,2.7). The integral in (12,2.8) is easily evaluated by using Laplace transforms. Employing eqn (C.11) we find that

$$\begin{aligned}\int_0^\infty & \langle R(t)\rangle_{11}e^{-i\omega t}dt \\ = & \left[\frac{kT}{IB}+i\omega\right]^{-1} + \gamma\left\{\left[\frac{kT}{IB}+i\omega\right]^{-1} - \left[\frac{kT}{IB}+B+i\omega\right]^{-1}\right\} \\ & +\gamma^2\left\{\frac{1}{2}\left[\frac{kT}{IB}+i\omega\right]^{-1} - \left[\frac{kT}{IB}+B+i\omega\right]^{-1} + \frac{1}{2}\left[\frac{kT}{IB}+2B+i\omega\right]^{-1}\right\} \\ & +\gamma^3\left\{\frac{1}{6}\left[\frac{kT}{IB}+i\omega\right]^{-1} - \frac{1}{2}\left[\frac{kT}{IB}+B+i\omega\right]^{-1}\right. \\ & \left.+\frac{1}{2}\left[\frac{kT}{IB}+2B+i\omega\right]^{-1} - \frac{1}{6}\left[\frac{kT}{IB}+3B+i\omega\right]^{-1}\right\}+\dots,\end{aligned}$$

and this equation simplifies to

$$\int_0^\infty \langle R(t)\rangle_{11} e^{-i\omega t}$$

$$= \left[\frac{kT}{IB} + i\omega\right]^{-1} + \gamma B\left[\frac{kT}{IB} + i\omega\right]^{-1}\left[\frac{kT}{IB} + B + i\omega\right]^{-1}$$

$$+ \gamma^2 B^2\left[\frac{kT}{IB} + i\omega\right]^{-1}\left[\frac{kT}{IB} + B + i\omega\right]^{-1}\left[\frac{kT}{IB} + 2B + i\omega\right]^{-1}$$

$$+ \gamma^3 B^3\left[\frac{kT}{IB} + i\omega\right]^{-1}\left[\frac{kT}{IB} + B + i\omega\right]^{-1}\left[\frac{kT}{IB} + 2B + i\omega\right]^{-1} \times$$

$$\left[\frac{kT}{IB} + 3B + i\omega\right]^{-1} + \ldots . \tag{12,2.9}$$

To express the complex polarizability in familiar notation we write τ for $IB/(kT)$, the relaxation time that appears in (1,1.14) for two-dimensional Debye theory. Then (12,2.8) and (12,2.9) yield

$$\frac{\alpha(\omega)}{\alpha_s} = \frac{1}{1 + i\omega\tau} - \frac{i\omega\tau}{(1 + i\omega\tau)(1 + B\tau + i\omega\tau)}\left\{1 + \frac{1}{1 + 2B\tau + i\omega\tau} \times\right.$$

$$\left.\left[1 + \frac{1}{1 + 3B\tau + i\omega\tau}(1 + \ldots)\right]\right\}. \tag{12,2.10}$$

If the series is terminated after the first term, we obtain an equation which agrees with the two-dimensional Debye result (1,3.12). On employing the relations

$$B\tau = \frac{1}{\gamma}, \quad B\tau_F = 1$$

we may identify (12,2.10) with (8,2.27). When we limit the series of (12,2.10) to the first two terms, we get

$$\frac{\alpha(\omega)}{\alpha_s} = \frac{1 + B\tau}{(1 + i\omega\tau)(1 + B\tau + i\omega\tau)}$$

$$= \frac{1 + \gamma}{(1 + i\omega\tau)(1 + \gamma + i\omega\tau_F)},$$

so that approximately

$$\frac{\alpha(\omega)}{\alpha_s} = \frac{1}{(1 + i\omega\tau)[1 + (1-\gamma)i\omega\tau_F]}. \qquad (12,2.11)$$

If we omit γ is the denominator, we have eqn (2,5.4) of Rocard's theory with τ_D replaced by the two-dimensional relaxation time. Equation (12,2.11) may therefore be interpreted as giving a correction to first order in γ to Rocard's theory applied to the two-dimensional rotational Brownian motion of the sphere.

12.3 Complex Polarizability for the Spherical Model

We shall first derive an expression for G that will be convenient for calculating polarizabilities. From eqn (9,2.23)–(9,2.26)

$$\epsilon^2 \Omega^{(2)}(t) = -J^2 \frac{kT}{I} \frac{dI^{(2)}(t)}{dt}, \qquad (12,3.1)$$

$$\epsilon^4 \Omega^{(4)}(t) = -J^2 \left(\frac{kT}{I}\right)^2 \frac{dI_2^{(4)}(t)}{dt}, \qquad (12,3.2)$$

$$\epsilon^6 \Omega^{(6)}(t) = -J^2 \left(\frac{kT}{I}\right)^3 \left(\frac{dI_3^{(6)}(t)}{dt} + 4\frac{dI_4^{(6)}(t)}{dt}\right), \qquad (12,3.3)$$

$$\epsilon^8 \Omega^{(8)}(t) = -J^2 \left(\frac{kT}{I}\right)^4 \left(\frac{dI_4^{(8)}(t)}{dt} + 8\frac{dI_5^{(8)}(t)}{dt} + [-4J^2 + 16\mathbf{I}] \times \frac{dI_7^{(8)}(t)}{dt} + [-10J^2 + 38\mathbf{I}]\frac{dI_8^{(8)}(t)}{dt}\right). \qquad (12,3.4)$$

We see from eqn (D.2) of Appendix D that the expressions for all the above $I_i^{(2n)}(t)$ are such that $B^{2n} I_i^{(2n)}(t)$ starts with a term proportional to Bt and that

$$\lim_{t\to\infty} \frac{dI_i^{(2n)}(t)}{dt} = \lim_{t\to\infty} \frac{I_i^{(2n)}(t)}{t}.$$

Hence from (12,1.5) and (D.2)

$$\epsilon^2 G^{(2)} = -\gamma J^2 B, \quad \epsilon^4 G^{(4)} = -\tfrac{1}{2}\gamma^2 J^2 B,$$
$$\epsilon^6 G^{(6)} = -\tfrac{7}{12}\gamma^3 J^2 B, \quad \epsilon^8 G^{(8)} = -(\tfrac{17}{18} - \tfrac{1}{8}J^2)\gamma^4 J^2 B, \qquad (12,3.5)$$

and so from (11,5.6)

$$G = \epsilon^2 G^{(2)} + \epsilon^4 G^{(4)} + \epsilon^6 G^{(6)} + \epsilon^8 G^{(8)} + \ldots$$
$$= -\gamma B\{1 + \tfrac{1}{2}\gamma + \tfrac{7}{12}\gamma^2 + (\tfrac{17}{18} - \tfrac{1}{8}J^2)\gamma^3 + \ldots\}J^2. \qquad (12,3.6)$$

We must now calculate $\epsilon^2 V^{(2)}(t)$, $\epsilon^4 V^{(4)}(t)$, $\epsilon^6 V^{(6)}(t)$ from (12,1.1)–(12,1.3), (12,3.1)–(12,3.5) and (D.2). On integrating (12,1.1) we obtain

$$\epsilon^2 V^{(2)}(t) = \epsilon^2 \int_0^t (\Omega^{(2)}(t_1) - G^{(2)})dt_1$$
$$= -J^2\gamma B^2(I^{(2)}(t) - I^{(2)}(0)) + J^2\gamma Bt,$$

and so

$$\epsilon^2 V^{(2)}(t) = J^2\gamma(1 - e^{-Bt}). \qquad (12,3.7)$$

We note that

$$\epsilon^2(\Omega^{(2)}(t) - G^{(2)}) = J^2\gamma B e^{-Bt},$$

and therefore

$$\epsilon^4 \int_0^t (\Omega^{(2)}(t_1) - G^{(2)})V^{(2)}(t_1)dt_1 = (J^2)^2\gamma^2(\tfrac{1}{2} - e^{-Bt} + \tfrac{1}{2}e^{-Bt}). \qquad (12,3.8)$$

Moreover

$$\epsilon^4 \int_0^t (\Omega^{(4)}(t_1) - G^{(4)})dt_1 = -J^2\gamma^2 B^4 I_2^{(4)}(t) + \tfrac{1}{2}J^2\gamma^2 Bt$$
$$= -J^2\gamma^2[-\tfrac{5}{4} + (Bt + 1)e^{-Bt} + \tfrac{1}{4}e^{-2Bt}], \qquad (12,3.9)$$

so from (12,1.2)

$$\epsilon^4 V^{(4)}(t)$$
$$= J^2\gamma^2[\tfrac{5}{4} - (Bt + 1)e^{-Bt} - \tfrac{1}{4}e^{-2Bt} + J^2(\tfrac{1}{2} - e^{-Bt} + \tfrac{1}{2}e^{-2Bt})]. \qquad (12,3.10)$$

In order to calculate $V^{(6)}(t)$ we use the above results to write (12,1.3) as

$$\epsilon^6 \frac{dV^{(6)}(t)}{dt}$$
$$= -J^2\gamma^3\left[B^6\left(\frac{dI_3^{(6)}(t)}{dt} + 4\frac{dI_4^{(6)}(t)}{dt}\right) - \frac{7}{12}B\right]$$
$$+ (J^2)^2\gamma^3 B[\tfrac{5}{4}e^{-Bt} + Bte^{-Bt} - \tfrac{1}{2}e^{-2Bt} - 2Bte^{-2Bt} - \tfrac{3}{4}e^{-3Bt}]$$
$$+ (J^2)^3\gamma^3 B[\tfrac{1}{2}e^{-Bt} - e^{-2Bt} + \tfrac{1}{2}e^{-3Bt}].$$

On integrating we find that

$$
\begin{aligned}
&\epsilon^6 V^{(6)}(t)\\
&= J^2\gamma^3[\tfrac{19}{9} - (\tfrac{1}{2}B^2t^2 + 2Bt + 1)e^{-Bt} - (\tfrac{3}{4}Bt + 1)e^{-2Bt} - \tfrac{1}{9}e^{-3Bt}]\\
&\quad + (J^2)^2\gamma^3[\tfrac{5}{4} - (Bt + \tfrac{9}{4})e^{-Bt} + (Bt + \tfrac{3}{4})e^{-2Bt} + \tfrac{1}{4}e^{-3Bt}]\\
&\quad + (J^2)^3\gamma^3[\tfrac{1}{6} - \tfrac{1}{2}e^{-Bt} + \tfrac{1}{2}e^{-2Bt} - \tfrac{1}{6}e^{-3Bt}]. \qquad (12,3.11)
\end{aligned}
$$

Since the γ^4 term of the series in (12,3.6) gives a correction to $-\gamma BJ^2$ of relative order γ^3 and since $\epsilon^6 V^{(6)}(t)$ gives in (11,5.7) a correction of the same order, our results will be correct to this order.

We deduce from (11,5.16), (12,3.6), (12,3.7), (12,3.10) and (12,3.11) that

$$
\begin{aligned}
&\langle R(t)\rangle\\
&= [\mathbf{I} + \gamma J^2(1 - e^{-Bt}) + \gamma^2\{J^2[\tfrac{5}{4} - (Bt + 1)e^{-Bt} - \tfrac{1}{4}e^{-2Bt}]\\
&\quad + (J^2)^2[\tfrac{1}{2} - e^{-Bt} + \tfrac{1}{2}e^{-2Bt}]\}\\
&\quad + \gamma^3\{J^2[\tfrac{19}{9} - (\tfrac{1}{2}B^2t^2 + 2Bt + 1)e^{-Bt} - (\tfrac{3}{4}Bt + 1)e^{-2Bt} - \tfrac{1}{9}e^{-3Bt}]\\
&\quad + (J^2)^2[\tfrac{5}{4} - (Bt + \tfrac{9}{4})e^{-Bt} + (Bt + \tfrac{3}{4})e^{-2Bt} + \tfrac{1}{4}e^{-3Bt}]\\
&\quad + (J^2)^3[\tfrac{1}{6} - \tfrac{1}{2}e^{-Bt} + \tfrac{1}{2}e^{-2Bt} - \tfrac{1}{6}e^{-3Bt}]\} + \ldots]\\
&\quad \times \exp[-\gamma B\{1 + \tfrac{1}{2}\gamma + \tfrac{7}{12}\gamma^2 + (\tfrac{17}{18} - \tfrac{1}{8}J^2)\gamma^3 + \ldots\}J^2 t]. \qquad (12,3.12)
\end{aligned}
$$

This is the required expression for $\langle R(t)\rangle$. Equation (12,3.12) is valid for all times, long or short. Comparing it with (9,3.1) for the Debye theory,

$$\langle R(t)\rangle = \exp[-\gamma BJ^2 t],$$

we see that for $Bt \gg 1$ we now have an exponential decay with a different time coefficient and in addition a multiplying factor that differs from unity. The different time coefficient did not appear in eqn (12,2.7) for the disc on account of the vanishing of $G^{(4)}$, $G^{(6)}$, etc. Comparing (12,3.12) with (9,2.28) we see from (9,2.35) that the normalized autocorrection function of $Y_{jm}(\beta(t), \alpha(t))$ is equal to the right-hand side of (12,3.12) with **I** replaced by unity and J^2 replaced throughout by $j(j + 1)$.

For the study of dielectric relaxation we require $\langle \cos\beta'(t)\rangle$, where $\beta'(t)$ is defined in Section 9.2. As in the derivation of (9,2.30), $\langle \cos\beta'(t)\rangle$ is obtained on replacing **I** by 1 and J^2 by 2 in the expression for $\langle R(t)\rangle$. We thus deduce from (12,3.12) that

$$\begin{aligned}&\langle\cos\beta'(t)\rangle\\&= [1 + \gamma(2 - 2e^{-Bt}) + \gamma^2\{\tfrac{9}{2} - 6e^{-Bt} - 2Bte^{-Bt} + \tfrac{3}{2}e^{-2Bt}\}\\&\quad + \gamma^3\{\tfrac{95}{9} - (B^2t^2 + 8Bt + 15)e^{-Bt} + (\tfrac{5}{2}Bt + 5)e^{-2Bt} - \tfrac{5}{9}e^{-3Bt}\}\\&\quad + \ldots]\exp[-2\gamma B\{1 + \tfrac{1}{2}\gamma + \tfrac{7}{12}\gamma^2 + \tfrac{25}{36}\gamma^3 + \ldots\}t]. \quad (12,3.13)\end{aligned}$$

We express the exponential as $\exp(-BG_1t)$ with

$$G_1 = 2\gamma(1 + \tfrac{1}{2}\gamma + \tfrac{7}{12}\gamma^2 + \tfrac{25}{36}\gamma^3 + \ldots), \qquad (12,3.14)$$

and we put $\omega = B\omega'$ as in (9,1.20). The dipole axis is taken along the third axis of the body frame, as it was in Section 9.1. Hence $\langle(\mathbf{n}(0)\cdot\mathbf{n}(t))\rangle$ in the Kubo relation (2,2.23) is $\langle\cos\beta'(t)\rangle$ and, from (12,3.13) and (12,3.14), the complex polarizability for orientational polarization satisfies

$$\begin{aligned}\frac{\alpha(\omega)}{\alpha_s} &= 1 - iB\omega'\int_0^\infty [1 + \gamma(2 - 2e^{-Bt})\\&\quad + \gamma^2\{\tfrac{9}{2} - 6e^{-Bt} - 2Bte^{-Bt} + \tfrac{3}{2}e^{-2Bt}\}\\&\quad + \gamma^3\{\tfrac{95}{9} - (B^2t^2 + 8Bt + 15)e^{-Bt} + (\tfrac{5}{2}Bt + 5)e^{-2Bt}\\&\quad - \tfrac{5}{9}e^{-3Bt}\} + \ldots]\exp[-B(G_1 + i\omega')t]\,dt. \quad (12,3.15)\end{aligned}$$

The value of the integral may be written down from (C.11). We thus obtain

$$\begin{aligned}\frac{\alpha(\omega)}{\alpha_s} &= \frac{G_1}{G_1 + i\omega'} - i\omega'\left\{\gamma\left[\frac{2}{G_1 + i\omega'} - \frac{2}{G_1 + 1 + i\omega'}\right]\right.\\&\quad + \gamma^2\left[\frac{\frac{9}{2}}{G_1 + i\omega'} - \frac{6}{G_1 + 1 + i\omega'} - \frac{2}{(G_1 + 1 + i\omega')^2} + \frac{\frac{3}{2}}{G_1 + 2 + i\omega'}\right]\\&\quad + \gamma^3\left[\frac{\frac{95}{9}}{G_1 + i\omega'} - \frac{15}{G_1 + 1 + i\omega'} - \frac{8}{(G_1 + 1 + i\omega')^2} - \frac{2}{(G_1 + 1 + i\omega')^3}\right.\\&\quad \left.\left. + \frac{5}{G_1 + 2 + i\omega'} + \frac{\frac{5}{2}}{(G_1 + 2 + i\omega')^2} - \frac{\frac{5}{9}}{G_1 + 3 + i\omega'}\right] + \ldots\right\}.\end{aligned}$$

(12,3.16)

Equation (12,3.16) expresses the complex polarizability in terms of γ both directly and also indirectly through G_1. A rather lengthy expression for $\alpha(\omega)$ in terms of γ alone was derived otherwise by McConnell (1979). If we expand γ in terms of G_1 by

$$\gamma = \tfrac{1}{2}G_1 + aG_1^2 + bG_1^3 + cG_1^4 + \ldots,$$

we find on substituting into (12,3.14) that

$$\gamma = \tfrac{1}{2}G_1 - \tfrac{1}{8}G_1^2 - \tfrac{1}{96}G_1^3 + \tfrac{5}{576}G_1^4 + \ldots.$$

Employing this result to eliminate γ in (12,3.16) it may be shown (Ford *et al.*, 1978a) that

$$\frac{\alpha(\omega)}{\alpha_s} = \frac{G_1(G_1+1)}{(G_1+i\omega')(G_1+1+i\omega')}$$
$$-\frac{i\omega' G_1^2\{\tfrac{5}{4}(G_1+i\omega')+\tfrac{7}{4}\}}{(G_1+i\omega')(G_1+1+i\omega')^2(G_1+2+i\omega')}$$
$$-\frac{i\omega' G_1^3\{(G_1+i\omega')^4+93(G_1+i\omega')^3+475(G_1+i\omega')^2+783(G_1+i\omega')+424\}}{48(G_1+i\omega')(G_1+1+i\omega')^3(G_1+2+i\omega')^2(G_1+3+i\omega')}$$
$$+\ldots. \qquad (12,3.17)$$

If we truncate the series after the first term and substitute for G_1 from (12,3.14), we obtain

$$\frac{\alpha(\omega)}{\alpha_s} = \frac{(2\gamma+\gamma^2+\ldots)(1+2\gamma+\ldots)}{(2\gamma+\gamma^2+i\omega'+\ldots)(1+2\gamma+i\omega'+\ldots)}$$
$$= \frac{1}{[1+i\omega'(1-\tfrac{1}{2}\gamma)/(2\gamma)+\ldots][1+i\omega'(1-2\gamma)+\ldots]},$$

that is approximately,

$$\frac{\alpha(\omega)}{\alpha_s} = \frac{1}{[1+(1-\tfrac{1}{2}\gamma)i\omega\tau_D][1+(1-2\gamma)i\omega\tau_F]}, \qquad (12,3.18)$$

where $\tau_D = IB/(2kT)$, $\tau_F = B^{-1}$. This equation gives a first order correction to (2,5.4) of Rocard's theory.

12.4 Complex Polarizability for the Linear Rotator Model

We treat the linear rotator model in much the same way as we treated the spherical model of a polar molecule. We calculate $\epsilon^2\Omega^{(2)}(t)$, $\epsilon^4\Omega^{(4)}(t)$, $\epsilon^6\Omega^{(6)}(t)$ by differentiating (10,2.11)–(10,2.13) and we put the J_3 operators equal to zero, since, as was explained in Section 10.2, they do not contribute to the final result. Thence we obtain

$$\epsilon^2 \Omega^{(2)}(t) = -J^2 \frac{kT}{I} \frac{dI^{(2)}(t)}{dt}, \tag{12,4.1}$$

$$\epsilon^4 \Omega^{(4)}(t) = -2J^2 \left(\frac{kT}{I}\right)^2 \frac{dI_2^{(4)}(t)}{dt}, \tag{12,4.2}$$

$$\epsilon^6 \Omega^{(6)}(t) = -J^2 \left(\frac{kT}{I}\right)^3 \left(4\frac{dI_3^{(6)}(t)}{dt} + 20\frac{dI_4^{(6)}(t)}{dt}\right).$$

From (12,1.5) and (D.2)

$$\epsilon^2 G^{(2)} = \epsilon^2 \Omega^{(2)}(\infty) = -\gamma J^2 B,$$
$$\epsilon^4 G^{(4)} = \epsilon^4 \Omega^{(4)}(\infty) = -\gamma^2 J^2 B,$$
$$\epsilon^6 G^{(6)} = \epsilon^6 \Omega^{(6)}(\infty) = -\tfrac{8}{3}\gamma^3 J^2 B,$$

and from (11,5.6)

$$G = \epsilon^2 G^{(2)} + \epsilon^4 G^{(4)} + \epsilon^6 G^{(6)} + \dots$$
$$= -\gamma(1 + \gamma + \tfrac{8}{3}\gamma^2 + \dots)BJ^2. \tag{12,4.3}$$

By comparing (12,4.1) and (12,3.1) we see that, as in (12,3.7),

$$\epsilon^2 V^{(2)}(t) = J^2 \gamma(1 - e^{-Bt}). \tag{12,4.4}$$

On comparing (12,4.2) with (12,3.2) we deduce from (12,3.8) and (12,3.9) that

$$\epsilon^4 V^{(4)}(t) = J^2 \gamma^2 [\tfrac{5}{2} - (2Bt + 2)e^{-Bt} - \tfrac{1}{2}e^{-2Bt}]$$
$$+ (J^2)^2 \gamma^2 [\tfrac{1}{2} - e^{-Bt} + \tfrac{1}{2}e^{-2Bt}]. \tag{12,4.5}$$

Since (12,4.3) gives a correction to $-\gamma BJ^2$ of relative order γ^2, which is as far as the calculations in Section 10.2 allow us to proceed, and since $\epsilon^4 V^{(4)}(t)$ gives a correction to the same order in (11,5.7), we do not need $\epsilon^6 V^{(6)}(t)$. We deduce from (11,5.3), (11,5.6), (11,5.7), (11,5.15) and (12,4.3)–(12,4.5) that

$$\langle R(t)\rangle = [\mathbf{I} + \gamma J^2 (1 - e^{-Bt}) + \gamma^2 \{J^2 [\tfrac{5}{2} - (2Bt + 2)e^{-Bt} - \tfrac{1}{2}e^{-2Bt}]$$
$$+ (J^2)^2 [\tfrac{1}{2} - e^{-Bt} + \tfrac{1}{2}e^{-2Bt}]\} + \dots]$$
$$\times \exp[-\gamma B(1 + \gamma + \tfrac{8}{3}\gamma^2 + \dots)J^2 t]. \tag{12,4.6}$$

This gives for all times a correction to the Debye result $\langle R(t)\rangle = \exp[-\gamma BJ^2 t]$, which is obtainable from (11,5.25) on putting J_3

equal to zero. The normalized autocorrelation function of $Y_{jm}(\beta(t), \alpha(t))$ is found by replacing **I** by unity and J^2 by $j(j+1)$ in the right-hand side of (12,4.6).

To calculate the complex polarizability we first put $j = 1$, $m = m' = 0$ in the matrix representative of $\langle R(t)\rangle$, as was done in the derivation of (10,2.17). Since then $J^2 = 2\mathbf{I}$, we obtain

$$\langle \cos \beta'(t)\rangle = [1 + \gamma(2 - 2e^{-Bt}) + \gamma^2(7 - 8e^{-Bt} - 4Bte^{-Bt} + e^{-2Bt}) + \ldots] \exp[-2\gamma B(1 + \gamma + \tfrac{8}{3}\gamma^2 + \ldots)t]. \qquad (12,4.7)$$

We write the exponential as $\exp(-BG_1 t)$, where now

$$G_1 = 2\gamma(1 + \gamma + \tfrac{8}{3}\gamma^2 + \ldots). \qquad (12,4.8)$$

On comparing (12,4.7) and (12,4.8) with (12,3.13) and (12,3.14) we write down from (12,3.15) and (12,3.16) the relations for the complex polarizability arising from dipolar polarization

$$\frac{\alpha(\omega)}{\alpha_s} = 1 - iB\omega' \int_0^\infty [1 + \gamma(2 - 2e^{-Bt}) + \gamma^2(7 - 8e^{-Bt} - 4Bte^{-Bt} + e^{-2Bt}) + \ldots] \times \exp[-B(G_1 + i\omega')t]\,dt,$$

$$\frac{\alpha(\omega)}{\alpha_s} = \frac{G_1}{G_1 + i\omega'} - i\omega'\left\{\gamma\left[\frac{2}{G_1 + i\omega'} - \frac{2}{G_1 + 1 + i\omega'}\right] + \gamma^2\left[\frac{7}{G_1 + i\omega'} - \frac{8}{G_1 + 1 + i\omega'} - \frac{4}{(G_1 + 1 + i\omega')^2} + \frac{1}{G_1 + 2 + i\omega'}\right] + \ldots\right\}, \qquad (12,4.9)$$

with $\omega' = \omega/B$.

Equation (12,4.8) gives

$$\gamma = \tfrac{1}{2}G_1 - \tfrac{1}{4}G_1^2 - \tfrac{1}{12}G_1^3 + \ldots,$$

and it may be shown (Ford *et al.*, 1978a) that

$$\frac{\alpha(\omega)}{\alpha_s} = \frac{G_1(G_1 + 1)}{(G_1 + i\omega')(G_1 + 1 + i\omega')} - \frac{i\omega' G_1^2\{\tfrac{3}{2}(G_1 + i\omega') + \tfrac{5}{2}\}}{(G_1 + i\omega')(G_1 + 1 + i\omega')^2(G_1 + 2 + i\omega')} + \ldots. \qquad (12,4.10)$$

The first term in this series is the same as that in (12,3.17), but G_1 now has a different value. If we truncate the series after the first term, we obtain

$$\frac{\alpha(\omega)}{\alpha_s} = \frac{1}{[1+(1-\gamma)i\omega\tau_D][1+(1-2\gamma)i\omega\tau_F]}. \quad (12,4.11)$$

This is again a first order correction in γ to (2,5.4), which was derived in Rocard's theory for the sphere. The correction differs from that found in (12,3.18) for the spherical model. The polarizability has also been expressed in terms of γ rather than G_1 (McConnell, 1979).

12.5 Complex Polarizability for the Asymmetric Rotator Model

In the two previous sections we examined dielectric relaxation by considering the rotational Brownian motion of a polar molecule, whose dipole axis lies along the radius of a sphere or along a linear rotator. To employ the Kubo relation we obtained $\langle \cos \beta'(t) \rangle$ in (12,3.13) and (12,4.7) by calculating the 00-element of the $j = 1$ representation of $\langle R(t) \rangle$. According to Section 7.3 the basis elements of this representation are $Y_{1,-1}$, $Y_{1,0}$, $Y_{1,1}$.

In the asymmetric rotator model of a polar molecule there is no obvious position for the dipole axis. Let us, however, suppose that it lies along a line through the centre of mass. We denote by n_1, n_2, n_3 the constant direction cosines of the dipole axis with respect to the body-frame coordinate system. In the Kubo relation (2,2.23) for the complex polarizability due to orientational polarization,

$$\frac{\alpha(\omega)}{\alpha_s} = 1 - i\omega \int_0^\infty \langle (\mathbf{n}(0) \cdot \mathbf{n}(t)) \rangle e^{-i\omega t} dt, \quad (12,5.1)$$

$\mathbf{n}(0)$ is the unit vector in the direction of the dipole axis at time zero and $\mathbf{n}(t)$ the corresponding vector at time t. Since we are concerned only with a scalar product, we can refer the components to the laboratory frame or to the body frame at a specified time.

Let us write $n(0)$ for the column vector $\begin{bmatrix} n_1(0) \\ n_2(0) \\ n_3(0) \end{bmatrix}$ whose elements are the components of $\mathbf{n}(0)$ referred to the body frame at time t,

$n(t)$ for the vector $\begin{bmatrix} n_1(t) \\ n_2(t) \\ n_3(t) \end{bmatrix}$ whose elements are the components of $\mathbf{n}(t)$ referred to the body frame at time t, so that $n(t) = \begin{bmatrix} n_1 \\ n_2 \\ n_3 \end{bmatrix}$, and let us write $n^{\mathrm{T}}(t)$ for the row vector $\overline{n_1, n_2, n_3}$. Then eqn (7,3.1) gives

$$n(0) = R^1(t)n(t), \tag{12,5.2}$$

where $R^1(t)$ is the three-dimensional rotation matrix that transforms $n(t)$ into $n(0)$. Our calculation of $R^1(t)$ is, by (7,3.8), based on the equation

$$\frac{dR^1(t)}{dt} = -i(\mathbf{J}\cdot\boldsymbol{\omega}(t))R^1(t). \tag{12,5.3}$$

The rotation operators in the present instance are, by an obvious generalization of (8,3.5), represented as follows:

$$J_1 = \begin{bmatrix} 0 & 0 & 0 \\ 0 & 0 & i \\ 0 & -i & 0 \end{bmatrix}, \quad J_2 = \begin{bmatrix} 0 & 0 & -i \\ 0 & 0 & 0 \\ i & 0 & 0 \end{bmatrix}, \quad J_3 = \begin{bmatrix} 0 & i & 0 \\ -i & 0 & 0 \\ 0 & 0 & 0 \end{bmatrix}. \tag{12,5.4}$$

These satisfy $J_2J_3 - J_3J_2 = -iJ_1$, etc., and they reduce to (8,3.5) for rotations in a coordinate plane. They provide a three-dimensional representation of the rotation operators, which is different from that with the basis $Y_{1,-1}$, $Y_{1,0}$, $Y_{1,1}$; indeed we see from (B.12), (B.13) and (B.15) that in the latter representation J_3 is a diagonal matrix with elements -1, 0, 1. The matrices in (12,5.4) are those which, when substituted into (12,5.3), yield the three-dimensional matrix $R^1(t)$ satisfying $R^1(0) = \mathbf{I}$ that produces the correct transformation (12,5.2) of $n(t)$.

We deduce from (12,5.4) that

$$J_1^2 = \begin{bmatrix} 0 & 0 & 0 \\ 0 & 1 & 0 \\ 0 & 0 & 1 \end{bmatrix}, \quad J_2^2 = \begin{bmatrix} 1 & 0 & 0 \\ 0 & 0 & 0 \\ 0 & 0 & 1 \end{bmatrix}, \quad J_3^2 = \begin{bmatrix} 1 & 0 & 0 \\ 0 & 1 & 0 \\ 0 & 0 & 0 \end{bmatrix}, \tag{12,5.5}$$

so that

$$J_2^2 J_3^2 - J_3^2 J_2^2 = 0, \text{ etc..}$$

Hence from (E.19) the operator P vanishes, and so (11,5.24) gives

$$\langle R^1(t)\rangle = \left\{ \mathbf{I} + \sum_{i=1}^{3} \frac{kT}{I_i B_i^2}(1 - e^{-B_i t})J_i^2 + \ldots \right\}$$

$$\times \exp\left\{\left(-\sum_{i=1}^{3} (D_i^{(1)} + D_i^{(2)})J_i^2 + \ldots \right) t\right\}. \qquad (12,5.6)$$

From (12,5.2)

$$(\mathbf{n}(0) \cdot \mathbf{n}(t)) = (\mathbf{n}(t) \cdot \mathbf{n}(0)) = n^{\mathrm{T}}(t)n(0)$$

$$= n^{\mathrm{T}}(t)R^1(t)n(t) = n^{\mathrm{T}}R^1(t)n, \qquad (12,5.7)$$

where the time argument has been omitted in n and n^{T} because in fact they are time independent. Since they are consequently non-stochastic, substitution into (12,5.1) gives

$$\frac{\alpha(\omega)}{\alpha_s} = n^{\mathrm{T}}\left(\mathbf{I} - i\omega \int_0^\infty \langle R^1(t)\rangle e^{-i\omega t} dt \right) n. \qquad (12,5.8)$$

The complex polarizability is obtained from this equation with $\langle R^1(t)\rangle$ given by (12,5.6), J_i^2 by (12,5.5) and $D_i^{(1)}$, $D_i^{(2)}$ by (11,5.18) and (11,5.23).

As an aid to future computation we write

$$D_l^{(1)} + D_l^{(2)} = D_l, \qquad (12,5.9)$$

$$\sum_{l=1}^{3} D_l J_l^2 = -G. \qquad (12,5.10)$$

This G is no longer a multiple of the identity as it was for the sphere and the linear rotator, so care must be taken when integrations are being performed, Thus in calculating $\int_0^\infty e^{Gt} e^{-i\omega t} dt$ for (12,5.8) we first write the integral as $\int_0^\infty \exp(Gt) \exp(-i\omega \mathbf{I} t) dt$. Since G commutes with $\mathbf{I}$, the integral is expressible as $\int_0^\infty \exp[(G - i\omega\mathbf{I})t]\, dt$. Now

$$\frac{d}{dt}\{(G - i\omega\mathbf{I})^{-1} \exp[(G - i\omega\mathbf{I})t]\}$$

$$= \frac{d}{dt}\left\{(G - i\omega\mathbf{I})^{-1} + t\mathbf{I} + \frac{t^2}{2!}(G - i\omega\mathbf{I}) + \frac{t^3}{3!}(G - i\omega\mathbf{I})^2 + \ldots\right\}$$

$$= \mathbf{I} + \frac{t}{1!}(G - i\omega\mathbf{I}) + \frac{t^2}{2!}(G - i\omega\mathbf{I})^2 + \dots$$
$$= \exp\,[(G - i\omega\mathbf{I})t]\,,$$

and so

$$\int_0^\infty \exp\,[(G - i\omega\mathbf{I})t]\,dt$$
$$= (G - i\omega\mathbf{I})^{-1} \lim_{t\to\infty} \exp\,[(G - i\omega\mathbf{I})t] - (G - i\omega\mathbf{I})^{-1}. \tag{12,5.11}$$

Since the absolute value of $e^{-i\omega t}$ is unity,

$$\lim_{t\to\infty} \exp\,[(G - i\omega\mathbf{I})t] = \lim_{t\to\infty} [e^{Gt}e^{-i\omega t}]$$
$$= \lim_{t\to\infty} e^{Gt} \lim_{t\to\infty} e^{-i\omega t} = 0,$$

by (11,5.28). Consequently (12,5.11) yields

$$\int_0^\infty e^{Gt}e^{-i\omega t}dt = \int_0^\infty \exp\,[(G - i\omega\mathbf{I})t]\,dt = (-G + i\omega\mathbf{I})^{-1}, \tag{12,5.12}$$

and we find similarly that

$$\int_0^\infty e^{Gt} \exp\,[-(B_l + i\omega)t]\,dt = (-G + [i\omega + B_l]\,\mathbf{I})^{-1}. \tag{12,5.13}$$

Since

$$(-G + a\mathbf{I})(-G + b\mathbf{I})^{-1}(-G + b\mathbf{I}) = -G + a\mathbf{I}$$
$$= (-G + b\mathbf{I})^{-1}(-G + b\mathbf{I})(-G + a\mathbf{I})$$
$$= (-G + b\mathbf{I})^{-1}(-G + a\mathbf{I})(-G + b\mathbf{I}),$$

we have on multiplying to the right by $(-G + b\mathbf{I})^{-1}$

$$(-G + a\mathbf{I})(-G + b\mathbf{I})^{-1} = (-G + b\mathbf{I})^{-1}(-G + a\mathbf{I}). \tag{12,5.14}$$

Now from (12,5.6), (12,5.10), (12,5.12) and (12,5.13)

$$\int_0^\infty \langle R^1(t)\rangle e^{-i\omega t}dt$$
$$= (-G + i\omega\mathbf{I})^{-1} + \sum_{l=1}^{3} \frac{kT}{I_l B_l^2} J_l^2\,[(-G + i\omega\mathbf{I})^{-1}$$
$$-(-G + [i\omega + B_l]\,\mathbf{I})^{-1}] + \dots. \tag{12,5.15}$$

On multiplying this square bracket to the left by $(-G + i\omega\mathbf{I})(-G + [i\omega + B_l]\mathbf{I})$ and employing (12,5.14) we deduce that

$$\int_0^\infty \langle R^1(t)\rangle e^{-i\omega t} dt$$
$$= (-G + i\omega\mathbf{I})^{-1} + \sum_{l=1} \frac{kT}{I_l B_l} J_l^2 (-G + i\omega\mathbf{I})^{-1} \times (-G + [i\omega + B_l]\mathbf{I})^{-1} + \ldots. \quad (12,5.16)$$

It follows that

$$\mathbf{I} - i\omega \int_0^\infty \langle R^1(t)\rangle e^{-i\omega t} dt$$
$$= -G(-G + i\omega\mathbf{I})^{-1} - i\omega \sum_{l=1}^{3} D_l^{(1)} J_l^2 (-G + i\omega\mathbf{I})^{-1} \times (-G + [i\omega + B_l]\mathbf{I})^{-1} + \ldots, \quad (12,5.17)$$

by (11,5.18).

To complete the calculations we employ (12,5.5) to write down the explicit representations of the operators occurring in (12,5.17). The matrices are all diagonal, so their reciprocals are diagonal. We have from (12,5.10)

$$-G + a\mathbf{I} = \begin{bmatrix} D_2 + D_3 + a & 0 & 0 \\ 0 & D_3 + D_1 + a & 0 \\ 0 & 0 & D_1 + D_2 + a \end{bmatrix}$$

$$(-G + a\mathbf{I})^{-1}$$
$$= \begin{bmatrix} (D_2 + D_3 + a)^{-1} & 0 & 0 \\ 0 & (D_3 + D_1 + a)^{-1} & 0 \\ 0 & 0 & (D_1 + D_2 + a)^{-1} \end{bmatrix}, \quad (12,5.18)$$

and so

$$-G(-G + i\omega\mathbf{I})^{-1}$$
$$= \begin{bmatrix} \dfrac{D_2 + D_3}{D_2 + D_3 + i\omega} & 0 & 0 \\ 0 & \dfrac{D_3 + D_1}{D_3 + D_1 + i\omega} & 0 \\ 0 & 0 & \dfrac{D_1 + D_2}{D_1 + D_2 + i\omega} \end{bmatrix}, \quad (12,5.19)$$

$$(-G+i\omega)^{-1}(-G+[i\omega+B_l]\mathbf{I})^{-1}$$

$$=\begin{bmatrix}(D_2+D_3+i\omega)^{-1}(D_2+D_3+B_l+i\omega)^{-1} & 0 & 0\\ 0 & (D_3+D_1+i\omega)^{-1}(D_3+D_1+B_l+i\omega)^{-1} & 0\\ 0 & 0 & (D_1+D_2+i\omega)^{-1}(D_1+D_2+B_l+i\omega)^{-1}\end{bmatrix},$$

$$\sum_{l=1}^{3} D_l^{(1)} J_l^2(-G+i\omega\mathbf{I})^{-1}(-G+[i\omega+B_l]\mathbf{I})^{-1}$$

$$=\begin{bmatrix}\frac{D_2^{(1)}}{(D_2+D_3+i\omega)(D_2+D_3+B_2+i\omega)}+\frac{D_3^{(1)}}{(D_2+D_3+i\omega)(D_2+D_3+B_3+i\omega)} & 0 & 0\\ 0 & \frac{D_3^{(1)}}{(D_3+D_1+i\omega)(D_3+D_1+B_3+i\omega)}+\frac{D_1^{(1)}}{(D_3+D_1+i\omega)(D_3+D_1+B_1+i\omega)} & 0\\ 0 & 0 & \frac{D_1^{(1)}}{(D_1+D_2+i\omega)(D_1+D_2+B_1+i\omega)}+\frac{D_2^{(1)}}{(D_1+D_2+i\omega)(D_1+D_2+B_2+i\omega)}\end{bmatrix}.$$

(12,5.20)

Hence, if we restrict (12,5.17) to the terms that have been written out explicitly, we see from (12,5.19) and (12,5.20) that the three-dimensional representation of $\mathbf{I}-i\omega\int_0^\infty\langle R^1(t)\rangle e^{-i\omega t}dt$ is a diagonal matrix whose 11-element is

$$\frac{1}{D_2+D_3+i\omega}\times$$

$$\left\{\frac{D_2(D_2+D_3+B_2)+i\omega D_2^{(2)}}{D_2+D_3+B_2+i\omega}+\frac{D_3(D_2+D_3+B_3)+i\omega D_3^{(2)}}{D_2+D_3+B_3+i\omega}\right\},$$

the other two non-vanishing elements being obtained by cyclic permutation. In deriving the last expression we have made use of (12,5.9). On substituting into (12,5.8) we finally obtain

$$\frac{\alpha(\omega)}{\alpha_s}=\frac{n_1^2}{D_2+D_3+i\omega}\times$$

$$\left\{\frac{D_2(D_2+D_3+B_2)+i\omega D_2^{(2)}}{D_2+D_3+B_2+i\omega}+\frac{D_3(D_2+D_3+B_3)+i\omega D_3^{(2)}}{D_2+D_3+B_3+i\omega}\right\}$$

$$+\frac{n_2^2}{D_3+D_1+i\omega}\times$$

$$\left\{\frac{D_3(D_3+D_1+B_3)+i\omega D_3^{(2)}}{D_3+D_1+B_3+i\omega}+\frac{D_1(D_3+D_1+B_1)+i\omega D_1^{(2)}}{D_3+D_1+B_1+i\omega}\right\}$$

$$+\frac{n_3^2}{D_1+D_2+i\omega}\times$$

$$\left\{\frac{D_1(D_1+D_2+B_1)+i\omega D_1^{(2)}}{D_1+D_2+B_1+i\omega}+\frac{D_2(D_1+D_2+B_2)+i\omega D_2^{(2)}}{D_1+D_2+B_2+i\omega}\right\}.$$

(12,5.21)

If the molecular dipole is in the direction of the third coordinate axis, $n_1 = n_2 = 0, n_3 = 1$ and (12,5.21) reduces to

$$\frac{\alpha(\omega)}{\alpha_s} = \frac{1}{D_1 + D_2 + i\omega} \times$$

$$\left\{\frac{D_1(D_1 + D_2 + B_1) + i\omega D_1^{(2)}}{D_1 + D_2 + B_1 + i\omega} + \frac{D_2(D_1 + D_2 + B_2) + i\omega D_2^{(2)}}{D_1 + D_2 + B_2 + i\omega}\right\}. \tag{12,5.22}$$

We see from (11,5.18) that $D^{(1)}$ is of order γB and from (11,5.23) that $D^{(2)}$ is of order $\gamma D^{(1)}$, so that $D_1(D_1 + D_2 + B_1)$ is of order $BD^{(1)}$ and $\omega D_1^{(2)}$ is of order $\gamma\omega D^{(1)}$. Hence the $i\omega D_1^{(2)}$ term will be of significance only if ω is at least of order B/γ. It will be found that the value B/γ of ω corresponds to a frequency where quantum theory has to be employed. Since our treatment has been entirely classical, we should for consistency omit the $i\omega D_l^{(2)}$ terms in (12,5.21) and (12,5.22).

When the polar molecule is a symmetric rotator, the dipole must lie along the axis of symmetry. Taking this to be the third coordinate axis we have $B_2 = B_1$, $D_2 = D_1$ and (12,5.22) with the $i\omega D_l^{(2)}$ terms omitted reduces to

$$\frac{\alpha(\omega)}{\alpha_s} = \frac{2D_1(2D_1 + B_1)}{(2D_1 + i\omega)(2D_1 + B_1 + i\omega)}.$$

If we approximate D_1 by $D_1^{(1)}$ and $2D_1 + B_1$ by B_1, this becomes

$$\frac{\alpha(\omega)}{\alpha_s} = \frac{1}{\left(1 + \dfrac{i\omega I_1 B_1}{2kT}\right)\left(1 + \dfrac{i\omega}{B_1}\right)},$$

which is eqn (2,5.3) for the Rocard theory.

When the symmetric rotator is a sphere, we put $B_1 = B_2 = B_3 = B$, $I_1 = I_2 = I_3 = I$, $D_1 = D_2 = D_3 = D$. From (11,5.18) and (11,5.23)

$$D^{(1)} = \frac{kT}{IB}, \quad D^{(2)} = \frac{1}{2}\gamma\frac{kT}{IB}, \tag{12,5.23}$$

$$D = \left(1 + \frac{1}{2}\gamma\right)\frac{kT}{IB}. \tag{12,5.24}$$

On substituting into (12,5.21) with the $i\omega D_i^{(2)}$ terms omitted, we get

$$\frac{\alpha(\omega)}{\alpha_s} = \frac{2D(2D+B)}{(2D+i\omega)(2D+B+i\omega)}.$$

From (12,5.24)

$$2D+B = (2+\gamma)\frac{kT}{IB} + B \doteqdot B(1+2\gamma)$$

and so

$$\frac{\alpha(\omega)}{\alpha_s} = \frac{1}{[1+(1-\frac{1}{2}\gamma)i\omega\tau_D][1+(1-2\gamma)i\omega\tau_F]},$$

which agrees with (12,3.18).

Let us finally calculate the complex polarizability for the asymmetric rotator in the Debye limit. Then according to (11,5.25)

$$\langle R^1(t)\rangle = e^{Gt}$$

with

$$G = -\sum_{l=1}^{3} \frac{kT}{I_l B_l} J_l^2. \qquad (12,5.25)$$

Equation (12,5.17) now reduces to

$$\mathbf{I} - i\omega \int_0^\infty \langle R^1(t)\rangle e^{-i\omega t} dt = -G(-G+i\omega\mathbf{I})^{-1}. \qquad (12,5.26)$$

We see from (12,5.19) and (12,5.25) that $-G(-G+i\omega\mathbf{I})^{-1}$ is represented by a diagonal matrix with leading element

$$\frac{\dfrac{kT}{I_2B_2} + \dfrac{kT}{I_3B_3}}{\dfrac{kT}{I_2B_2} + \dfrac{kT}{I_3B_3} + i\omega}.$$

We express this as

$$\frac{1}{1+i\omega\tau_{D,1}},$$

where

$$\tau_{D,1} = \left(\frac{kT}{I_2B_2} + \frac{kT}{I_3B_3}\right)^{-1},$$

as previously defined in (11,6.2). The other non-vanishing elements are found by cyclic permutation. On multiplying both sides of (12,5.26) to the left by n^{T} and to the right by n we deduce from

(12,5.8) that

$$\frac{\alpha(\omega)}{\alpha_s} = n^{\mathrm{T}}[-G(-G+i\omega\mathbf{I})^{-1}]n$$

$$= \frac{n_1^2}{1+i\omega\tau_{D,1}} + \frac{n_2^2}{1+i\omega\tau_{D,2}} + \frac{n_3^2}{1+i\omega\tau_{D,3}}.$$

This is just eqn (11,6.3). Hence the Debye limit of the theory of the present section is equivalent to Perrin's theory of the rotational Brownian motion of an ellipsoid.

13

Correlation Times and Spectral Densities for Rotational Brownian Motion

13.1 Angular Velocity and Orientational Correlation Times

At the end of Section 4.4 the correlation time for a random variable was defined as the time integral from 0 to ∞ of its normalized autocorrelation function. Later during the discussion of the rotational Brownian motion of a rigid body it was shown in eqn (9,4.2) that, if $\alpha(t)$, $\beta(t)$ are the Euler angles with reference to the laboratory system of the third body fixed axis, the normalized autocorrelation function of the spherical harmonic $Y_{jm}(\beta(t), \alpha(t))$ is $\langle R^j_{mm}(t)\rangle^*$, where $R^j_{mm}(t)$ is the mm element of the $(2j+1)$-dimensional representation of the rotation operator that brings the body coordinate system at time 0 to the body system at time t, the basis of the representation being the harmonics Y_{js}. Hence the correlation time τ_{jm} for $Y_{jm}(\beta(t), \alpha(t))$ is given by

$$\tau_{jm} = \int_0^\infty \langle R^j_{mm}(t)\rangle^* \, dt. \tag{13,1.1}$$

In the special case of $j = 1$, $m = 0$ we see from (7,3.15) that

$$Y_{10}(\beta(t), \alpha(t)) = \left(\frac{3}{4\pi}\right)^{1/2} \cos \beta(t).$$

Thus the normalized autocorrelation function of $\cos \beta(t)$ is the same as that of $Y_{10}(\beta(t), \alpha(t))$, and from (13,1.1) their correlation time

$$\tau_{10} = \int_0^\infty \langle R^1_{00}(t)\rangle^* \, dt.$$

Alternatively since $\cos \beta(t) = (\mathbf{n}(0) \cdot \mathbf{n}(t))$, where $\mathbf{n}(t)$ is the unit vector in the direction of the third body fixed axis, we have

$$\tau_{10} = \lim_{\omega \to 0} \int_0^\infty \langle (\mathbf{n}(0) \cdot \mathbf{n}(t)) \rangle e^{-i\omega t} dt. \qquad (13,1.2)$$

The integral has been evaluated for different models of a Brownian particle in the previous chapter. Since the evaluation of the integral did not presuppose that the particle was polar, we can employ (13,1.2) to find τ_{10} for any Brownian particle.

Most of the investigations in Chapters 8–11 were devoted to calculating orientational correlation functions. When these investigations were based on stochastic differential equations, the results were deduced from values of angular velocity correlation functions. These also lead to correlation times. We shall therefore return to expressions for correlation functions, both angular velocity and orientational, derived for the disc and for the spherical, linear and asymmetric rotators and deduce the corresponding correlation times. Since the angular velocity correlation times may be found rather quickly, we shall deal with them for all the models of a Brownian particle in the next section. In the subsequent four sections we shall find orientational correlation times for each model. Then in the final section of this chapter we shall generalize these results in order to calculate power spectra of spherical harmonics.

13.2 Angular Velocity Correlation Times for a Brownian Particle

In the disc model the angular velocity has only one component and eqn (8,3.7),

$$\langle \omega(t_1)\omega(t_2) \rangle = \frac{kT}{I} e^{-B|t_1 - t_2|},$$

gives

$$\langle \omega(0)\omega(t) \rangle = \frac{kT}{I} e^{-Bt} \qquad (t \geqslant 0).$$

The normalized autocorrelation function,

$$\frac{\langle \omega(0)\omega(t) \rangle}{\langle \omega(0)\omega(0) \rangle} = e^{-Bt}.$$

Hence the angular velocity correlation time

$$\tau_\omega = \int_0^\infty e^{-Bt}dt = \frac{1}{B},$$

which is just the friction time τ_F.

For the spherical model we write (9,2.2) as

$$\langle \omega_i(t_k)\omega_j(t_l)\rangle = \delta_{ij}\frac{kT}{I}e^{-B|t_k-t_l|},$$

since the ensemble average over angular velocity space is equal to that over both angular velocity and configuration space. The last equation gives

$$\frac{\langle \omega_i(0)\omega_i(t)\rangle}{\langle \omega_i(0)\omega_i(0)\rangle} = e^{-Bt} \qquad (i = 1, 2, 3; t \geqslant 0).$$

Hence the correlation time for any component of angular velocity is τ_F.

Equation (10,2.2) for the linear rotator expressed as

$$\langle \omega_i(t_k)\omega_j(t_l)\rangle = \delta_{ij}\frac{kT}{I}\, e^{-B|t_k-t_l|} \qquad (i, j = 1, 2)$$

gives $e^{-B|t|}$ for the normalized autocorrelation function of $\omega_1(t)$ or $\omega_2(t)$, the component of angular velocity along the axis of the rotator being zero. The correlation times for $\omega_1(t)$ and $\omega_2(t)$ are again just the friction time.

Finally for the asymmetric rotator we have from (11,4.5) for $t \geqslant 0$

$$\frac{\langle \omega_i(0)\omega_i(t)\rangle}{\langle \omega_i(0)\omega_i(0)\rangle} = e^{-B_i t}$$
$$+ \frac{(I_j - I_k)^2 kT}{I_1 I_2 I_3}\,\frac{e^{-B_i t}\{1-(B_j+B_k-B_i)t - e^{-(B_j+B_k-B_i)t}\}}{(B_j+B_k-B_i)^2}$$
$$+ \dots,$$

where i, j, k is a cyclic permutation of 1, 2, 3. On integrating with respect to t from 0 to ∞ we obtain the correlation time for $\omega_i(t)$

$$\tau_{\omega,i} = \frac{1}{B_i}$$
$$- \frac{(I_j - I_k)^2}{I_1 I_2 I_3} \frac{kT}{(B_j + B_k - B_i)^2} \left\{ \frac{1}{B_i} - \frac{B_j + B_k - B_i}{B_i^2} - \frac{1}{B_j + B_k} \right\}$$
$$+ \ldots .$$

This simplifies to

$$\tau_{\omega,i} = \frac{1}{B_i} - \frac{(I_j - I_k)^2}{I_1 I_2 I_3} \frac{kT}{B_i^2 (B_j + B_k)} + \ldots,$$

which agrees with the result of Hubbard (1977, eqn (6.5)).

13.3 Orientational Correlation Times for the Disc Model

When studying the rotational Brownian motion of a disc we took a unit vector $\mathbf{n}(t)$ fixed in the disc and perpendicular to the axis of rotation. The normalized autocorrelation function of $\mathbf{n}(t)$ is $\langle(\mathbf{n}(0) \cdot \mathbf{n}(t))\rangle$ and according to (8,3.15)

$$\langle(\mathbf{n}(0) \cdot \mathbf{n}(t))\rangle = \exp\left[-\gamma(Bt - 1 + e^{-Bt})\right]. \qquad (13,3.1)$$

In order to integrate this from 0 to ∞ we appeal to results of Section 12.2. In the notation of this section

$$\langle(\mathbf{n}(0) \cdot \mathbf{n}(t))\rangle = \langle R_{11}(t)\rangle,$$

so the correlation time τ_1 for $\mathbf{n}(t)$ is given by

$$\tau_1 = \int_0^\infty \langle R_{11}(t)\rangle dt.$$

This is obtained by putting $\omega = 0$ in (12,2.9), and so

$$\tau_1 = \frac{IB}{kT}\left\{1 + \frac{\gamma}{1+\gamma}\left[1 + \frac{\gamma}{2+\gamma}\left(1 + \frac{\gamma}{3+\gamma}\,[1 + \ldots]\right)\right]\right\}. \qquad (13,3.2)$$

If $\theta(t)$ is the angle between $\mathbf{n}(0)$ and $\mathbf{n}(t)$, so that $\theta(0) = 0$, we see from (8,3.18) that the autocorrelation function of $\cos m\theta(t)$,

$$\langle \cos[m\theta(0)] \cos[m\theta(t)] \rangle = \exp\left[-m^2\gamma(Bt - 1 + e^{-Bt})\right].$$

This may be obtained from (13,3.1) simply by replacing γ by $m^2\gamma$. Making this replacement in (13,3.2) we have the corresponding correlation time

$$\tau_m = \frac{IB}{m^2 kT}\left\{1 + \frac{m^2\gamma}{1+m^2\gamma}\left[1 + \frac{m^2\gamma}{2+m^2\gamma}\left(1 + \frac{m^2\gamma}{3+m^2\gamma}[1+\ldots]\right)\right]\right\}.$$

13.4 Orientational Correlation Times for the Sphere

We shall calculate the correlation time in the case of the sphere for $Y_{jm}(\beta(t), \alpha(t))$, where $\alpha(t)$, $\beta(t)$ are the Euler angles with reference to the laboratory frame of a radius fixed in the sphere. According to (12,3.12) the mean value of the rotation operator $R(t)$ depends only on $\mathbf{I}$ and on J^2 which is $j(j+1)\mathbf{I}$ in the $(2j+1)$-dimensional representation. Thus $\langle R^j_{mm}(t)\rangle$ is independent of m, and replacing $\mathbf{I}$ by unity and J^2 by $j(j+1)$ in (12,3.12) we have

$$\begin{aligned}
\langle R^j_{mm}(t)\rangle = \Big[& 1 + \gamma j(j+1)(1-e^{-Bt}) \\
& + \gamma^2\left\{ j(j+1)\left[\frac{5}{4} - (1+Bt)e^{-Bt} - \frac{1}{4}e^{-2Bt}\right]\right. \\
& \left. + j^2(j+1)^2\left[\frac{1}{2} - e^{-Bt} + \frac{1}{2}e^{-2Bt}\right]\right\} \\
& + \gamma^3\left\{ j(j+1)\left[\frac{19}{9} - (1+2Bt+\frac{1}{2}B^2t^2)e^{-Bt}\right.\right. \\
& \left. - (1+\frac{3}{4}Bt)e^{-2Bt} - \frac{1}{9}e^{-3Bt}\right] + j^2(j+1)^2 \times \\
& \left[\frac{5}{4} - \left(\frac{9}{4}+Bt\right)e^{-Bt} + \left(\frac{3}{4}+Bt\right)e^{-2Bt} + \frac{1}{4}e^{-3Bt}\right] \\
& \left.\left. + j^3(j+1)^3\left[\frac{1}{6} - \frac{1}{2}e^{-Bt} + \frac{1}{2}e^{-2Bt} - \frac{1}{6}e^{-3Bt}\right]\right\} + \ldots \Big]
\end{aligned} \qquad (13,4.1)$$

$$\times \exp\left[-j(j+1)B\gamma\left\{1+\frac{1}{2}\gamma+\frac{7}{12}\gamma^2+\left(\frac{17}{18}-\frac{j(j+1)}{8}\right)\gamma^3 + \ldots\right\}t\right],$$

which is real. We write the exponential as exp $(-BG_jt)$, where

$$G_j = j(j+1)\gamma\left\{1+\frac{1}{2}\gamma+\frac{7}{12}\gamma^2+\left(\frac{17}{18}-\frac{j(j+1)}{8}\right)\gamma^3+\ldots\right\}. \tag{13,4.2}$$

The correlation time is independent of m and writing it as τ_j we have from (13,1.1), (13,4.1) and (C.11) with $\omega = 0$

$$\tau_j = \frac{1}{BG_j} + \gamma\frac{j(j+1)}{B}\left[\frac{1}{G_j}-\frac{1}{1+G_j}\right]$$

$$+\frac{\gamma^2}{B}\left\{j(j+1)\left[\frac{\frac{5}{4}}{G_j}-\frac{1}{1+G_j}-\frac{1}{(1+G_j)^2}-\frac{\frac{1}{4}}{2+G_j}\right]\right.$$

$$\left.+j^2(j+1)^2\left[\frac{\frac{1}{2}}{G_j}-\frac{1}{1+G_j}+\frac{\frac{1}{2}}{2+G_j}\right]\right\}$$

$$+\frac{\gamma^3}{B}\left\{j(j+1)\left[\frac{\frac{19}{9}}{G_j}-\frac{1}{1+G_j}-\frac{2}{(1+G_j)^2}-\frac{1}{(1+G_j)^3}\right.\right.$$

$$\left.-\frac{1}{2+G_j}-\frac{\frac{3}{4}}{(2+G_j)^2}-\frac{\frac{1}{9}}{3+G_j}\right]$$

$$+j^2(j+1)^2\left[\frac{\frac{5}{4}}{G_j}-\frac{\frac{9}{4}}{1+G_j}-\frac{1}{(1+G_j)^2}+\frac{\frac{3}{4}}{2+G_j}\right.$$

$$\left.+\frac{1}{(2+G_j)^2}+\frac{\frac{1}{4}}{3+G_j}\right]$$

$$\left.+j^3(j+1)^3\left[\frac{\frac{1}{6}}{G_j}-\frac{\frac{1}{2}}{1+G_j}+\frac{\frac{1}{2}}{2+G_j}-\frac{\frac{1}{6}}{3+G_j}\right]\right\}+\ldots$$

On substitution for G_j from (13,4.2) and expanding the quantities in the square brackets as powers of γ we find after some calculation that

$$\tau_j = \frac{IB}{j(j+1)kT} \times$$

$$\left\{1 + \left[j(j+1) - \frac{1}{2}\right]\gamma - \left[\frac{1}{2}j^2(j+1)^2 - \frac{3}{4}j(j+1) + \frac{1}{3}\right]\gamma^2\right.$$

$$\left. + \left[\frac{5}{12}j^3(j+1)^3 - \frac{9}{8}j^2(j+1)^2 + \frac{23}{18}j(j+1) - \frac{35}{72}\right]\gamma^3 + \ldots\right\}. \tag{13,4.3}$$

Hence the correlation time for $Y_{1m}(\beta(t), \alpha(t))$,

$$\tau_1 = \frac{IB}{2kT}\left\{1 + \frac{3}{2}\gamma - \frac{5}{6}\gamma^2 + \frac{65}{72}\gamma^3 + \ldots\right\}, \tag{13,4.4}$$

and for $Y_{2m}(\beta(t), \alpha(t))$,

$$\tau_2 = \frac{IB}{6kT}\left\{1 + \frac{11}{2}\gamma - \frac{83}{6}\gamma^2 + 56\,\frac{49}{72}\gamma^3 + \ldots\right\}. \tag{13,4.5}$$

This value of τ_2 was found by Hubbard (1973, eqn (6.3)) as far as the term $-\frac{83}{6}\gamma^2$ in the braces.

For the Debye theory $\langle R(t)\rangle$ is given by (9,3.1):

$$\langle R(t)\rangle = \exp\{-J^2\gamma Bt\}. \tag{13,4.6}$$

Then

$$\langle R^j_{mm}(t)\rangle = \exp\{-j(j+1)\gamma Bt\}$$

and

$$\tau_j = \frac{1}{j(j+1)\gamma B}. \tag{13,4.7}$$

This is the first term in the series of (13,4.3); the remaining terms give the correction due to inertial effects.

Since it was shown in Section 13.2 that the correlation time for any component of angular velocity is B^{-1}, we now write τ_ω for B^{-1}. We then deduce from (13,4.7) that

$$\tau_\omega \tau_j = \frac{I}{j(j+1)kT}, \tag{13,4.8}$$

so that for $j = 2$

$$\tau_\omega \tau_2 = \frac{I}{6kT}. \tag{13,4.9}$$

This is known as the *Hubbard relation* (Hubbard, 1963, eqn (5.18)). It is valid only if inertial effects are neglected.

13.5 Orientational Correlation Times for the Linear Rotator

We wish to calculate the correlation time for $Y_{jm}(\beta(t), \alpha(t))$ in the linear model of a Brownian particle. The $\alpha(t), \beta(t)$ are the Euler angles of the axis of the rotator referred to the laboratory frame. To calculate the correlation time we proceed along the lines of the previous section. According to (12,4.6)

$$\begin{aligned}\langle R^j_{mm}(t)\rangle = {} & [1 + \gamma j(j+1)(1 - e^{-Bt}) \\ & + \gamma^2 \{j(j+1)\,[\tfrac{5}{2} - (2+2Bt)e^{-Bt} - \tfrac{1}{2}e^{-2Bt}] \\ & + j^2(j+1)^2[\tfrac{1}{2} - e^{-Bt} + \tfrac{1}{2}e^{-2Bt}]\} + \ldots]\exp(-BG_j t),\end{aligned} \tag{13,5.1}$$

where now

$$G_j = j(j+1)\gamma\{1 + \gamma + \tfrac{8}{3}\gamma^2 + \ldots\}. \tag{13,5.2}$$

We see that $\langle R^j_{mm}(t)\rangle$ is real and independent of m. Hence from (13.1.1), (13,5.1) and (C.11) the correlation time

$$\begin{aligned}\tau_j = {} & \frac{1}{BG_j} + \gamma\frac{j(j+1)}{B}\left[\frac{1}{G_j} - \frac{1}{1+G_j}\right] \\ & + \frac{\gamma^2}{B}\left\{j(j+1)\left[\frac{\tfrac{5}{2}}{G_j} - \frac{2}{1+G_j} - \frac{2}{(1+G_j)^2} - \frac{\tfrac{1}{2}}{2+G_j}\right]\right. \\ & \left. + j^2(j+1)^2\left[\frac{\tfrac{1}{2}}{G_j} - \frac{1}{1+G_j} + \frac{\tfrac{1}{2}}{2+G_j}\right]\right\} + \ldots\end{aligned}$$

Using (13,5.2) to expand in powers of γ the expressions in the square brackets we find that

$$\tau_j = \frac{IB}{j(j+1)kT}\left\{1 + [j(j+1) - 1]\gamma - \left[\frac{1}{2}j^2(j+1)^2 - \frac{3}{2}j(j+1) + \frac{5}{3}\right]\gamma^2 + \ldots\right\}.$$

From this it is deduced that

$$\tau_1 = \frac{IB}{2kT}\left\{1 + \gamma - \frac{2}{3}\gamma^2 + \ldots\right\},$$

$$\tau_2 = \frac{IB}{6kT}\left\{1 + 5\gamma - \frac{32}{3}\gamma^2 + \ldots\right\}. \qquad (13,5.4)$$

The Debye approximation (11,5.25) yields for the linear rotator

$$\langle R(t)\rangle = \exp\left[-\frac{kT}{IB}(J^2 - J_3^2)t\right].$$

This will lead to

$$\langle R^j_{m'm}(t)\rangle = \langle D^j_{m'm}(\alpha'(t), \beta'(t), \gamma'(t))\rangle$$
$$= \delta_{m'm}\exp[-j(j+1)\gamma Bt],$$

as in eqn (9,3.2) for the sphere. The resulting correlation times will therefore continue to satisfy (13,4.8) and (13,4.9).

13.6 Orientational Correlation Times for the Asymmetric Rotator

In investigating the orientational correlation times for the asymmetric rotator we first consider the case of $j = 1$, which was discussed in Section 12.5. Indeed it was found that the ensemble average of the rotation matrix $R^1(t)$ assumes a simple form given in (12,5.6):

$$\langle R^1(t)\rangle = \left\{\mathbf{I} + \sum_{i=1}^{3}\frac{kT}{I_iB_i^2}(1 - e^{-B_it})J_i^2 + \ldots\right\}$$
$$\times \exp\left\{\left(-\sum_{i=1}^{3}(D_i^{(1)} + D_i^{(2)})J_i^2 + \ldots\right)t\right\}.$$

We may then deduce from (12,5.16) that approximately

$$\int_0^\infty \langle R^1(t)\rangle dt = (-G)^{-1} + \sum_{l=1}^{3} \frac{kT}{I_l B_l} J_l^2 (-G)^{-1}(-G + B_l \mathbf{I})^{-1}.$$

On referring to (12,5.18) and (12,5.20) we see that to this approximation

$$\int_0^\infty \langle R^1(t)\rangle dt \tag{13,6.1}$$

$$= \begin{bmatrix} \frac{1}{D_2 + D_3}\left[1 + \frac{D_2^{(1)}}{D_2 + D_3 + B_2} + \frac{D_3^{(1)}}{D_2 + D_3 + B_3}\right] & 0 & 0 \\ 0 & \frac{1}{D_3 + D_1}\left[1 + \frac{D_3^{(1)}}{D_3 + D_1 + B_3} + \frac{D_1^{(1)}}{D_3 + D_1 + B_1}\right] & 0 \\ 0 & 0 & \frac{1}{D_1 + D_2}\left[1 + \frac{D_1^{(1)}}{D_1 + D_2 + B_1} + \frac{D_2^{(1)}}{D_1 + D_2 + B_2}\right] \end{bmatrix}$$

We now calculate the correlation time τ_{10} corresponding to the correlation function $\langle(\mathbf{n}(0)\cdot\mathbf{n}(t))\rangle$, where $\mathbf{n}(t)$ is a unit vector from the centre of mass fixed in the body with constant components n_1, n_2, n_3 referred to the body-frame axes. Then from (12,5.7)

$$(\mathbf{n}(0)\cdot\mathbf{n}(t)) = n^{\mathrm{T}} R^1(t) n,$$

and from (13,1.2)

$$\tau_{10} = n^{\mathrm{T}} \int_0^\infty \langle R^1(t)\rangle n \, dt.$$

On substituting from (13,6.1) we obtain

$$\begin{aligned} \tau_{10} = {} & \frac{n_1^2}{D_2 + D_3}\left[1 + \frac{D_2^{(1)}}{D_2 + D_3 + B_2} + \frac{D_3^{(1)}}{D_2 + D_3 + B_3}\right] \\ & + \frac{n_2^2}{D_3 + D_1}\left[1 + \frac{D_3^{(1)}}{D_3 + D_1 + B_3} + \frac{D_1^{(1)}}{D_3 + D_1 + B_1}\right] \\ & + \frac{n_3^2}{D_1 + D_2}\left[1 + \frac{D_1^{(1)}}{D_1 + D_2 + B_1} + \frac{D_2^{(1)}}{D_1 + D_2 + B_2}\right]. \end{aligned} \tag{13,6.2}$$

To check this result we apply it to the case of the sphere. Then from (12,5.23) and (12,5.24)

$$D_1^{(1)} = D_2^{(1)} = D_3^{(1)} = \frac{kT}{IB},$$

$$D_1 = D_2 = D_3 = D = (1 + \frac{1}{2}\gamma)\frac{kT}{IB}.$$

Hence

$$\tau_{10} = \frac{1}{2D}\left(1 + \frac{2D^{(1)}}{B + 2D}\right)$$

$$= \frac{IB}{2kT(1 + \frac{1}{2}\gamma)}\left(1 + \frac{2\gamma}{1 + 2\gamma(1 + \frac{1}{2}\gamma)}\right)$$

$$= \frac{IB}{2kT}\left(1 + \frac{3}{2}\gamma + \ldots\right)$$

in agreement with (13,4.4) in the approximation to which we have taken calculations for the asymmetric rotator.

We now approach the general problem of finding the correlation time for $Y_{jm}(\beta(t), \alpha(t))$. We define an operator τ by

$$\tau = \int_0^\infty \langle R(t)\rangle dt, \tag{13,6.3}$$

so that from (13,1.1)

$$\tau_{jm} = \tau_{mm}^{j*}, \tag{13,6.4}$$

the complex conjugate of the diagonal mm-element of the $(2j + 1)$-dimensional representation of τ, when the Y_{js} constitute the basis of the representation. From (11,5.24)

$$\tau = \int_0^\infty dt\left\{\mathbf{I} + \sum_{l=1}^{3} \frac{kT}{I_l B_l^2}(1 - e^{-B_l t})J_l^2 + \ldots\right\} e^{Gt}, \tag{13,6.5}$$

where

$$G = \sum_{i=1}^{3} [-(D_i^{(1)} + D_i^{(2)})J_i^2] + \frac{i}{3}P\frac{(kT)^2}{I_1 I_2 I_3}\sum_{i=1}^{3} I_i \frac{B_j - B_k}{B_j^2 B_k^2} + \ldots, \tag{13,6.6}$$

i, j, k being a cyclic permutation of 1, 2, 3 and $D_i^{(1)}$, $D_i^{(2)}$, P being defined in (11,5.18), (11,5.23) and (11,3.7). On comparing (13,6.3) with (12,5.16) we get

$$\tau = (-G)^{-1} + \sum_{l=1}^{3} \frac{kT}{I_l B_l} J_l^2 (-G)^{-1}(-G + B_l \mathbf{I})^{-1}. \qquad (13,6.7)$$

We now derive an approximate form of (13,6.7) that will facilitate future computation. We see from (13,6.6) that

$$G = G_0 + \frac{i}{3} AP + \ldots \qquad (13,6.8)$$

with

$$G_0 = -\sum_{i=1}^{3} (D_i^{(1)} + D_i^{(2)}) J_i^2 = -\sum_{i=1}^{3} D_i J_i^2, \qquad (13,6.9)$$

$$A = \frac{(kT)^2}{I_1 I_2 I_3} \sum_{i=1}^{3} I_i \frac{B_j - B_k}{B_j^2 B_k^2}. \qquad (13,6.10)$$

Considering order of magnitude we omit subscripts of the I's and B's, and we put $kT/(IB^2) = \gamma$. From (11,5.18) and (11,5.23), $D_i^{(1)}$ is of order $kT/(IB)$, $D_i^{(2)}$ is of order $\gamma kT/(IB)$ and so, from (13,6.9), D_i is of order $kT/(IB)$. From (13,6.10), A is of order $\gamma kT/(IB)$. Hence, since J_i^2 is of order unity for the not very large values of j with which we shall be concerned (Edmonds, 1968, p. 17), G_0 and G are of order $kT/(IB)$. Expanding to first order of approximation in γ we obtain from (13,6.8)

$$(-G)^{-1} = (-G_0 - \frac{i}{3} AP)^{-1}$$

$$= (-G_0)^{-1} + \frac{i}{3} A(-G_0)^{-1} P(-G_0)^{-1}, \qquad (13,6.11)$$

as can easily be verified by multiplying the equations to the left by $-G_0 - (i/3)AP$.

Returning to (13,6.7) we note that, since $-G/B_l$ is of order γ, the summation term is of order γ in comparison with $(-G)^{-1}$. Now

$$(B_l \mathbf{I} - G)^{-1} = (B_l \mathbf{I})^{-1} - (B_l \mathbf{I})^{-1}(-G)(B_l \mathbf{I})^{-1} + \ldots$$
$$= B_l^{-1} \mathbf{I} - B_l^{-2}(-G) + \ldots$$

Since we have throughout neglected corrections of order γ^2, we now approximate in the summation term $(-G + B_l\mathbf{I})^{-1}$ by $B_l^{-1}\mathbf{I}$ and $(-G)^{-1}$ by $(-G_0)^{-1}$, so that (13,6.7) becomes

$$\begin{aligned}\tau &= (-G)^{-1} + \sum_{l=1}^{3} \frac{D_l^{(1)}}{B_l} J_l^2 (-G_0)^{-1} \\ &= (-G_0)^{-1} + \frac{i}{3} A(-G_0)^{-1} P(-G_0)^{-1} \\ &\quad + \sum_{l=1}^{3} \frac{D_l^{(1)}}{B_l} J_l^2 (-G_0)^{-1},\end{aligned}$$

by (13,6.11). Equation (E.20) may be expressed in our present notation as

$$-\frac{2i}{3} AP = \left[\sum_{i=1}^{3} D_i^{(1)} J_i^2, \sum_{l=1}^{3} \frac{D_l^{(1)}}{B_l} J_l^2\right],$$

and therefore

$$\begin{aligned}\tau = (-G_0)^{-1} &- \frac{1}{2}(-G_0)^{-1} \left\{\sum_{i=1}^{3} D_i^{(1)} J_i^2 \sum_{l=1}^{3} \frac{D_l^{(1)}}{B_l} J_l^2\right. \\ &\left. - \sum_{l=1}^{3} \frac{D_l^{(1)}}{B_l} J_l^2 \sum_{i=1}^{3} D_i^{(1)} J_i^2\right\} (-G_0)^{-1} + \sum_{l=1}^{3} \frac{D_l^{(1)}}{B_l} J_l^2 (-G_0)^{-1}.\end{aligned}$$

Since the second term on the right-hand side is of order γ in comparison with the first, we approximate $\sum_{i=1}^{3} D_i^{(1)} J_i^2$ in it by $-G_0$ and obtain

$$\begin{aligned}\tau = (-G_0)^{-1} &+ \frac{1}{2} \sum_{l=1}^{3} \frac{D_l^{(1)}}{B_l} J_l^2 (-G_0)^{-1} \\ &+ \frac{1}{2}(-G_0) \sum_{l=1}^{3} \frac{D_l^{(1)}}{B_l} J_l^2.\end{aligned}$$

In the last two terms we may replace $D_l^{(1)}$ by D_l to our order of approximation, and then

$$\tau = (-G_0)^{-1} + \frac{1}{2} \sum_{l=1}^{3} \frac{D_l}{B_l} J_l^2 (-G_0)^{-1} + \frac{1}{2}(-G_0)^{-1} \sum_{l=1}^{3} \frac{D_l}{B_l} J_l^2. \tag{13,6.12}$$

This is the equation on which we shall base specific calculations of correlation times. It could be used to give an alternative treatment of the $j = 1$ case. Of greater interest, however, is the $j = 2$ case.

Let us find the correlation time for $Y_{2m}(\beta(t), \alpha(t))$. The matrix elements for angular momentum operators M_x, M_y, M_z obeying

$$M_y M_z - M_z M_y = i\hbar M_x, \text{ etc.,}$$

in the representation with basis elements Y_{js} are well known (Edmonds, 1968, p. 17). To deduce the matrix elements for our J_1, J_2, J_3 which obey

$$J_2 J_3 - J_3 J_2 = -iJ_1, \text{ etc.,}$$

we make the substitutions

$$M_x = -\hbar J_1, M_y = -\hbar J_2, M_z = -\hbar J_3.$$

Putting $j = 2$ and labelling the rows and columns by the values of m progressively from -2 to $+2$ we have

$$J_1 = -\frac{1}{2}\begin{bmatrix} 0 & 2 & 0 & 0 & 0 \\ 2 & 0 & \sqrt{6} & 0 & 0 \\ 0 & \sqrt{6} & 0 & \sqrt{6} & 0 \\ 0 & 0 & \sqrt{6} & 0 & 2 \\ 0 & 0 & 0 & 2 & 0 \end{bmatrix}, \quad J_2 = \frac{-i}{2}\begin{bmatrix} 0 & 2 & 0 & 0 & 0 \\ -2 & 0 & \sqrt{6} & 0 & 0 \\ 0 & -\sqrt{6} & 0 & \sqrt{6} & 0 \\ 0 & 0 & -\sqrt{6} & 0 & 2 \\ 0 & 0 & 0 & -2 & 0 \end{bmatrix}, \quad J_3 = \begin{bmatrix} 2 & 0 & 0 & 0 & 0 \\ 0 & 1 & 0 & 0 & 0 \\ 0 & 0 & 0 & 0 & 0 \\ 0 & 0 & 0 & -1 & 0 \\ 0 & 0 & 0 & 0 & -2 \end{bmatrix}$$

$$J_1^2 = \tfrac{1}{2}\begin{bmatrix} 2 & 0 & \sqrt{6} & 0 & 0 \\ 0 & 5 & 0 & 3 & 0 \\ \sqrt{6} & 0 & 6 & 0 & \sqrt{6} \\ 0 & 3 & 0 & 5 & 0 \\ 0 & 0 & \sqrt{6} & 0 & 0 \end{bmatrix}, \quad J_2^2 = \tfrac{1}{2}\begin{bmatrix} 2 & 0 & -\sqrt{6} & 0 & 0 \\ 0 & 5 & 0 & -3 & 0 \\ -\sqrt{6} & 0 & 6 & 0 & -\sqrt{6} \\ 0 & -3 & 0 & 5 & 0 \\ 0 & 0 & -\sqrt{6} & 0 & 2 \end{bmatrix}, \quad J_3^2 = \begin{bmatrix} 4 & 0 & 0 & 0 & 0 \\ 0 & 1 & 0 & 0 & 0 \\ 0 & 0 & 0 & 0 & 0 \\ 0 & 0 & 0 & 1 & 0 \\ 0 & 0 & 0 & 0 & 4 \end{bmatrix}$$

(13,6.13)

From (13,6.9) we obtain the matrix representative of G_0. Putting

$$D_1 + D_2 + 4D_3 = a, \quad \frac{5}{2}(D_1 + D_2) + D_3 = b,$$

$$3(D_1 + D_2) = c, \quad \frac{\sqrt{6}}{2}(D_1 - D_2) = d, \qquad (13,6.14)$$

$$\frac{3}{2}(D_1 - D_2) = e,$$

we find that

$$-G_0 = \begin{bmatrix} a & 0 & d & 0 & 0 \\ 0 & b & 0 & e & 0 \\ d & 0 & c & 0 & d \\ 0 & e & 0 & b & 0 \\ 0 & 0 & d & 0 & a \end{bmatrix}, \qquad (13,6.15)$$

$$(-G_0)^{-1} = \begin{bmatrix} \frac{ac-d^2}{a(ac-2d^2)} & 0 & -\frac{ad}{a(ac-2d^2)} & 0 & \frac{d^2}{a(ac-2d^2)} \\ 0 & \frac{b}{b^2-e^2} & 0 & -\frac{e}{b^2-e^2} & 0 \\ -\frac{ad}{a(ac-2d^2)} & 0 & \frac{a^2}{a(ac-2d^2)} & 0 & -\frac{ad}{a(ac-2d^2)} \\ 0 & -\frac{e}{b^2-e^2} & 0 & \frac{b}{b^2-e^2} & 0 \\ \frac{d^2}{a(ac-2d^2)} & 0 & -\frac{ad}{a(ac-2d^2)} & 0 & \frac{ac-d^2}{a(ac-2d^2)} \end{bmatrix}.$$

(13,6.16)

We also have

$$\sum_{l=1}^{3} \frac{D_l}{B_l} J_l^2 = \begin{bmatrix} a' & 0 & d' & 0 & 0 \\ 0 & b' & 0 & e' & 0 \\ d' & 0 & c' & 0 & d' \\ 0 & e' & 0 & b' & 0 \\ 0 & 0 & d' & 0 & a' \end{bmatrix} \qquad (13,6.17)$$

with

$$a' = \frac{D_1}{B_1} + \frac{D_2}{B_2} + \frac{4D_3}{B_3}, \quad b' = \frac{5}{2}\left(\frac{D_1}{B_1} + \frac{D_2}{B_2}\right) + \frac{D_3}{B_3},$$

$$c' = 3\left(\frac{D_1}{B_1} + \frac{D_2}{B_2}\right), \qquad d' = \frac{\sqrt{6}}{2}\left(\frac{D_1}{B_1} - \frac{D_2}{B_2}\right), \tag{13,6.18}$$

$$e' = \frac{3}{2}\left(\frac{D_1}{B_1} - \frac{D_2}{B_2}\right),$$

and hence

$$\sum_{l=1}^{3} \frac{D_l}{B_l} J_l^2 (-G_0)^{-1} + (-G_0)^{-1} \sum_{l=1}^{3} \frac{D_l}{B_l} J_l^2 \tag{13,6.19}$$

$$= \begin{bmatrix} \frac{2(aa'c - a'd^2 - add')}{a(ac - 2d^2)} & 0 & \frac{acd' - ac'd - aa'd + a^2d'}{a(ac - 2d^2)} & 0 & \frac{2(a'd^2 - add')}{a(ac - 2d^2)} \\ 0 & \frac{2(bb' - ee')}{b^2 - e^2} & 0 & \frac{2(be' - b'e)}{b^2 - e^2} & 0 \\ \frac{acd' - ac'd - aa'd + a^2d'}{a(ac - 2d^2)} & 0 & \frac{2(ac' - 2dd')}{ac - 2d^2} & 0 & \frac{acd' - ac'd - aa'd + a^2d'}{a(ac - 2d^2)} \\ 0 & \frac{2(be' - b'e)}{b^2 - e^2} & 0 & \frac{2(bb' - ee')}{b^2 - e^2} & 0 \\ \frac{2(a'd^2 - add')}{a(ac - 2d^2)} & 0 & \frac{acd' - ac'd - aa'd + a^2d'}{a(ac - 2d^2)} & 0 & \frac{2(aa'c - a'd^2 - add')}{a(ac - 2d^2)} \end{bmatrix}.$$

On substituting from (13,6.16) and (13,6.19) into (13,6.12) we can write down the five-dimensional representation of τ. It is clearly a real matrix.

We have now all the information required to calculate the correlation time τ_{2m} for $Y_{2m}(\beta(t), \alpha(t))$. Since τ is represented by a real matrix, it follows from (13,6.4) that τ_{2m} is just the mm-element of the matrix. The correlation times are therefore the diagonal elements and it is seen from (13,6.16) and (13,6.19) that $\tau_{2,-m}$ is equal to $\tau_{2,m}$. It is readily found that

$$\tau_{20} = \frac{a + ac' - 2dd'}{ac - 2d^2}, \tag{13,6.20}$$

$$\tau_{21} = \frac{b + bb' - ee'}{b^2 - e^2}, \tag{13,6.21}$$

$$\tau_{22} = \frac{ac - d^2 + a'(ac - d^2) - add'}{a(ac - 2d^2)}. \tag{13,6.22}$$

Since from (7,3.15)

$$Y_{20}(\beta(t), \alpha(t)) = \left(\frac{5}{4\pi}\right)^{1/2} P_2(\cos\beta(t)),$$

τ_{20} is the correlation time for $P_2(\cos\beta(t))$. On substituting from (13,6.14) and (13,6.18) into (13,6.20)–(13,6.22) we obtain

$$\tau_{20} = \frac{D_1 + D_2 + 4D_3 + 6D_1D_2\left(\dfrac{1}{B_1} + \dfrac{1}{B_2}\right) + \dfrac{12D_2D_3}{B_2} + \dfrac{12D_3D_1}{B_1}}{12(D_2D_3 + D_3D_1 + D_1D_2)}, \tag{13,6.23}$$

$$\tau_{21} = \frac{\begin{aligned}&\frac{5}{2}(D_1 + D_2) + D_3 + \frac{4D_1^2}{B_1} + \frac{4D_2^2}{B_2} + \frac{D_3^2}{B_3}\\ &+ \frac{5}{2}D_2D_3\left(\frac{1}{B_2} + \frac{1}{B_3}\right) + \frac{5}{2}D_3D_1\left(\frac{1}{B_3} + \frac{1}{B_1}\right)\\ &\qquad + \frac{17}{2}D_1D_2\left(\frac{1}{B_1} + \frac{1}{B_2}\right)\end{aligned}}{4D_1^2 + 4D_2^2 + D_3^2 + 5(D_2D_3 + D_3D_1) + 17D_1D_2}, \tag{13,6.24}$$

$$\tau_{22} = \frac{\begin{aligned}&\frac{3}{2}D_1^2 + \frac{3}{2}D_2^2 + 9D_1D_2 + 12(D_2D_3 + D_3D_1)\\ &+ D_1^2D_2\left(\frac{9}{B_1} + \frac{3}{B_2}\right) + D_1D_2^2\left(\frac{3}{B_1} + \frac{9}{B_2}\right) + \frac{48(D_1 + D_2)D_3^2}{B_3}\\ &+ 6D_2^2D_3\left(\frac{1}{B_2} + \frac{1}{B_3}\right)\\ &+ 6D_1^2D_3\left(\frac{1}{B_3} + \frac{1}{B_1}\right) + 18D_1D_2D_3\left(\frac{1}{B_1} + \frac{1}{B_2} + \frac{2}{B_3}\right)\end{aligned}}{12(D_1 + D_2 + 4D_3)(D_2D_3 + D_3D_1 + D_1D_2)}. \tag{13,6.25}$$

As a check on (13,6.23)–(13,6.25) we apply these equations to the sphere. Then

$$D_1 = D_2 = D_3 = D, \quad B_1 = B_2 = B_3 = B,$$

and for every equation we obtain

$$\tau_{2m} = \frac{1 + \dfrac{6D}{B}}{6D}.$$

Now from (12,5.24)

$$D = \left(1 + \frac{1}{2}\gamma\right)\frac{kT}{IB},$$

and hence

$$\tau_{2m} = \frac{IB[1 + 6\gamma(1 + \frac{1}{2}\gamma)]}{6kT(1 + \frac{1}{2}\gamma)} = \frac{IB}{6kT}\left(1 + \frac{11}{2}\gamma + \ldots\right),$$

which in the approximation of the present section agrees with (13,4.5).

13.7 Spectral Densities of Spherical Harmonics

For future applications to nuclear magnetic phenomena it will be helpful to define by

$$\tau(\omega) = \int e^{-i\omega t}\langle R(t)\rangle dt \qquad (13,7.1)$$

a generalization $\tau(\omega)$ of the operator τ given in (13,6.3). We see that

$$\tau = \tau(0). \qquad (13,7.2)$$

The operator $\tau(\omega)$ may be used to calculate power spectra, or spectral densities, which we met in Sections 2.2 and 4.4. Thus from (4,4.6), generalized for complex $X(t)$, the power spectrum $\mathscr{Y}_{jm}(\omega)$ of $Y_{jm}(\beta(t), \alpha(t))$ is given by

$$\mathscr{Y}_{jm}(\omega) = \int_0^\infty e^{-i\omega t}\langle Y^*_{jm}(\beta(0), \alpha(0))\, Y_{jm}(\beta(t), \alpha(t))\rangle dt. \quad (13,7.3)$$

Then from (9,4.1) and (5,4.18)

$$\langle Y^*_{jm}(\beta(0), \alpha(0))\, Y_{jm}(\beta(t), \alpha(t))\rangle = \frac{1}{4\pi}\langle R(t)\rangle^{j*}_{mm}, \qquad (13,7.4)$$

so that, by (4,4.2),

$$\begin{aligned}\langle R(-t)\rangle^{j*}_{mm} &= 4\pi\langle Y^*_{jm}(\beta(0), \alpha(0))\, Y_{jm}(\beta(-t), \alpha(-t))\rangle^* \\ &= 4\pi\langle Y_{jm}(\beta(t), \alpha(t))\, Y^*_{jm}(\beta(0), \alpha(0))\rangle \\ &= \langle R(t)\rangle^{j}_{mm}.\end{aligned}$$

Hence, from (13,7.3) and (13,7.4)

$$\begin{aligned}\mathscr{Y}_{jm}(\omega) &= \frac{1}{4\pi}\int_0^\infty e^{-i\omega t}\langle R(t)\rangle^{j*}_{mm}\, dt + \frac{1}{4\pi}\int_0^\infty e^{i\omega t}\langle R(-t)\rangle^{j*}_{mm}\, dt \\ &= \frac{1}{4\pi}\int_0^\infty e^{i\omega t}\langle R(t)\rangle^{j}_{mm}\, dt + \frac{1}{4\pi}\int_0^\infty e^{-i\omega t}\langle R(t)\rangle^{j*}_{mm}\, dt,\end{aligned} \qquad (13,7.5)$$

so that from (13,7.1)

$$\mathscr{Y}_{jm}(\omega) = \frac{1}{4\pi}(\tau(-\omega)^{j}_{mm} + \tau(-\omega)^{j*}_{mm}).$$

It follows from (13,7.2) that

$$\mathscr{Y}_{jm}(0) = \frac{1}{4\pi}(\tau^j_{mm} + \tau^{j*}_{mm}).$$

We shall be concerned chiefly with the representation where $j = 2$. For this case eqns (11,5.24), (11,5.20), and (13,6.13) show that $\langle R(t)\rangle^j_{mm}$ is a real matrix. Then from (13,7.5) and (13,7.2)

$$\mathscr{Y}_{2m}(\omega) = \frac{1}{4\pi}(\tau(\omega)^2_{mm} + \tau(-\omega)^2_{mm}), \tag{13,7.6}$$

$$\mathscr{Y}_{2m}(0) = \frac{\tau^2_{mm}}{2\pi}. \tag{13,7.7}$$

If we wish to calculate $\tau(\omega)$ for an asymmetric rotator explicitly from (13,7.1), we proceed along the lines of the deduction of (13,6.12) from (13,6.3). For values of ω not greater in order of magnitude than $kT/(IB)$ it is found that

$$\tau(\omega) = (-G_0 + i\omega\mathbf{I})^{-1} + \frac{1}{2}\sum_{l=1}^{3}\frac{D_l}{B_l}J_l^2(-G_0 + i\omega\mathbf{I})^{-1} + \frac{1}{2}(-G_0 + i\omega\mathbf{I})^{-1}\sum_{l=1}^{3}\frac{D_l}{B_l}J_l^2$$

(Ford *et al.*, 1979). When this is applied to the five-dimensional representation, it will be seen that

$$\tau(\omega)^2_{-1,-1} = \tau(\omega)^2_{11}, \quad \tau(\omega)^2_{-2,-2} = \tau(\omega)^2_{22}$$

and that the values of $\tau(\omega)^2_{00}$, $\tau(\omega)^2_{11}$, $\tau(\omega)^2_{22}$ are given by (13,6.20), (13,6.21), (13,6.22), respectively, with the replacements

$$a \to a + i\omega, \quad b \to b + i\omega, \quad c \to c + i\omega, \tag{13,7.8}$$

the other variables being left unaltered. The expression for $\mathscr{Y}_{2m}(\omega)$ may then be written down from (13,7.6).

Let us calculate $\mathscr{Y}_{2m}(\omega)$ when the rotator becomes a sphere. Then from (13,6.14), (13,6.18) and (13,7.8) a, b, c, become $6D + i\omega$, a', b', c' become $6D/B$ and d, e, d', e' vanish. It is easily deduced from (13,7.6) that

$$\mathscr{Y}_{2m}(\omega) = \frac{3D}{\pi} \frac{1 + \dfrac{6D}{B}}{36D^2 + \omega^2}. \qquad (13,7.9)$$

If we now approximate D by $D^{(1)}$, that is $kT/(IB)$, this relation becomes

$$\mathscr{Y}_{2m}(\omega) = \frac{(1 + 6\gamma)IB}{12\pi kT} \frac{1}{1 + \left(\dfrac{IB\omega}{6kT}\right)^2}.$$

Since in the above approximation we have neglected small quantities of order γ, we replace the last equation by

$$\mathscr{Y}_{2m}(\omega) = \frac{IB}{12\pi kT} \frac{1}{1 + \left(\dfrac{IB\omega}{6kT}\right)^2}. \qquad (13,7.10)$$

When $\omega \ll kT/(IB)$, we see from (13,6.14) and (13,7.8) that $\tau(\omega)^2_{00}$, $\tau(\omega)^2_{11}$, $\tau(\omega)^2_{22}$ become approximately equal to τ_{20}, τ_{21}, τ_{22}, respectively. Such values of ω are said to give the *extreme narrowing case* (Abragam, 1961, p. 279). Hence from (13,7.7)

$$\mathscr{Y}_{2m}(\omega) = \frac{\tau_{2m}}{2\pi} \quad \left(\omega \ll \frac{kT}{IB}\right). \qquad (13,7.11)$$

Thus we need only correlation times for the evaluation of spectral densities for $j = 2$ in the extreme narrowing case. Equation (13,7.11) is, of course, also true for the sphere and linear rotator models of a Brownian particle and then τ_{2m} is independent of m, as we saw when deriving (13,4.5) and (13,5.4).

14

Rotational Brownian Motion and Relaxation Experiments

14.1 The Rocard Relation for Complex Permittivity

In the previous two chapters the theory of rotational Brownian motion was employed to derive results for complex polarizability, correlation times and spectral densities. In this final chapter we shall consider how these results may be related to experiments on dielectric and nuclear magnetic relaxation phenomena. We shall first attempt to express permittivities in a form that will be convenient for making comparison with experimental figures.

During the discussion of (12,5.22) it was asserted that the value B/γ of ω is beyond the limits of classical statistical mechanics. This is something that will have to be verified for each experiment. When this has been done, it will allow the suppression of the $i\omega D_l^{(2)}$ terms in eqn (12,5.21) for dipolar polarization. When displacement polarization is also considered and we are dealing with a gas dielectric or a very dilute solution of a liquid dielectric in a nonpolar solvent, we replace $\alpha(\omega)/\alpha_s$ by $(\epsilon(\omega)-\epsilon_\infty)/(\epsilon_s-\epsilon_\infty)$, as was explained in Section 2.3. We than obtain from (12,5.21)

$$\frac{\epsilon(\omega)-\epsilon_\infty}{\epsilon_s-\epsilon_\infty}$$

$$= \frac{n_1^2}{D_2+D_3+i\omega}\left\{\frac{D_2(D_2+D_3+B_2)}{D_2+D_3+B_2+i\omega}+\frac{D_3(D_2+D_3+B_3)}{D_2+D_3+B_3+i\omega}\right\}$$

$$+\frac{n_2^2}{D_3+D_1+i\omega}\left\{\frac{D_3(D_3+D_1+B_3)}{D_3+D_1+B_3+i\omega}+\frac{D_1(D_3+D_1+B_1)}{D_3+D_1+B_1+i\omega}\right\}$$

$$+\frac{n_3^2}{D_1+D_2+i\omega}\left\{\frac{D_1(D_1+D_2+B_1)}{D_1+D_2+B_1+i\omega}+\frac{D_2(D_1+D_2+B_2)}{D_1+D_2+B_2+i\omega}\right\},$$

(14,1.1)

where n_1, n_2, n_3 are the direction cosines of the dipole axis and D_1, D_2, D_3 are given by (12,5.9), (11,5.18) and (11,5.23).

We apply this result to a polar molecule which has the configuration of a symmetric rotator. The dipole must lie along the axis of symmetry, which we take to be the third coordinate axis, and so

$$n_1 = n_2 = 0, \quad n_3 = 1,$$

$$I_2 = I_1, \quad B_2 = B_1, \quad D_2 = D_1.$$

Thence (14,1.1) reduces to

$$\frac{\epsilon(\omega)-\epsilon_\infty}{\epsilon_s-\epsilon_\infty}=\frac{1}{\left(1+\dfrac{i\omega}{2D_1}\right)\left(1+\dfrac{i\omega}{B_1+2D_1}\right)}. \tag{14,1.2}$$

If we neglect small quantities of relative order γ, so that D_1 is approximated by $D_1^{(1)}$, eqn (14,1.2) becomes

$$\frac{\epsilon(\omega)-\epsilon_\infty}{\epsilon_s-\epsilon_\infty}=\frac{1}{\left(1+\dfrac{i\omega I_1 B_1}{2kT}\right)\left(1+\dfrac{i\omega}{B_1}\right)}. \tag{14,1.3}$$

This is eqn (2,5.3), if the above replacement of complex polarizability by complex permittivity is made. We shall refer to an equation of the form (14,1.3) as a *Rocard equation* or a *Rocard relation.* Equation (14,1.2) gives corrections of order γ to the ω-dependent terms of (14,1.3), but we see from (11,5.23) that in order to write down these corrections we need to know I_3 and B_3.

When a polar molecule is spherical or approximately spherical in shape, we may employ (12,3.17). We shall assume in future that ω is at most of order B, so that ω', or $\omega\tau_F$, is at most of order unity: $\omega\tau_F \leqslant 0(1)$. Since by (12,3.14), G_1 and γ are of the same order, the second term on the right-hand side of (12,3.17) is at most of order G_1 or γ relative to the first. Truncating the series after the first term and replacing polarizability by permittivity as before we deduce from (12,3.18) that

$$\frac{\epsilon(\omega)-\epsilon_\infty}{\epsilon_s-\epsilon_\infty}=\frac{1}{[1+(1-\frac{1}{2}\gamma)i\omega\tau_D][1+(1-2\gamma)i\omega\tau_F]}. \tag{14,1.4}$$

In the same way we deduce

$$\frac{\epsilon(\omega)-\epsilon_\infty}{\epsilon_s-\epsilon_\infty}=\frac{1}{[1+(1-\gamma)i\omega\tau_D][1+(1-2\gamma)i\omega\tau_F]} \tag{14,1.5}$$

from (12,4.11) for the linear rotator, and

$$\frac{\epsilon(\omega)-\epsilon_\infty}{\epsilon_s-\epsilon_\infty}=\frac{1}{[1+i\omega\tau][1+(1-\gamma)i\omega\tau_F]} \tag{14,1.6}$$

from (12,2.11) for the disc model.

We conclude from (14,1.3)–(14,1.6) that, when $\omega\tau_F$ does not exceed unity in order of magnitude, the complex permittivity is given in first approximation for the disc and symmetric rotator models of a polar molecule, including the spherical and linear rotator models, by a Rocard equation

$$\frac{\epsilon(\omega)-\epsilon_\infty}{\epsilon_s-\epsilon_\infty}=\frac{1}{(1+i\omega\tau_D)(1+i\omega\tau_F)}, \tag{14,1.7}$$

where for the symmetric rotator

$$\tau_D=\frac{I_1B_1}{2kT}, \quad \tau_F=\frac{1}{B_1}. \tag{14,1.8}$$

When the molecule is completely asymmetric, the permittivity is given by (14,1.1) and this will not reduce in first approximation to a Rocard equation even when the dipole is in the direction of a principal axis of inertia.

In the limit of infinitely large friction τ_F becomes vanishingly small and (14,1.7) reduces to

$$\frac{\epsilon(\omega)-\epsilon_\infty}{\epsilon_s-\epsilon_\infty}=\frac{1}{1+i\omega\tau_D}, \tag{14,1.9}$$

which is eqn (2,4.1) of the Debye theory. However, on account of the quadratic dependence on ω of the denominator of the right-hand side of (14,1.7) this equation is structurally different from (14,1.9): eqn (14,1.7) is not just a correction of order γ to (14,1.9). This situation may be contrasted with what happens when we include inertial effects in the calculation of correlation times. Thus we see from eqn (13,4.3) for a rotating spherical Brownian particle that the correlation time is expressible as a power series in γ, of which the first term is that which occurs in the Debye theory. The difference in structure of (14,1.7) and (14,1.9) has profound experimental implications, as we shall see below.

The conclusion of greatest relevance to experiments on dielectric relaxation that we have reached as a result of the study of the theory of rotational Brownian motion is that, if a polar molecule has axial symmetry, the complex permittivity is given to a high degree of accuracy by the Rocard relation (14,1.7). On equating real and imaginary parts this equation yields

$$\epsilon'(\omega) = \epsilon_\infty + \frac{(\epsilon_s - \epsilon_\infty)(1 - \omega^2 \tau_D \tau_F)}{(1 + \omega^2 \tau_D^2)(1 + \omega^2 \tau_F^2)} \qquad (\omega\tau_F \leqslant 0(1)), \tag{14,1.10}$$

$$\epsilon''(\omega) = \frac{(\epsilon_s - \epsilon_\infty)(\tau_D + \tau_F)\omega}{(1 + \omega^2 \tau_D^2)(1 + \omega^2 \tau_F^2)}, \tag{14,1.11}$$

which are very different from (2,4.2) and (2,4.3) for the Debye theory. On substituting the values of $\epsilon'(\omega)$ and $\epsilon''(\omega)$ into (2,3.14) and (2,3.16) we obtain the index of refraction $n(\omega)$ and the absorption coefficient $a(\omega)$. However, on account of the restriction on ω we cannot proceed as in the discussion of (2,4.2) and (2,4.3) by allowing $\omega\tau_D$ to become indefinitely great.

By taking $\omega^2\tau_D^2$ as the independent variable and writing $\tau_F = 2\gamma\tau_D$ with $\gamma = kT/(I_1 B_1^2)$ from (14,1.8), it is easily deduced from (14,1.10) that the minimum value of $\epsilon'(\omega)$ occurs when

$$\omega\tau_D = \frac{1}{8^{1/4}\gamma^{3/4}}(1 + \tfrac{1}{2}\gamma^{1/2} + \tfrac{3}{4}\gamma + \ldots). \tag{14,1.12}$$

Then

$$\omega\tau_F = 2\gamma\omega\tau_D = 2^{1/4}\gamma^{1/4}(1 + \tfrac{1}{2}\gamma^{1/2} + \tfrac{3}{4}\gamma + \ldots),$$

so that the minimum occurs in the region where the condition $\omega\tau_F \leqslant 0(1)$ is satisfied. The minimum value of $\epsilon'(\omega)$,

$$\epsilon'_{\min} = \epsilon_\infty - (\epsilon_s - \epsilon_\infty)2\gamma(1 - 2\sqrt{2}\gamma^{1/2} + \ldots), \tag{14,1.13}$$

which is less than ϵ_∞. When ω satisfies (14,1.12), the Cole–Cole plot corresponding to the Rocard relation (14,1.7) bends backwards and, as we deduce from (14,1.11),

$$\epsilon''(\omega) = (\epsilon_s - \epsilon_\infty)8^{1/4}\gamma^{3/4}(1 - \tfrac{3}{2}\sqrt{2}\gamma^{1/2} + \ldots). \tag{14,1.14}$$

Likewise it is found from (14,1.11) that $\epsilon''(\omega)$ attains its maximum when $\omega\tau_D = 1 - 16\gamma^2 + \ldots$, which is extremely close to unity. Thus we may say that the highest point of the Cole–Cole plot occurs when $\omega\tau_D = 1$. The maximum value of $\epsilon''(\omega)$,

$$\epsilon''_{\max} = \tfrac{1}{2}(1 + 2\gamma)(\epsilon_s - \epsilon_\infty),$$

which is slightly greater than the Debye value $\frac{1}{2}(\epsilon_s - \epsilon_\infty)$ obtained from Fig. 2.4. For $\omega\tau_D = 1$ eqn (14,1.10) yields

$$\epsilon'(\omega) = \tfrac{1}{2}(\epsilon_s + \epsilon_\infty) - \gamma(\epsilon_s - \epsilon_\infty).$$

The maximum point thus occurs slightly above and to the left of the maximum point in Fig. 2.4.

14.2 Dielectric Relaxation Experiments

We next consider how the implications of the theory of dielectric relaxation which we have been developing may be compared with the results of experiments. Unfortunately there is no abundance of experimental information that will provide a reliable check on the theory. In establishing our equations for complex permittivity we treated each polar molecule as being undisturbed by the dipole fields of other molecules. To make this possible we assumed that we were dealing with a dielectric in a gaseous state or in a liquid state as a very dilute solution in a nonpolar solvent. On the other hand many of the experiments are performed with pure polar liquids, where individual polar molecules must inevitably be influenced by the neighbouring polar molecules.

We saw in the previous section that, when the polar molecules have axial symmetry, the complex permittivity is given to a high degree of accuracy by a Rocard equation (14,1.7), if $\omega\tau_F$ does not exceed unity in order of magnitude. In this equation there are four parameters: ϵ_s, ϵ_∞, τ_D, τ_F. If an experimental Cole–Cole plot is given, then from the lower frequency part it will be possible to determine the semi-circular plot that would have resulted, if the Debye theory were applicable; indeed, within the limits of experimental error, the semicircle seems to persist until after the plot has reached its maximum (Leroy *et al.*, 1967–68; Goulon *et al.*, 1973; Chantry, 1977). The values of ϵ_s and ϵ_∞ are read off from the intersections of the semicircle and the ϵ'-axis. Since the maximum point of the semicircle is reached when $\omega\tau_D = 1$, the value of τ_D is the reciprocal of 2π times the value of the frequency when the semicircular plot attains its maximum. Having τ_D we immediately deduce B_1 from $B_1 = 2kT\tau_D/I_1$, by (14,1.8), where I_1 is the moment of inertia of the molecule about a line through its centre of mass and perpendicular to the axis of symmetry. Finally $\tau_F = B_1^{-1}$ and, for completeness, $\gamma = \frac{1}{2}\tau_F/\tau_D$.

As an example of a polar molecule let us consider the chloroform molecule $CHCl_3$, which is roughly spherical in shape with bond lengths r(C–H) = 1.07 Å, r(C–Cl) = 1.77 Å and with ∠ClCCl = 110.4° (Sutton, 1958, M.106). Its moment of inertia I is given in terms of its rotational constant B_0 for the ground vibrational state by

$$I = \frac{h}{8\pi^2 B_0}$$

(Herzberg, 1966, p. 670). On substituting $B_0 = 3129.5 \times 10^6\ \text{sec}^{-1}$ (Townes and Schawlow, 1975, p. 618), $h = 6.626 \times 10^{-27}$ erg sec we deduce that $I = 2.7 \times 10^{-38}\ \text{g cm}^2$.

Goulon *et al.* (1973) have provided rather good data on the dielectric properties of pure liquid chloroform for a temperature of 25°C, and we shall use their results to illustrate how theory and experiment may be compared. According to these authors

$$\tau_D = 6.36 \times 10^{-12}\ \text{sec}. \tag{14,2.1}$$

Hence from (1,2.7),

$$B = \frac{2kT\tau_D}{I} = \frac{2 \times 1.38 \times (273 + 25) \times 6.36 \times 10^{38}}{10^{16} \times 10^{12} \times 2.7}\ \text{sec}^{-1}$$

$$= 1.93 \times 10^{13}\ \text{sec}^{-1},$$

which corresponds to a frequency of 3.07 THz. It follows that

$$\tau_F = B^{-1} = 5.18 \times 10^{-14}\ \text{sec}, \tag{14,2.2}$$

$$\gamma = \frac{\tau_F}{2\tau_D} = 4.07 \times 10^{-3}, \tag{14,2.3}$$

$$\frac{B}{\gamma} = 4.74 \times 10^{15}\ \text{sec}^{-1}. \tag{14,2.4}$$

The *wave number* $\bar{\nu}$ is defined as the reciprocal of the wave length for electromagnetic rays *in vacuo*, and is therefore expressed in terms of an angular velocity ω by

$$\bar{\nu} = \frac{\omega}{2\pi c}. \tag{14,2.5}$$

Classical statistical mechanics is valid only in the region where $\hbar\omega \ll kT$ (Landau and Lifschitz, 1958, Chap. 4). We deduce from (14,2.5) that for temperature 25°C this is the region where $\bar{\nu} \ll 207\ \mathrm{cm}^{-1}$. On the other hand we see from (14,2.4) and (14,2.5) that for the chloroform molecule at this temperature $\bar{\nu} = 2.53 \times 10^4\ \mathrm{cm}^{-1}$, if $\omega = B/\gamma$. This justifies in the present instance the assertion made in Section 12.5 that, when ω is equal to B/γ, classical statistical mechanics is no longer valid.

We are therefore entitled to accept the Rocard equation (14,1.7) but in order to employ it we still need to know the values of ϵ_s and ϵ_∞. The value of ϵ_s is 4.72, the value of $\epsilon'(\omega)$ when $\epsilon''(\omega)$ attains its maximum value is approximately 3.45, and so continuing the semicircular Cole–Cole plot we would get a value about 2.18 for ϵ_∞. With these values of ϵ_s and ϵ_∞ and with the values of τ_D and τ_F coming from (14,2.1) and (14,2.2) we can deduce $\epsilon'(\omega)$ and $\epsilon''(\omega)$ from (14,1.10) and (14,1.11) for frequencies not greater in order of magnitude than 3 THz. The corresponding limit on the order of magnitude of the wave number is $103\ \mathrm{cm}^{-1}$, which is about as far as the experiments on chloroform have been taken. On substituting for γ from (14,2.3) we see from (14,1.13) and (14,1.14) that

$$\epsilon'_{\min} = \epsilon_\infty - 0.02 = 2.16$$

and that then $\epsilon''(\omega) = 0.07$. These may be compared with the experimental figures 2.08 and 0.1, respectively. The discrepancies are well within the limits of uncertainty in reading off values of $\epsilon'(\omega)$ and $\epsilon''(\omega)$ from the experimental Cole–Cole plot. The experimental plot bends backwards at a wave number about $40\ \mathrm{cm}^{-1}$. On calculating the theoretical value of the wave number from (14,2.5), (14,1.12), (14,2.1) and (14,2.3) we obtain $31\ \mathrm{cm}^{-1}$. When one considers not only that the theory was developed for a dilute solution of a dielectric but also the uncertainties in the values of ϵ_∞ and τ_D as deduced from the semicircular Cole–Cole plot, it may be claimed that the Rocard relation provides a reasonably good explanation of the experimental Cole–Cole plot for pure chloroform at 25°C.

When one calculates numerically the absorption coefficient $a(\omega)$

from (2,3.16), (14,1.10) and (14,1.11), and plots $a(\omega)$ expressed in neper cm^{-1} as a function of wave number, one obtains a curve which rises from zero to a maximum of about 9 neper cm^{-1} and then decrease steadily with increasing wave number. This curve is drawn in Fig. 14.1 together with the experimental curve and, for comparison, the absorption curve deduced from eqns (2,4.2) and (2,4.3) of the Debye theory. We see first of all that the inclusion of inertial effects in the study of rotational Brownian motion, which led to the

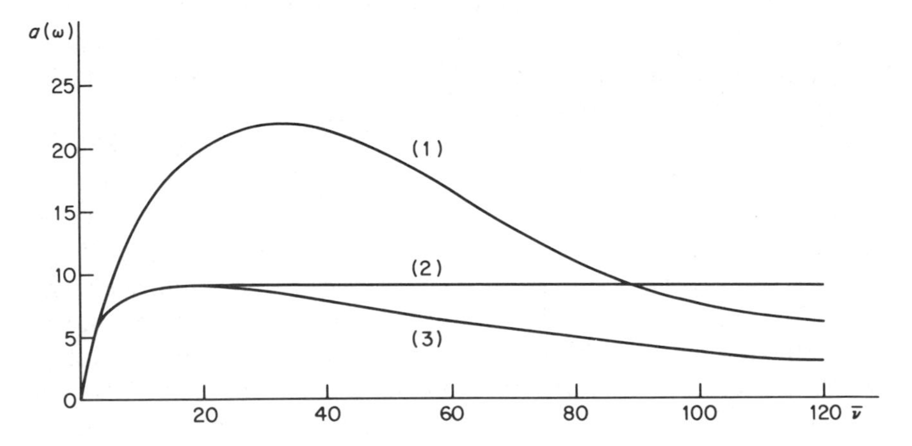

Fig. 14.1. The absorption coefficient $a(\omega)$ in neper cm^{-1} for pure chloroform at 25°C as a function of wave number in cm^{-1}: (1) experimental curve (Goulon *et al.*, 1973); (2) theoretical curve in the Debye theory; (3) theoretical curve according to the Rocard relation (14,1.7).

Rocard equation, eliminates the difficulty of the Debye plateau, to which attention was drawn in Section 2.4. The second result is that the theory of rotational Brownian motion as developed in the previous chapters leads to an absorption that is appreciably less than that obtained experimentally for wave numbers above 10 cm^{-1}. It is also to be noted that the experimental maximum is well above the Debye plateau. The extra absorption in the submillimetre region is found commonly in experiments on polar liquids (Chantry, 1971, Chap. 5). It is known as *Poley absorption*, since its existence was first suggested by Poley (1955) as a result of observation of deviations from semicircular Cole–Cole plots in the microwave region on a variety of polar liquids.

The Poley absorption process also manifests itself in dispersion effects. If, for example, we calculate from (14,1.10) and (14,1.11) the values of $\epsilon'(\omega)$ and $\epsilon''(\omega)$ for $\bar{\nu} = 120\,\text{cm}^{-1}$, we shall find that to a very close approximation $\epsilon'(\omega) = \epsilon_\infty$ and $\epsilon''(\omega)$ is negligible. The same is found to be true, if we take $\epsilon'(\omega)$ and $\epsilon''(\omega)$ from eqns (2,4.2) and (2,4.3) of the Debye theory. Hence we deduce from (2,3.14) that in both cases

$$n(\omega) = \epsilon_\infty^{1/2} = 2.18^{1/2} = 1.48,$$

which is greater than the experimental value 1.46 of the refractive index (Goulon *et al.*, 1973). This shows that the permittivity ϵ_∞ that appears in (14,1.7) and (14,1.9) exceeds the square of the experimental refractive index at high frequencies.

Measurements of $\epsilon'(\omega)$, $\epsilon''(\omega)$, $n(\omega)$, $a(\omega)$ for liquid chlorobenzene, both pure and in solutions of varying concentrations in cyclohexane, have been made in Leiden, Nancy, Brussels, Mainz and Teddington for wave numbers ranging from $0.02\,\text{cm}^{-1}$ to $125\,\text{cm}^{-1}$. The experimental absorption curve for pure chlorobenzene at 25°C is qualitatively much the same as that in Fig. 14.1 for chloroform but the Cole–Cole plot does not bend backwards (Chantry, 1977). The absorption curve has a maximum value of about 18 neper cm^{-1} when $\bar{\nu}$ is about $40\,\text{cm}^{-1}$. The chlorobenzene molecule C_6H_5Cl has the configuration of a plate and is therefore an asymmetric rotator. The complex permittivity is given theoretically by (14,1.1) and, since there is no obvious way to determine the unknowns B_1, B_2, B_3, the theory in its present state cannot be compared with experiment.

Gerschel *et al.* (1976) investigated absorption and dispersion for liquid methyl chloride, methyl fluoride and fluoroform whose molecules CH_3Cl, CH_3F, CHF_3, respectively, are highly polar symmetric rotators. Absorption curves were given for a wide range of temperatures going from the solid to the gaseous states for CH_3F and CHF_3 and from the solid 129 K to the liquid 270 K for CH_3Cl, the wave numbers varying from about 20 to $175\,\text{cm}^{-1}$. The absorption curves for the liquids are qualitatively similar to those for chloroform and chlorobenzene. An unexpected result for fluoroform in the gaseous state at 293 K is that the theoretical absorption curve based on our investigations appears to be higher than the experimental curve, as was first noticed by Evans and Evans (1978).

A somewhat similar result was found by Leroy *et al.* (1967–68) working with a 14% solution of trichloroethane in hexane at 25°C. The molecule $C_2H_3Cl_3$ has axial symmetry and, according to these authors, the maximum value of the experimental absorption coefficient is decidedly below the Debye plateau. They also claim that the experimental Cole–Cole plot comes inside the semicircular plot at high frequencies.

The weight of experimental evidence favours the existence of a Poley absorption, the cause of which is something other than orientational polarization. In Section 1.3 we introduced displacement polarization as that arising from the displacement of positive and negative charges in a molecule by an external electric field. The action of the field on the electrons is to induce an oscillating dipole moment which gives rise to optical polarization, the frequencies of the oscillations being in the visible or ultraviolet region. There are also oscillations of the nuclei in the molecule, and the frequencies of these are usually in the infrared region (Fröhlich, 1968, Section 14). We refer to these oscillations of electrons and nuclei within a molecule generically as *intramolecular vibrations.* These vibrations have discrete energy levels and therefore lead to absorption peaks, which are however far removed from the broad absorption maxima arising from Poley absorption; for example, the lowest resonance absorption peak for liquid chlorobenzene is at 195 cm^{-1} as contrasted with 40 cm^{-1} for the previous absorption maximum. Thus the intramolecular vibrations do not explain Poley absorption. Likewise difference bands of the form $\nu_i - \nu_j$ (Herzberg,1945, p. 266), although they undoubtedly contribute to the absorption profile of some liquids, cannot explain the general phenomena and in particular cannot explain the lack of temperature sensitivity (Chantry, 1971, p. 183).

One of the attempts to explain Poley absorption is based on the itinerant oscillator model, which supposes that every polar molecule is surrounded by a cage of neighbouring molecules (Hill, 1963; Wyllie, 1971; Calderwood and Coffey, 1977). It is assumed that each polar molecule executes damped oscillations with respect to its cage, and that the cage itself undergoes rotational Brownian motion due to the thermal motion of its environment. The equations have been solved in two-dimensional approximation and their solution gives two absorption peaks, which could be identified with the experimental broad absorption peak and the resonance peak in the 200 cm^{-1} region.

Lassier and Brot (1968) observed that in plastic crystals and in many liquids with a quasi-crystalline structure the rotational motion of molecules is performed by jumps from one set of discrete orientations to another. The orientations behave as potential wells which give rise to resonant absorption, and the jumps lead to the relaxation components of the absorption. Chantry (1977) developed these ideas into a liquid lattice theory, where the Poley curve is obtained from a few broadened Lorentz peaks. He found good agreement with experiment up to wave numbers of about $100\,\text{cm}^{-1}$.

Darmon *et al.* (1971) examined far infrared absorption in the polar carbonyl sulphide and the nonpolar carbon disulphide, the respective molecules OCS and CS_2 being linear rotators. Poley absorption was detected for both substances in the liquid phase. They concluded that for both polar and nonpolar substances the extra absorption is due to transient induced moments.

A detailed description of the processes that cause Poley absorption is obviously a matter that requires further investigation.

14.3 Nuclear Magnetic Relaxation

We shall now illustrate how the results of our calculations on rotational Brownian motion may be applied to problems in the theory of nuclear magnetic relaxation processes. These processes should be studied quantum-mechanically but a semiclassical theory, in which the motion of molecules is treated classically, is adequate. This allows us to employ our classical theory. In particular we shall make use of expressions derived in Chapter 13 for orientational correlation times and for spectral densities of spherical harmonics. Our approach will be to examine the simple case of isotropic motion where any line in a molecule rotates in a random manner, to compare the result of our calculations with experiment and so to forecast what relevance our previous results may have in the study of anisotropic motion.

Focusing our attention on nuclear spins we understand by *lattice* in this section the environment of these spins. Let us suppose that the spins in the presence of a constant magnetic field H_0 in a fixed direction are in thermal equilibrium with the lattice. If $\hbar\mathbf{J}$ is the spin angular momentum of a system of identical nuclei, the system has a magnetic moment $\Gamma\hbar\mathbf{J}$, where Γ is the *gyromagnetic ratio* of a nucleus, and the angular velocity of the Larmor precession ω_0 is ΓH_0. When a radiofrequency magnetic field with angular velocity ω_0 is suitably applied, the spin system will absorb energy and the previous state of equilibrium will be disturbed. The spin system returns to a new state of equilibrium and the approach to this state will be characterized by a time T_1, called the *spin-lattice relaxation time.* The theory shows that thermal motion plays a fundamental role in this relaxation process (Bloch, 1946).

Let us consider how T_1 may be expressed in terms of spectral densities that were encountered in Section 13.7. For simplicity we confine our attention to two identical nuclei in a molecule at a fixed distance r apart, and we let the polar angles of one relative to the other be $\beta(t)$, $\alpha(t)$ referred to the laboratory frame. We then define $F^{(0)}(t)$, $F^{(1)}(t)$, $F^{(2)}(t)$ by (Abragam, 1961, p. 289)

$$F^{(0)}(t) = \frac{1-3\cos^2\beta(t)}{r^3}, \quad F^{(1)}(t) = \frac{\sin\beta(t)\cos\beta(t)e^{-i\alpha(t)}}{r^3},$$

$$F^{(2)}(t) = \frac{\sin^2\beta(t)e^{-2i\alpha(t)}}{r^3},$$

so that

$$F^{(0)}(t)^* = -\frac{1}{r^3}\left(\frac{16\pi}{5}\right)^{1/2} Y_{20}(\beta(t),\alpha(t)),$$

$$F^{(1)}(t)^* = -\frac{1}{r^3}\left(\frac{8\pi}{15}\right)^{1/2} Y_{21}(\beta(t),\alpha(t)),$$

$$F^{(2)}(t)^* = -\frac{1}{r^3}\left(\frac{32\pi}{15}\right)^{1/2} Y_{22}(\beta(t),\alpha(t)), \qquad (14,3.1)$$

by (7,3.10). We next put

$$\mathscr{J}^{(q)}(\omega) = \int_{-\infty}^{\infty} e^{-i\omega t}\langle F^{(q)}(0)F^{(q)}(t)^*\rangle dt$$

$$= \int_{-\infty}^{\infty} e^{-i\omega t}\langle F^{q}(0)^{**}F^{(q)}(t)^*\rangle dt, \qquad (14,3.2)$$

where the ensemble average is for the rotational Brownian motion of the molecule which contains the nuclei. Hence from (14,3.1) and (13,7.3)

$$\mathscr{J}^{(0)}(\omega) = \frac{16\pi}{5r^6}\mathscr{Y}_{20}(\omega), \quad \mathscr{J}^{(1)}(\omega) = \frac{8\pi}{15r^6}\mathscr{Y}_{21}(\omega),$$

$$\mathscr{J}^{(2)}(\omega) = \frac{32\pi}{15r^6}\mathscr{Y}_{22}(\omega). \tag{14,3.3}$$

Now the value of T_1 is given by (Abragam, 1961, p. 291)

$$\frac{1}{T_1} = \frac{3}{2}\mathscr{I}(\mathscr{I}+1)\Gamma^4\hbar^2\{\mathscr{J}^{(1)}(\omega_0) + \mathscr{J}^{(2)}(2\omega_0)\}, \tag{14,3.4}$$

where $\mathscr{I}$ is the spin quantum number of a nucleus. Employing (14,3.3) we express this result as

$$\frac{1}{T_1} = \frac{4\pi}{5r^6}\mathscr{I}(\mathscr{I}+1)\Gamma^4\hbar^2\{\mathscr{Y}_{21}(\omega_0) + 4\mathscr{Y}_{22}(2\omega_0)\}, \tag{14,3.5}$$

which is what we set out to do.

This result is independent of the shape of the molecule. For the most general case of the asymmetric rotator model the method of calculating the spectral densities has been described in Section 13.7. The same method, with due simplifications, can be applied to the spherical and linear rotator models. In the extreme narrowing case of $\omega_0 \ll kT/(IB)$ we have from (14,3.5) and (13,7.11)

$$\frac{1}{T_1} = \frac{2\mathscr{I}(\mathscr{I}+1)\Gamma^4\hbar^2(\tau_{21} + 4\tau_{22})}{5r^6}, \tag{14,3.6}$$

which involves two correlation times. For a spherical molecule this becomes

$$\frac{1}{T_1} = \frac{2\mathscr{I}(\mathscr{I}+1)\Gamma^4\hbar^2\tau_2}{r^6}. \tag{14,3.7}$$

Then (13,4.5) gives

$$\frac{1}{T_1} = \frac{IB\mathscr{I}(\mathscr{I}+1)\Gamma^4\hbar^2}{3kTr^6}\{1 + \tfrac{11}{2}\gamma - \tfrac{83}{6}\gamma^2 + 56\tfrac{49}{72}\gamma^3 + \ldots\}$$

$$\left(\gamma = \frac{kT}{IB^2}\right). \tag{14,3.8}$$

We apply these results to calculate the spin-lattice relaxation time of protons in water. Let us first consider the intramolecular effect in the H_2O molecule of a proton on its neighbouring proton, denoting the relaxation time for this by $(T_1)_p$. Since a proton has spin $\frac{1}{2}$, we express (14,3.5) and (14,3.6) as

$$\left(\frac{1}{T_1}\right)_p = \frac{3\pi}{5r^6}\Gamma^4\hbar^2\{\mathscr{Y}_{21}(\omega_0) + 4\mathscr{Y}_{22}(2\omega_0)\}, \quad (14,3.9)$$

$$\left(\frac{1}{T_1}\right)_p = \frac{3}{10r^6}\Gamma^4\hbar^2(\tau_{21} + 4\tau_{22}) \qquad \left(\omega_0 \ll \frac{kT}{IB}\right).$$

For a spherical model of the water molecule (14,3.9) and (13,7.10) yield

$$\left(\frac{1}{T_1}\right)_p = \frac{IB\Gamma^4\hbar^2}{20kTr^6}\left\{\frac{1}{1+\left(\dfrac{IB\omega_0}{6kT}\right)^2} + \frac{4}{1+\left(\dfrac{IB\omega_0}{3kT}\right)^2}\right\}. \quad (14,3.10)$$

This is equivalent to eqn (46) of Bloembergen *et al.* (1948), if their τ_c is replaced by its value $IB/(6kT)$ and their eqn (34) is corrected so as to agree with (14,3.4). However, our method provides a closer approximation, if we employ (13,7.9) rather than (13,7.10) and take

$$D = (1 + \tfrac{1}{2}\gamma)\frac{kT}{IB}$$

from (12,5.24). In the extreme narrowing case eqn (14,3.8) yields for the spherical model of the water molecule

$$\left(\frac{1}{T_1}\right)_p = \frac{IB\Gamma^4\hbar^2}{4kTr^6}\left\{1 + \frac{11}{2}\gamma - \frac{83}{6}\gamma^2 + 56\frac{49}{72}\gamma^3 + \ldots\right\}. \quad (14,3.11)$$

We see from (13,4.7) that in the Debye limit (14,3.11) becomes

$$\left(\frac{1}{T_1}\right)_p = \frac{IB\Gamma^4\hbar^2}{4kTr^6}. \quad (14,3.12)$$

Let us find the numerical value of $\left(\frac{1}{T}\right)_p$ at 20°C. In the H_2O molecule the distance between the two hydrogen nuclei is 1.513×10^{-8} cm (Herzberg, 1945, p. 489). The Debye time τ_D is 9.3×10^{-12} s (Hasted, 1973, p. 47), so from (1,2.7)

$$\frac{kT}{IB} = \frac{1}{2\tau_D} = 5.37 \times 10^{10}\ \mathrm{s}^{-1}. \quad (14,3.13)$$

The gyromagnetic ratio Γ of the proton is $2.66 \times 10^4 \ s^{-1} G^{-1}$. The magnetic fields used in these experiments are of order 10^4 gauss. Taking H_0 equal to 10^4 G we have

$$\omega_0 = \Gamma H_0 = 2.66 \times 10^8 \ s^{-1},$$

which on comparison with (14,3.13) shows that we are well within the limits of the extreme narrowing case. Then from (14,3.12)

$$\left(\frac{1}{T_1}\right)_p = \frac{\tau_D \Gamma^4 \hbar^2}{2r^6} = 0.224 \ s^{-1}. \qquad (14,3.14)$$

This compares with the value $0.26 \ s^{-1}$ obtained by using the Rayleigh–Stokes equation (1,1.1) between viscosity and frictional constant (Pople *et al.*, 1959, p. 202). A correction to (14,3.14) could be obtained by employing (14,3.11) rather than (14,3.12).

The intermolecular effect on the spin-relaxation time due to neighbouring molecules in the water has been estimated also by Bloembergen *et al.* (1948). When their eqn (34) is corrected as explained above, it is found that the relaxation rate due to the neighbours,

$$\left(\frac{1}{T_1}\right)_n = \frac{3\pi^2 \Gamma^4 \hbar^2 \eta N_0}{kT}, \qquad (14,3.15)$$

where η is the viscosity and N_0 the number of molecules per unit volume. The relaxation time T_1 satisfies

$$\frac{1}{T_1} = \left(\frac{1}{T_1}\right)_p + \left(\frac{1}{T_1}\right)_n,$$

and on substituting the numerical values of the quantities in (14,3.15) it is deduced from (14,3.14) and (14,3.15) that the value of T_1 for water at 20°C is 2.7 s. This compares favourably with the experimental figure 2.3 ± 0.5 s (Bloembergen *et al.*, 1948, p. 703). However, the agreement is probably fortuitous because the theoretical model is rather crude and the evaluation of $(T_1)_n$ is based on the Rayleigh–Stokes equation. Nevertheless even an agreement to an order of magnitude is regarded as a confirmation of the general correctness of the nuclear magnetic relaxation model.

A second important relaxation time takes account of the interaction between nuclear spins. This *spin–spin relaxation time* is denoted by T_2. Spin–spin relaxation is largely responsible for the observation of broad lines in spectra from solids. Motions on a time scale greater than $10^{-5}-10^{-4}$ s progressively narrow the lines (McBrierty, 1974). The relaxation time T_2 is given by (Abragam, 1961, p. 292)

$$\frac{1}{T_2} = \mathscr{I}(\mathscr{I}+1)\Gamma^4\hbar^2\{\tfrac{3}{8}\mathscr{J}^{(2)}(2\omega_0) + \tfrac{15}{4}\mathscr{J}^{(1)}(\omega_0) + \tfrac{3}{8}\mathscr{J}^{(0)}(0)\},$$

where $\mathscr{J}^{(q)}$ is defined by (14,3.1) and (14,3.2). Then from (14,3.3)

$$\frac{1}{T_2} = \frac{2\pi}{5r^6}\mathscr{I}(\mathscr{I}+1)\Gamma^4\hbar^2\{2\mathscr{Y}_{22}(2\omega_0) + 5\mathscr{Y}_{21}(\omega_0) + 3\mathscr{Y}_{20}(0)\},$$

and in the extreme narrowing case, by (13,7.11),

$$\frac{1}{T_2} = \frac{\mathscr{I}(\mathscr{I}+1)\Gamma^4\hbar^2}{5r^6}(2\tau_{22} + 5\tau_{21} + 3\tau_{20}).$$

For a spherical molecule this becomes

$$\frac{1}{T_2} = \frac{2\mathscr{I}(\mathscr{I}+1)\Gamma^4\hbar^2\tau_2}{r^6},$$

which is equal to the $1/T_1$ of (14,3.7).

Brownian motion investigations of nuclear magnetic relaxation have been made by methods other than those referred to above. Hubbard (1961, 1963, 1974) examined nuclear magnetic relaxation in spherical molecules by dipole–dipole, spin-rotational and quadrupole interactions. He compared the results of his calculation of spin-rotational time with τ_2 for carbon tetrachloride and chloryl fluoride, and found reasonable agreement with experiment. Huntress (1968) developed a comprehensive theory of nuclear magnetic relaxation in liquids for molecules of arbitrary shape. This theory was based on the investigations of Favro (1960) and so did not take account of inertial effects.

It appears from our calculations on isotropic motion that the theory elaborated in the previous chapters is applicable to nuclear magnetic relaxation, and that it leads in some cases to results that are more accurate than those commonly found in the literature. Moreover the expressions derived in Sections 13.6 and 13.7 for correlation times and spectral densities in the case of an asymmetric rotator, when inertial effects are included, are available for the calculation of T_1 and T_2 when the motion is not isotropic (Woessner, 1962).

14.4 Proposals for Future Theoretical and Experimental Studies

The theory of rotational Brownian motion that we have developed and applied to relaxation experiments is based on the Langevin equation, which in its simplest form is

$$\frac{d\omega}{dt} = -B\omega(t) + \frac{dW(t)}{dt}. \tag{14,4.1}$$

Historically the term $-B\omega(t)$ came from Debye's assumption that the problem is a hydrodynamical one, so that a frictional couple proportional to the angular velocity is acting on the polar molecule. However we try to apply (14,4.1) to situations where the conditions of the hydrodynamical model are not fulfilled. We saw in Section 5.3 that this procedure is not without justification, but further investigation of microscopic friction would be valuable in order to clarify this matter and also to attempt to provide the values of B_1, B_2, B_3 in (14,1.1).

In generalizing (14,4.1) to the Euler–Langevin equations (7,1.1) it was assumed that the couple acting on an asymmetric body has components about each principal axis of inertia proportional to the component of angular velocity about that axis. This assumption was made in order to perform the complicated calculations of Chapter 11. It would be worthwhile to investigate the conditions under which the assumption could be justified.

For the theoretical study of dielectric relaxation we have to examine the conditions for the validity of both sides of a relation like (14,1.1) and (14,1.4). Such equations are consequences of (2,3.24), namely,

$$\frac{\epsilon(\omega) - \epsilon_\infty}{\epsilon_s - \epsilon_\infty} = 1 - i\omega \int_0^\infty \langle(\mathbf{n}(0) \cdot \mathbf{n}(t)\rangle e^{-i\omega t} dt, \tag{14,4.2}$$

and the left-hand side will have a different form, if the dielectric is neither a gas nor a very dilute solution of a liquid dielectric in a nonpolar solvent. In Section 14.2 we have followed the common practice of reading off the value of ϵ_∞ from the Debye extrapolation of the low-frequency Cole–Cole plot. Since in fact we discard the Debye theory at very high frequencies, the justification for taking this value of ϵ_∞ needs to be considered further.

The right-hand side of (14,4.2) has been given to a high degree of accuracy for a spherical molecule by the right-hand side of (12,3.17). Since for $\omega\tau_F \leqslant 0(1)$ the second term in the right-hand side is of order γ relative to the first, eqn (14,1.4) obtained by truncation after the first term does not give the complete first order correction to (14,1.7). Similarly (14,1.5) does not give the complete first order correction for the linear rotator; to obtain the complete correction we would have to include a contribution from the second term on the right-hand side of (12,4.10). In the same way we see that (14,1.2) does not give the complete first order correction to (14,1.3); in order to get this we would need to take the calculations of Chapter 11, simplified on account of the axial symmetry of the molecule, to the next order of approximation. This is possible but it would be extremely tedious. Hence it would be desirable to find an alternative method of calculation that would at least be applicable to the symmetric rotator.

With regard to nuclear magnetic relaxation, there is considerable interest nowadays in anisotropic motion. The results in Chapter 11 for the Brownian motion of an asymmetric rotator, which, as we saw in Section 12.5, lead in the Debye approximation to those of Perrin for an ellipsoid, could well provide a basis for a fresh study of anisotropic processes.

We remarked in Section 14.2 that the experiments on dielectric relaxation which would provide a check on the theory of rotational Brownian motion with inclusion of inertial effects are rather few. It may therefore be permissible to list a few suggestions for ideal experiments that might be performed for this purpose:

1. The dielectric to be examined should be in the liquid phase at room temperatures, and it should be constituted of molecules that are rigid, highly polar and axially symmetric.
2. It should have associated with it a nonpolar liquid whose molecules have approximately the same mass and configuration as the polar ones. This would provide an estimate of the part of the Poley absorption that is due to orientational polarization alone.
3. The polar substance should be soluble in a nonpolar solvent, whose molecules have linear dimensions appreciably smaller than those of the molecules of the solute.
4. The concentration of the solution should be varied from about 5% to 100%. This would show whether the absorption is

proportional to the concentration (Davies *et al.*, 1968). It might also throw some light on the applicability of the Langevin equation at different concentrations and on the origin of the Poley absorption.

5. Measurements of the relative permittivity $\epsilon'(\omega)$ and the loss factor $\epsilon''(\omega)$ should be made at room temperatures for a range of wave numbers 0.01–200 cm^{-1} approximately. The lower limit is of the order of magnitude of wave numbers used in chlorobenzene experiments at Leiden and the upper limit is in the region where classical statistical mechanics becomes inapplicable, as we saw in Section 14.2.
6. The intervals at which these measurements are made should be sufficiently fine to provide not only accurate absorption and dispersion curves but also accurate Cole–Cole plots.
7. The temperature and pressure should be varied so that, if possible, the experiments could be performed on the pure polar substance from the solid through the liquid up to the gaseous phase. The behaviour in the gaseous phase could indicate whether a classical theory of rotational Brownian motion is adequate for the description of dielectric relaxation phenomena in gases below a certain density.

When adequate experimental information is available and has been evaluated, we shall be in a better position to assess not only the relevance of the theory of rotational Brownian motion for dielectric and nuclear magnetic resonance theory but also the value of the concept of Brownian motion as a physical picture of what occurs in the microscopic world of molecules.

Appendix A
Representation of Linear Operators

We say that an operator A is a *linear operator*, if given g_1 and g_2 on which A operates and two constants c_1 and c_2

$$A(c_1 g_1 + c_2 g_2) = c_1 A g_1 + c_2 A g_2.$$

An elementary example of a linear operator is the differential operator d/dx, since it satisfies

$$\frac{d}{dx}[c_1 g_1(x) + c_2 g_2(x)] = c_1 \frac{dg_1(x)}{dx} + c_2 \frac{dg_2(x)}{dx}.$$

If the effect of A operating on g is to multiply it by a constant k:

$$Ag = kg, \tag{A.1}$$

we say that g is an *eigenfunction* of A and that k is the corresponding *eigenvalue* of A.

Let us suppose that A is a function of one or more real variables q and of derivatives with respect to these variables, and that it operates on functions of q. The *adjoint operator* A^+ of A is defined by

$$\int f_1^* A f_2 \, dq = \int (A^+ f_1)^* f_2 \, dq, \tag{A.2}$$

where f_1 and f_2 are arbitrary functions, the asterisk denotes complex conjugate and dq denotes the volume element in q-space. We deduce from (A.2) that

$$\begin{aligned} \int f_1^* A f_2 \, dq &= \left(\int f_2^* A^+ f_1 \, dq\right)^* = \left(\int (A^{++} f_2)^* f_1 \, dq\right)^* \\ &= \int f_1^* A^{++} f_2 \, dq, \end{aligned}$$

so that

$$\int f_1^*(A^{++}-A)f_2dq = 0.$$

Since f_1 and f_2 are arbitrary, we see that

$$A^{++} = A. \tag{A.3}$$

Moreover, if A and B are any two operators, then on employing (A.2) and (A.3)

$$\int f_1^*(AB)^+f_2dq = \int (ABf_1)^*f_2dq = \int (A^{++}Bf_1)^*f_2dq$$

$$= \int (Bf_1)^*A^+f_2dq = \int f_1^*B^+A^+f_2dq,$$

and so

$$(AB)^+ = B^+A^+. \tag{A.4}$$

When $A^+ = A$, we say that A is a *self-adjoint operator.* If we regard a real linear coordinate q_l as an operator,

$$\int f_1^*q_l^+f_2dq = \int (q_lf_1)^*f_2dq = \int q_lf_1^*f_2dq$$

$$= \int f_1^*q_lf_2dq,$$

so that $q_l^+ = q_l$ and q_l is a self-adjoint operator. The same is true of any real quantity regarded as an operator. It may likewise be shown that the quantum-mechanical linear momentum operator p_l defined by

$$p_l = -i\hbar\frac{\partial}{\partial q_l}, \tag{A.5}$$

where $\hbar$ is the Planck constant h divided by 2π, is self-adjoint (McConnell, 1960, p. 46).

If A is self-adjoint, we deduce from (A.1) that

$$k\int g^*gdq = \int g^*Agdq = \int (A^+g)^*gdq$$

$$= \int (Ag)^*gdq = k^*\int g^*gdq,$$

and so eigenvalues of a self-adjoint operator are real. If f_1, f_2 are two different eigenfunctions of a self-adjoint A with distinct eigenvalues k_1 and k_2, respectively,

$$(k_1 - k_2) \int f_1^* f_2 dq = \int k_1^* f_1^* f_2 dq - \int f_1^* k_2 f_2 dq$$

$$= \int (Af_1)^* f_2 dq - \int f_1^* A f_2 dq = 0,$$

so that $\int f_1^* f_2 dq$ vanishes. Multiplying the eigenfunctions f_1, f_2, $f_3, \ldots$ of the self-adjoint operator A by suitable constants we can satisfy the relations

$$\int f_i^* f_k dq = \delta_{ik}, \tag{A.6}$$

where δ_{ik} is the Kronecker delta defined in (3,4.16) as being equal to unity for $i = k$ and zero otherwise. We then say that the eigenfunctions constitute an *orthonormal set.* We shall assume that the orthonormal set is *complete*, that is to say, that an arbitrary function $F(q)$ is expressible as a linear combination of f_1, f_2, etc. A complete orthonormal set is called a *basic set*, or a *basis.* If then we write

$$F(q) = a_1 f_1 + a_2 f_2 + a_3 f_3 + \ldots \tag{A.7}$$

and assume that we can integrate the series term-by-term, we deduce from (A.6) that

$$a_k = \int f_k^* F(q) dq. \tag{A.8}$$

Let an operator P act on f_i and let us expand Pf_i as in (A.7):

$$Pf_i = \sum_k P_{ki} f_k.$$

Then from (A.8)

$$P_{ki} = \int f_k^* P f_i dq. \tag{A.9}$$

We see that

$$(P^+)_{ki} = \int f_k^* P^+ f_i dq = \int (Pf_k)^* f_i dq$$

$$= \left(\int f_i^* P f_k dq \right)^* = P_{ik}^*, \tag{A.10}$$

and that for two operators P and Q

$$(PQ)_{ki} = \int f_k^* PQ f_i \, dq = \int (P^+ f_k)^* Q f_i \, dq$$

$$= \int \sum_l (P_{lk}^+ f_l)^* \sum_m Q_{mi} f_m \, dq$$

$$= \sum_l \sum_m P_{kl} Q_{mi} \int f_l^* f_m \, dq,$$

by (A.10). Employing the relation (A.6) we have

$$(PQ)_{ki} = \sum_l P_{kl} Q_{li}. \tag{A.11}$$

If in (A.9) we put $k = 1, 2, \ldots$ and $i = 1, 2, \ldots$, we obtain a two-dimensional array, and (A.11) shows that two such arrays multiply together like two matrices. Since P and Q are linear operators,

$$(P + Q)_{ki} = \int f_k^* (P + Q) f_i \, dq$$

$$= P_{ki} + Q_{ki},$$

which shows that the two arrays add together like two matrices. We therefore speak of (A.9) as providing a *matrix representation* of the operator P. We call the matrix with elements given by (A.9) the *matrix representative* of P in the representation with basis elements f_1, f_2, etc.

We define the *identity operator* **I** as that which leaves unaltered any function on which it operates. We see from (A.9) that the matrix representative has 1 for its diagonal elements and 0 for its off-diagonal elements. This is the *unit matrix*, which we also denote by **I**; it will always be clear from the context whether **I** is an operator or a matrix. If corresponding to an operator A there exists an operator A^{-1} such that

$$A^{-1} A = A A^{-1} = \mathbf{I},$$

we say that A^{-1} is the *inverse operator* of A. We see from (A.10) that the matrix representative of P^+ is obtained from that of P by changing rows into columns and taking the complex conjugate of the elements. This is just the *Hermitian conjugate* of the representative of P. For this reason we shall denote the Hermitian conjugate of a matrix S by S^+. If P is self-adjoint, its representative is a *Hermitian matrix*, that is, a matrix which is identical with its Hermitian conjugate.

Corresponding to different bases we have different matrix representatives of an operator P. Suppose that the basis $g_1, g_2, \ldots$ is related to the basis $f_1, f_2, \ldots$ by

$$g_m = \sum_l U_{lm} f_l. \tag{A.12}$$

Applying (A.6) to $g_1, g_2, \ldots$ we have, by (A.6) and (A.10),

$$\begin{aligned}
\delta_{ik} &= \int g_i^* g_k \, dq = \sum_l \sum_n \int U_{li}^* f_l^* U_{nk} f_n \, dq \\
&= \sum_l \sum_n U_{li}^* U_{nk} \delta_{ln} = \sum_l U_{li}^* U_{lk} \\
&= \sum_l (U^+)_{il} U_{lk} = (U^+U)_{ik}.
\end{aligned}$$

Thus the matrix representative of U^+U is the unit matrix, so by definition the matrix with elements U_{lm} is a *unitary matrix*. Also by definition an operator U satisfying

$$U^+U = UU^+ = \mathbf{I}$$

is an *unitary operator*. With reference to the basis $g_1, g_2, \ldots$ the representative P'_{ik} of P is given by

$$\begin{aligned}
P'_{ik} &= \int g_i^* P g_k \, dq = \sum_l \sum_m U_{li}^* f_l^* P U_{mk} f_m \, dq \\
&= \sum_l \sum_m U_{il}^+ P_{lm} U_{mk} = (U^+PU)_{ik}.
\end{aligned} \tag{A.13}$$

This is, by definition, a *unitary transformation* of the matrix with *ik*-element P_{ik}, the unitary transforming matrix being that in (A.12) for the change of basis. Similarly the unitary transformation of an operator P with respect to a unitary operator U is defined by

$$P' = U^+PU.$$

We see from (A.13) that the unit matrix will transform to the unit matrix. Hence from (A.9) the operator P is the identity operator, if its representative in any representation is the unit matrix.

The *zero operator* is that which produces zero when it operates on any function. Its matrix representative has, from (A.9), zero for all its elements; it is the *zero matrix.* If in any representation an operator is represented by the zero matrix, then by (A.13) it is represented by the zero matrix in all other representations. We deduce from (A.9) that the operator is the zero operator.

Appendix B
Angular Momentum and Rotation Operators

The theory of rotation operators is closely related to the quantum theory of angular momentum. In classical mechanics the angular momentum $\mathbf{M}(M_x, M_y, M_z)$ of a particle is the vector product of the position vector $\mathbf{r}(x, y, z)$ and the linear momentum vector $\mathbf{p}(p_x, p_y, p_z)$, so that

$$M_x = yp_z - zp_y, \quad M_y = zp_x - xp_z, \quad M_z = xp_y - yp_x. \tag{B.1}$$

In quantum mechanics the components of linear momentum $\mathbf{p}$ are defined by

$$p_x = -i\hbar \frac{\partial}{\partial x}, \quad p_y = -i\hbar \frac{\partial}{\partial y}, \quad p_z = -i\hbar \frac{\partial}{\partial z}, \tag{B.2}$$

as we had in (A.5). Then for a function $f(x, y, z)$

$$(xp_x - p_x x)f = -i\hbar \left\{ x\frac{\partial f}{\partial x} - \frac{\partial}{\partial x}(xf) \right\} = ihf,$$

$$(xp_y - p_y x)f = -i\hbar \left\{ x\frac{\partial f}{\partial y} - \frac{\partial}{\partial y}(xf) \right\} = 0.$$

Hencc wc may rcgard

$$x, y, z, p_x, p_y, p_z \tag{B.3}$$

as operators satisfying

$$xp_x - p_x x = i\hbar \mathbf{I}, \quad xp_y - p_y x = 0, \text{ etc.}, \tag{B.4}$$

where $\mathbf{I}$ is the identity and 0 the zero operator. According to the discussion in Appendix A each member of (B.3) is a self-adjoint operator. In quantum mechanics we continue to define the angular momentum of a particle by (B.1). Then, by (B.4) and (A.4),

$$M_x^+ = p_z y - p_y z = y p_z - z p_y = M_x, \text{ etc.},$$

so that M_x, M_y, M_z are self-adjoint operators. On defining the *commutator* $[A, B]$ of A and B by

$$[A, B] = AB - BA \tag{B.5}$$

we immediately deduce from (B.1) and (B.4) that

$$[M_y, M_z] = i\hbar M_x, \quad [M_z, M_x] = i\hbar M_y, \quad [M_x, M_y] = i\hbar M_z. \tag{B.6}$$

We now generalize the notion of angular momentum of a single particle to that of any quantum-mechanical system. We define the *angular momentum operator* as any vector $\mathbf{M}$, whose components M_x, M_y, M_z are self-adjoint operators obeying (B.6). We define the *total angular momentum operator* M by the relation

$$M^2 = M_x^2 + M_y^2 + M_z^2. \tag{B.7}$$

We confine our attention to the case of a single particle, which is all that we require for our discussion of rotational Brownian motion. Introducing spherical polar coordinates by

$$x = r \sin\theta \cos\phi, \quad y = r \sin\theta \sin\phi, \quad z = r\cos\theta$$

we find from (B.1), (B.2) and (B.7)

$$M_z = -i\hbar \frac{\partial}{\partial \phi},$$

$$M^2 = -\hbar^2 \left\{ \frac{1}{\sin\theta} \frac{\partial}{\partial\theta} \left(\sin\theta \frac{\partial}{\partial\theta} \right) + \frac{1}{\sin^2\theta} \frac{\partial^2}{\partial\phi^2} \right\} \tag{B.8}$$

(McConnell, 1960, Section 21). The eigenfunctions of M_z obtained from

$$M_z g = -i\hbar \frac{\partial g}{\partial\phi} = kg$$

are $e^{mi\phi}$ multiplied by a function of r and θ, where m is zero or an integer, positive or negative, so that g is unaltered when ϕ is increased by 2π. The corresponding eigenvalue of M_z is $m\hbar$. The eigenvalues of M^2 are known from the theory of the hydrogen atom to be

$j(j+1)\hbar^2$, where $j = 0, 1, 2, \ldots$ (McConnell, 1960, Section 12). Hence the eigenfunctions g of M^2 are the solutions of the equation

$$\frac{1}{\sin\theta}\frac{\partial}{\partial\theta}\left(\sin\theta\frac{\partial g}{\partial\theta}\right)+\frac{1}{\sin^2\theta}\frac{\partial^2 g}{\partial\phi^2}+j(j+1)g = 0. \qquad \text{(B.9)}$$

The solutions of (B.9) are the spherical harmonics $Y_{jm}(\theta, \phi)$ defined by

$$Y_{jm}(\theta,\phi) = \left[\frac{2j+1}{4\pi}\frac{(j-m)!}{(j+m)!}\right]^{1/2}\frac{e^{im\phi}(-\sin\theta)^m}{2^j j!}\times$$

$$\left[\frac{d}{d(\cos\theta)}\right]^{j+m}(\cos^2\theta-1)^j, \qquad \text{(B.10)}$$

the constant multiplier being chosen such that

$$\int_0^{2\pi} d\phi \int_0^{\pi} \sin\theta d\theta Y^*_{jm}(\theta,\phi)Y_{j'm'}(\theta,\phi) = \delta_{jj'}\delta_{mm'} \qquad \text{(B.11)}$$

(Edmonds, 1968, p. 21). It is clear that $Y_{jm}(\theta, \phi)$ is an eigenfunction of M_z with eigenvalue $m\hbar$. We see from (B.10) that m has only the values $-j, -j+1, \ldots j-1, j$. Since $\sin\theta d\theta d\phi$ is the volume element in (θ, ϕ)-space, eqn (B.11) with $j' = j$ expresses the condition (A.6) that

$$Y_{j,-j}(\theta,\phi), Y_{j,-j+1}(\theta,\phi), \ldots Y_{j,j-1}(\theta,\phi), Y_{j,j}(\theta,\phi) \qquad \text{(B.12)}$$

constitute an orthonormal set in the space of the angular variables. We shall assume that the set is complete, so that (B.12) is a basis in terms of which an arbitrary function of θ and ϕ may be expanded. With respect to the basis (B.12) the operator M^2 is represented by $j(h+1)\hbar^2$ times the $(2j+1)$-dimensional unit matrix and M_z is represented by the diagonal matrix with diagonal elements

$$-j\hbar, (-j+1)\hbar, \ldots (j-1)\hbar, j\hbar. \qquad \text{(B.13)}$$

According to a theorem established near the end of Appendix A, the operator M^2 is just $j(j+1)\hbar^2\mathbf{I}$, where $j = 0, 1, 2, \ldots$.

We next relate angular momentum operators to rotations in three-dimensional space. If we rotate the coordinate axes about the z-axis through a small angle $\delta\alpha$ the angle ϕ is altered to $\phi - \delta\alpha$ while r and θ are unaltered. Let us take a function f of the coordinates of a point fixed in space. The rotation of the coordinate axes will give an increase in the function

$$\begin{aligned}\delta f &= f(r,\theta,\phi-\delta\alpha)-f(r,\theta,\phi)\\ &= -\frac{\partial f}{\partial\phi}\delta\alpha = -\frac{i}{\hbar}M_z f\delta\alpha,\end{aligned} \tag{B.14}$$

by (B.8). We put

$$\mathbf{M} = \hbar\mathbf{J}, \tag{B.15}$$

so that J_x, J_y, J_z are self-adjoint operators which, by (B.6), obey

$$[J_y, J_z] = iJ_x, \quad [J_z, J_x] = iJ_y, \quad [J_x, J_y] = iJ_z. \tag{B.16}$$

We shall refer to **J** as a *rotation operator.* Then $Y_{jm}(\theta,\phi)$ is an eigenfunction of J_z with eigenvalue m and an eigenfunction of J^2 with eigenvalue $j(j+1)$. The operator J^2 is $j(j+1)\mathbf{I}$, so any $(2j+1)$-dimensional representative of J^2 will be $j(j+1)$ times the unit matrix. Equations (B.14) give

$$\delta f = -iJ_z f\delta\alpha. \tag{B.17}$$

Let us now suppose that the infinitesimal rotation of coordinate axes is about the direction specified by the unit vector **n**. The component of the rotation operator in this direction is the scalar product $(\mathbf{n}\cdot\mathbf{J})$. If the axis of rotation is taken as the axis of polar cordinates, we deduce from (B.17) that now

$$\begin{aligned}\delta f &= -i(\mathbf{n}\cdot\mathbf{J})f\delta\alpha,\\ f+\delta f &= [\mathbf{I}-i\delta\alpha(\mathbf{n}\cdot\mathbf{J})]f.\end{aligned} \tag{B.18}$$

When the rotation about **n** is through a finite angle χ, we write $\delta\alpha = \chi/s$, where s is a large positive integer. We denote by Rf the value of f when the coordinate axes have been subjected to the rotation χ about **n**. Then Rf is given by a succession of infinitesimal rotations χ/s, so that from (B.18)

$$Rf = \lim_{s\to\infty}\left[\mathbf{I}-\frac{i\chi(\mathbf{n}\cdot\mathbf{J})}{s}\right]^s f. \tag{B.19}$$

The exponential e^A of an operator A is defined by

$$e^A = \mathbf{I}+A+\frac{A^2}{2!}+\frac{A^3}{3!}+\dots.$$

A well-known theorem in elementary algebra gives

$$\lim_{s \to \infty} \left(\mathbf{I} + \frac{A}{s}\right)^s = e^A ,$$

so (B.19) is expressible as

$$Rf = \exp\left[-i\chi(\mathbf{n} \cdot \mathbf{J})\right] f. \tag{B.20}$$

In place of a single function f there may be a set of functions u_1, $u_2, \ldots u_p$. We may then retain (B.20), if we agree to understand by f the column vector with elements $u_1, u_2, \ldots u_p$ so that

$$R \begin{bmatrix} u_1 \\ u_2 \\ \cdot \\ \cdot \\ \cdot \\ u_p \end{bmatrix} = \exp\left[-i\chi(\mathbf{n} \cdot \mathbf{J})\right] \begin{bmatrix} u_1 \\ u_2 \\ \cdot \\ \cdot \\ \cdot \\ u_p \end{bmatrix} .$$

In this equation J and R must be interpreted as p-dimensional representatives of these operators. It is important to remember that, while the rotations are three-dimensional and that consequently we have three rotation operators J_x, J_y, J_z, the representations of $\mathbf{J}$ and R may be 1, 2, 3, 4, . . . -dimensional. This situation is familiar in the theory of the representation of angular momentum operators. We shall be concerned with the three-dimensional representation where the u's are Cartesian coordinates and the $2j + 1$-dimensional representation where the u's are the set (B.12). Equation (B.20) may also be applied to the case of the two-dimensional representation (McConnell, 1962).

Appendix C
Laplace Transforms

The Laplace transform $L_s\{f(t)\}$ of a function $f(t)$ is defined for s real or complex by

$$L_s\{f(t)\} = \int_{0+}^{\infty} e^{-st} f(t)\, dt, \tag{C.1}$$

when the integral exists. It is a function of s but not of t and so may be written $L_s\{f\}$, though it is often more convenient to retain the argument of the function f. Thus for a real and positive

$$L_s\{t^{a-1} e^{ct}\} = \int_{0+}^{\infty} e^{-(s-c)t} t^{a-1}\, dt$$

$$= (s - c) \quad \int_{0+}^{\infty} e^{-w} w^{a-1}\, dw.$$

The integral is the gamma function of a, and therefore

$$L_s\{t^{a-1} e^{ct}\} = \frac{\Gamma(a)}{(s-c)^a}\,. \tag{C.2}$$

Since for positive integral values of a, $\Gamma(a) = (a-1)!$, we have from (C.2)

$$L_s\{1\} = \frac{1}{s},\; L_s\{t\} = \frac{1}{s^2},\; L_s\{t^2\} = \frac{2}{s^3},$$

$$L_s\{e^{-bt}\} = \frac{1}{s+b},\; L_s\{te^{-bt}\} = \frac{1}{(s+b)^2},$$

$$L_s\{t^2 e^{-bt}\} = \frac{2}{(s+b)^3},\; L_s\{t^3 e^{-bt}\} = \frac{6}{(s+b)^4}. \tag{C.3}$$

The *convolution* $f_1 * f_2$ of two functions f_1 and f_2 is defined by

$$(f_1 * f_2)(t) = \int_{0+}^{t-} f_1(u) f_2(t-u) du. \tag{C.4}$$

By making the transformation $u = t - v$ it is immediately deduced that

$$f_2 * f_1 = f_1 * f_2. \tag{C.5}$$

Since there will be no convergence difficulties at the upper or lower limits of integration for the integrands with which we shall be concerned, we shall omit the + and − signs at the limits in (C.1) and (C.4). The importance of the convolution is due to the theorem that the Laplace transform of the convolution of two functions is equal to the product of their Laplace transforms:

$$L_s\{f_1 * f_2\} = L_s\{f_1\}L_s\{f_2\} \tag{C.6}$$

(Smith, 1966, p. 43).

For three functions f_1, f_2, f_3

$$((f_1 * f_2) * f_3)(t) = \int_0^t f_3(t-v) dv \int_0^v f_1(u) f_2(v-u) du. \tag{C.7}$$

By changing the integration variables it may be shown that

$$f_1 * (f_2 * f_3) = (f_1 * f_2) * f_3,$$

so that $f_1 * f_2 * f_3$ has a well-defined meaning. Combining this result with (C.5) we see that

$$f_1 * f_2 * f_3 = f_2 * f_3 * f_1 = f_3 * f_2 * f_1, \text{ etc.}$$

On interchanging f_1 and f_3 in (C.7) we have

$$(f_1 * f_2 * f_3)(t) = \int_0^t dt_1 \int_0^{t_1} dt_2 f_1(t-t_1) f_2(t_1 - t_2) f_3(t_2).$$

Then by two applications of (C.6)

$$L_s\{f_1 * f_2 * f_3\} = L_s\{f_1\}L_s\{f_2\}L_s\{f_3\}.$$

These relations may be generalized to

$$(f_1 * f_2 * \ldots * f_n)(t) = \int_0^t dt_1 \int_0^{t_1} dt_2 \ldots \int_0^{t_{n-2}} dt_{n-1} f_1(t-t_1) \times f_2(t_1^1 - t_2) \ldots f_{n-1}(t_{n-2} - t_{n-1}) f_n(t_{n-1}) \tag{C.8}$$

and

$$L_s\{f_1 * f_2 * \ldots * f_n\} = L_s\{f_1\}L_s\{f_2\}\ldots L_s\{f_n\}. \tag{C.9}$$

It can be proved that $L_s\{f(t)\}$ determines $f(t)$ in (C.1) uniquely (Smith, 1966, p. 31). For the functions with which we shall be concerned we may quote the result (Bateman, 1954, p. 232):

If

$$\int_0^\infty e^{-st} f(t)dt = \frac{Q(s)}{P(s)},$$

where

$$P(s) = (s - \alpha_1)^{m_1}(s - \alpha_2)^{m_2} \ldots (s - \alpha_n)^{m_n},$$

with $\alpha_1, \alpha_2, \ldots \alpha_n$ distinct and $Q(s)$ a polynomial of degree less than $m_1 + m_2 + \ldots + m_n - 1$, then

$$f(t) = \sum_{k=1}^{n} \sum_{l=1}^{m_k} \frac{\Phi_{kl}(\alpha_k)t^{m_k - l}e^{\alpha_k t}}{(m_k - l)!(l-1)!}, \tag{C.10}$$

where

$$\Phi_{kl}(s) = \frac{d^{l-1}}{ds^{l-1}}\left[\frac{Q(s)(s-\alpha_k)^{m_k}}{P(s)}\right].$$

For applications to the theory of dielectric relaxation we are interested in the integral $\int_0^\infty e^{-i\omega t} f(t)dt$. This is just $L_{i\omega}\{f\}$, and (C.3) yield

$$L_{i\omega}\{1\} = \frac{1}{i\omega}, L_{i\omega}\{t\} = \frac{1}{(i\omega)^2}, L_{i\omega}\{t^2\} = \frac{2}{(i\omega)^3},$$

$$L_{i\omega}\{e^{-bt}\} = \frac{1}{b + i\omega}, L_{i\omega}\{te^{-bt}\} = \frac{1}{(b+i\omega)^2},$$

$$L_{i\omega}\{t^2 e^{-bt}\} = \frac{2}{(b+i\omega)^3}, L_{i\omega}\{t^3 e^{-bt}\} = \frac{6}{(b+i\omega)^4}. \tag{C.11}$$

Appendix D
Calculation of the Functions $I_i^{(2n)}(t)$

In the study of the rotational Brownian motion of the sphere and of the linear rotator there occur functions of the time defined as follows:

$$I^{(2)}(t) = \int_0^t dt_1 \int_0^{t_1} dt_2 e^{-B(t_1 - t_2)} ,$$

$$I_1^{(4)}(t) = \int_0^t dt_1 \int_0^{t_1} dt_2 \int_0^{t_2} dt_3 \int_0^{t_3} dt_4 e^{-B(t_1 - t_2 + t_3 - t_4)} ,$$

$$I_2^{(4)}(t) = \int_0^t dt_1 \ldots \int_0^{t_3} dt_4 e^{-B(t_1 + t_2 - t_3 - t_4)} ,$$

$$I_1^{(6)}(t) = \int_0^t dt_1 \ldots \int_0^{t_5} dt_6 e^{-B(t_1 - t_2 + t_3 - t_4 + t_5 - t_6)} ,$$

$$I_2^{(6)}(t) = \int_0^t dt_1 \ldots \int_0^{t_5} dt_6 e^{-B(t_1 + t_2 - t_3 - t_4 + t_5 - t_6)}$$

$$= \int_0^t dt_1 \ldots \int_0^{t_5} dt_6 e^{-B(t_1 - t_2 + t_3 + t_4 - t_5 - t_6)} ,$$

$$I_3^{(6)}(t) = \int_0^t dt_1 \ldots \int_0^{t_5} dt_6 e^{-B(t_1 + t_2 - t_3 + t_4 - t_5 - t_6)} ,$$

$$I_4^{(6)}(t) = \int_0^t dt_1 \ldots \int_0^{t_5} dt_6 e^{-B(t_1 + t_2 + t_3 - t_4 - t_5 - t_6)} ,$$

$$I_1^{(8)}(t) = \int_0^t dt_1 \ldots \int_0^{t_7} dt_8 e^{-B(t_1 - t_2 + t_3 - t_4 + t_5 - t_6 + t_7 - t_8)} ,$$

$$I_2^{(8)}(t) = \int_0^t dt_1 \ldots \int_0^{t_7} dt_8 e^{-B(t_1 + t_2 - t_3 - t_4 + t_5 - t_6 + t_7 - t_8)}$$

$$= \int_0^t dt_1 \ldots \int_0^{t_7} dt_8 e^{-B(t_1 - t_2 + t_3 + t_4 - t_5 - t_6 + t_7 - t_8)}$$

$$= \int_0^t dt_1 \ldots \int_0^{t_7} dt_8 e^{-B(t_1 - t_2 + t_3 - t_4 + t_5 + t_6 - t_7 - t_8)} ,$$

$$I_3^{(8)}(t) = \int_0^t dt_1 \ldots \int_0^{t_7} dt_8 e^{-B(t_1 + t_2 - t_3 + t_4 - t_5 - t_6 + t_7 - t_8)}$$

$$= \int_0^t dt_1 \ldots \int_0^{t_7} dt_8 e^{-B(t_1 - t_2 + t_3 + t_4 - t_5 + t_6 - t_7 - t_8)}$$

$$= \int_0^t dt_1 \ldots \int_0^{t_7} dt_8 e^{-B(t_1 + t_2 - t_3 - t_4 + t_5 + t_6 - t_7 - t_8)} ,$$

$$I_4^{(8)}(t) = \int_0^t dt_1 \ldots \int_0^{t_7} dt_8 e^{-B(t_1 + t_2 - t_3 + t_4 - t_5 + t_6 - t_7 - t_8)} ,$$

$$I_5^{(8)}(t) = \int_0^t dt_1 \ldots \int_0^{t_7} dt_8 e^{-B(t_1 + t_2 + t_3 - t_4 - t_5 + t_6 - t_7 - t_8)}$$

$$= \int_0^t dt_1 \ldots \int_0^{t_7} dt_8 e^{-B(t_1 + t_2 - t_3 + t_4 + t_5 - t_6 - t_7 - t_8)} ,$$

$$I_6^{(8)}(t) = \int_0^t dt_1 \ldots \int_0^{t_7} dt_8 e^{-B(t_1 + t_2 + t_3 - t_4 - t_5 - t_6 + t_7 - t_8)}$$

$$= \int_0^t dt_1 \ldots \int_0^{t_7} dt_8 e^{-B(t_1 - t_2 + t_3 + t_4 + t_5 - t_6 - t_7 - t_8)} ,$$

$$I_7^{(8)}(t) = \int_0^t dt_1 \ldots \int_0^{t_7} dt_8 e^{-B(t_1 + t_2 + t_3 - t_4 + t_5 - t_6 - t_7 - t_8)}$$

$$I_8^{(8)}(t) = \int_0^t dt_1 \ldots \int_0^{t_7} dt_8 e^{-B(t_1 + t_2 + t_3 + t_4 - t_5 - t_6 - t_7 - t_8)} . \tag{D.1}$$

Referring to Appendix C, we note from (C.8) that all the above integrals are convolutions. We may therefore employ (C.9) to find their Laplace transforms. We can then use (C.10) to invert the Laplace transforms and so evaluate the integrals. Thus, for example,

$$I_1^{(6)}(t) = \int_0^t dt_1 \ldots \int^{t_5} dt_6 e^{O(t - t_1)} e^{-B(t_1 - t_2)} e^{O(t_2 - t_3)} e^{-B(t_3 - t_4)}$$

$$\times e^{O(t_4 - t_5)} e^{-B(t_5 - t_6)} e^{Ot_6}$$

$$= 1 * e^{-Bt} * 1 * e^{-Bt} * 1 * e^{-Bt} * 1,$$

so that

$$L_s\{I_1^{(6)}(t)\} = (L_s\{1\})^4 (L_s\{e^{-Bt}\})^3 = \frac{1}{s^4(s+B)^3},$$

by (C.3). Then (C.10) gives

$$I_1^{(6)}(t) = B^{-6}\{\tfrac{1}{6}(Bt)^3 - \tfrac{3}{2}(Bt)^2 + 6Bt - 10 + [\tfrac{1}{2}(Bt)^2 + 4Bt + 10]\, e^{-Bt}\}.$$

We list the values of $I^{(2)}(t), \ldots I_8^{(8)}(t)$ and of their Laplace transforms:

$$I^{(2)}(t) = B^{-2}\{Bt - 1 + e^{-Bt}\},$$

$$L_s\{I^{(2)}(t)\} = [s^2(s+B)]^{-1},$$

$$I_1^{(4)}(t) = B^{-4}\{\tfrac{1}{2}(Bt)^2 - 2Bt + 3 - [Bt + 3]\, e^{-Bt}\},$$

$$L_s\{I_1^{(4)}(t)\} = [s^3(s+B)^2]^{-1},$$

$$I_2^{(4)}(t) = B^{-4}\{\tfrac{1}{2}Bt - \tfrac{5}{4} + [Bt + 1]\, e^{-Bt} + \tfrac{1}{4}e^{-2Bt}\},$$

$$L_s\{I_2^{(4)}(t)\} = [s^2(s+B)^2(s+2B)]^{-1},$$

$$I_1^{(6)}(t) = B^{-6}\{\tfrac{1}{6}(Bt)^3 - \tfrac{3}{2}(Bt)^2 + 6Bt - 10 + [\tfrac{1}{2}(Bt)^2 + 4Bt + 10]\, e^{-Bt}\},$$

$$L_s\{I_1^{(6)}(t)\} = [s^4(s+B)^3]^{-1},$$

$$I_2^{(6)}(t) = B^{-6}\{\tfrac{1}{4}(Bt)^2 - \tfrac{7}{4}Bt + \tfrac{31}{8} - [\tfrac{1}{2}(Bt)^2 + 2Bt + 4]\, e^{-Bt} + \tfrac{1}{8}e^{-2Bt}\},$$

$$L_s\{I_2^{(6)}(t)\} = [s^3(s+B)^3(s+2B)]^{-1},$$

$$I_3^{(6)}(t) = B^{-6}\{\tfrac{1}{4}Bt - 1 + [\tfrac{1}{2}(Bt)^2 + 2]\, e^{-Bt} - [\tfrac{1}{4}Bt + 1]\, e^{-2Bt}\},$$

$$L_s\{I_3^{(6)}(t)\} = [s^2(s+B)^3(s+2B)^2]^{-1},$$

$$I_4^{(6)}(t) = B^{-6}\{\tfrac{1}{12}Bt - \tfrac{5}{18} + [\tfrac{1}{2}Bt - \tfrac{1}{4}]\, e^{-Bt} + [\tfrac{1}{4}Bt + \tfrac{1}{2}]\, e^{-2Bt} + \tfrac{1}{36}e^{-3Bt}\},$$

$$L_s\{I_4^{(6)}(t)\} = [s^2(s+B)^2(s+2B)^2(s+3B)]^{-1},$$

$$I_1^{(8)}(t) = B^{-8}\{\tfrac{1}{24}(Bt)^4 - \tfrac{2}{3}(Bt)^3 + 5(Bt)^2 - 20Bt + 35 - [\tfrac{1}{6}(Bt)^3 + \tfrac{5}{2}(Bt)^2 + 15Bt + 35]\, e^{-Bt}\},$$

$$
\begin{aligned}
L_s\{I_1^{(8)}(t)\} &= [s^5(s+B)^4]^{-1}, \\
I_2^{(8)}(t) &= B^{-8}\{\tfrac{1}{12}(Bt)^3 - \tfrac{9}{8}(Bt)^2 + \tfrac{49}{8}Bt - \tfrac{209}{16} \\
&\quad + [\tfrac{1}{6}(Bt)^3 + \tfrac{3}{2}(Bt)^2 + 7Bt + 13]\,e^{-Bt} + \tfrac{1}{16}e^{-2Bt}\}, \\
L_s\{I_2^{(8)}(t)\} &= [s^4(s+B)^4(s+2B)]^{-1}, \\
I_3^{(8)}(t) &= B^{-8}\{\tfrac{1}{8}(Bt)^2 - \tfrac{5}{4}Bt + \tfrac{59}{16} - [\tfrac{1}{6}(Bt)^3 + \tfrac{1}{2}(Bt)^2 \\
&\quad + 3Bt + 3]\,e^{-Bt} - [\tfrac{1}{8}Bt + \tfrac{11}{16}]\,e^{-2Bt}\}, \\
L_s\{I_3^{(8)}(t)\} &= [s^3(s+B)^4(s+2B)^2]^{-1}, \\
I_4^{(8)}(t) &= B^{-8}\{\tfrac{1}{8}Bt - \tfrac{11}{16} + [\tfrac{1}{6}(Bt)^3 - \tfrac{1}{2}(Bt)^2 + 3Bt - 3]\,e^{-Bt} \\
&\quad + [\tfrac{1}{8}(Bt)^2 + \tfrac{5}{4}Bt + \tfrac{59}{16}]\,e^{-2Bt}\}, \\
L_s\{I_4^{(8)}(t)\} &= [s^2(s+B)^4(s+2B)^3]^{-1}, \\
I_5^{(8)}(t) &= B^{-8}\{\tfrac{1}{24}Bt - \tfrac{29}{144} + [\tfrac{1}{4}(Bt)^2 - \tfrac{3}{4}Bt + \tfrac{15}{8}]\,e^{-Bt} \\
&\quad - [\tfrac{1}{8}(Bt)^2 + \tfrac{3}{4}Bt + \tfrac{27}{16}]\,e^{-2Bt} + \tfrac{1}{72}e^{-3Bt}\}, \\
L_s\{I_5^{(8)}(t)\} &= [s^2(s+B)^3(s+2B)^3(s+3B)]^{-1}, \\
I_6^{(8)}(t) &= B^{-8}\{\tfrac{1}{24}(Bt)^2 - \tfrac{13}{36}Bt + \tfrac{403}{432} - [\tfrac{1}{4}(Bt)^2 + \tfrac{1}{4}Bt + \tfrac{11}{8}]\,e^{-Bt} \\
&\quad + [\tfrac{1}{8}Bt + \tfrac{7}{16}]\,e^{-2Bt} + \tfrac{1}{216}e^{-3Bt}\}, \\
L_s\{I_6^{(8)}(t)\} &= [s^3(s+B)^3(s+2B)^2(s+3B)]^{-1}, \\
I_7^{(8)}(t) &= B^{-8}\{\tfrac{1}{72}Bt - \tfrac{25}{432} + [\tfrac{1}{4}Bt - \tfrac{1}{2}]\,e^{-Bt} \\
&\quad + [\tfrac{1}{8}(Bt)^2 + \tfrac{1}{4}Bt + \tfrac{11}{16}]\,e^{-2Bt} - [\tfrac{1}{36}Bt + \tfrac{7}{54}]\,e^{-3Bt}\}, \\
L_s\{I_7^{(8)}(t)\} &= [s^2(s+B)^2(s+2B)^3(s+3B)^2]^{-1}, \\
I_8^{(8)}(t) &= B^{-8}\{\tfrac{1}{144}Bt - \tfrac{47}{1728} + [\tfrac{1}{12}Bt - \tfrac{1}{9}]\,e^{-Bt} \\
&\quad + [\tfrac{1}{8}Bt + \tfrac{1}{16}]\,e^{-2Bt} + [\tfrac{1}{36}Bt + \tfrac{2}{27}]\,e^{-3Bt} + \tfrac{1}{576}e^{-4Bt}\}, \\
L_s\{I_8^{(8)}(t)\} &= [s^2(s+B)^2(s+2B)^2(s+3B)^2(s+4B)]^{-1}.
\end{aligned}
\tag{D.2}
$$

Each $B^{2r}I_i^{(2r)}(t)$ is a power series in Bt. It is found that

$$\lim_{Bt\to 0} B^{2r}I_i^{(2r)}(t) = \frac{(Bt)^{2r}}{(2r)!}.$$

The values of $\lim_{Bt\to\infty} B^{2r}I_i^{(2r)}(t)$ may be read off immediately from the above expressions for $I_i^{(2r)}(t)$.

The product of two or more $I_i^{(2n)}(t)$'s is expressible as a linear combination of the $I_i^{(2n)}(t)$'s. This has been done by applying combinatorics to the sequences of + and − signs that occur in the multipliers of $-B$ in the exponents of (D.1) (Ford *et al.*, 1976, Appendix). We give the results that are required for our calculations:

$$
\begin{aligned}
(I^{(2)}(t))^2 &= 2I_1^{(4)}(t) + 4I_2^{(4)}(t)\,, \\
I^{(2)}(t)I_1^{(4)}(t) &= 3I_1^{(6)}(t) + 8I_2^{(6)}(t) + 4I_3^{(6)}(t)\,, \\
I^{(2)}(t)I_2^{(4)}(t) &= 2I_2^{(6)}(t) + 4I_3^{(6)}(t) + 9I_4^{(6)}(t)\,, \\
(I^{(2)}(t))^3 &= 6I_1^{(6)}(t) + 24I_2^{(6)}(t) + 24I_3^{(6)}(t) + 36I_4^{(6)}(t)\,, \\
I^{(2)}(t)I_1^{(6)}(t) &= 4I_1^{(8)}(t) + 12I_2^{(8)}(t) + 8I_3^{(8)}(t) + 4I_4^{(8)}(t), \\
I^{(2)}(t)I_2^{(6)}(t) &= 3I_2^{(8)}(t) + 8I_3^{(8)}(t) + 4I_4^{(8)}(t), + 6I_5^{(8)}(t) + 9I_6^{(8)}(t)\,, \\
I^{(2)}(t)I_3^{(6)}(t) &= 2I_3^{(8)}(t) + 5I_4^{(8)}(t) + 12I_5^{(8)}(t) + 9I_7^{(8)}(t)\,, \\
I^{(2)}(t)I_4^{(6)}(t) &= 4I_5^{(8)}(t) + 2I_6^{(8)}(t) + 6I_7^{(8)}(t) + 16I_8^{(8)}(t), \\
(I_2^{(4)}(t))^2 &= 2I_3^{(8)}(t) + 2I_4^{(8)}(t) + 12I_5^{(8)}(t) + 18I_7^{(8)}(t) \\
&\quad + 36I_8^{(8)}(t)\,, \\
(I^{(2)}(t))^4 &= 24I_1^{(8)}(t) + 144I_2^{(8)}(t) + 288I_3^{(8)}(t) + 192I_4^{(8)}(t) \\
&\quad + 576I_5^{(8)}(t) + 288I_6^{(8)}(t) + 432I_7^{(8)}(t) \\
&\quad + 576I_8^{(8)}(t), \\
(I^{(2)}(t))^2 I_2^{(4)}(t) &= 6I_2^{(8)}(t) + 24I_3^{(8)}(t) + 24I_4^{(8)}(t) + 96I_5^{(8)}(t) \\
&\quad + 36I_6^{(8)}(t) + 90I_7^{(8)}(t) + 144I_8^{(8)}(t).
\end{aligned}
\tag{D.3}
$$

Appendix E
Calculation of $\Omega^{(4)}(t)$ for the Asymmetric Rotator

When performing subsequent calculations it will be helpful to note the relation

$$\frac{\lambda_2}{I_3 I_1} + \frac{\lambda_3}{I_1 I_2} = -\frac{\lambda_1}{I_2 I_3}, \tag{E.1}$$

which is an immediate consequence of (11,1.7).

According to (11,3.2), $\Omega^{(4)}(t_1)$ is the sum of $\Omega_2^{(4)}(t_1)$, $\Omega_3^{(4)}(t_1)$, $\Omega_4^{(4)}(t_1)$ given by (11,3.3)–(11,3.5). We deduce from (11,1.12), (11,1.16) and (11,1.17) that

$$\epsilon^4 \langle \omega_1^{(1)}(t_1) \omega_1^{(3)}(t_2) \rangle$$

$$= \lambda_1 \lambda_3 \epsilon^4 \int_{-\infty}^{t_2} dt_3 \int_{-\infty}^{t_3} dt_4 \exp\left[-B_1(t_2 - t_3) - B_3(t_3 - t_4)\right] \times$$

$$\langle \omega_1^{(1)}(t_1) \omega_2^{(1)}(t_3) \omega_1^{(1)}(t_4) \omega_2^{(1)}(t_4) \rangle$$

$$+ \lambda_1 \lambda_2 \epsilon^4 \int_{-\infty}^{t_2} dt_3 \int_{-\infty}^{t_3} dt_4 \exp\left[-B_1(t_2 - t_3) - B_2(t_3 - t_4)\right] \times$$

$$\langle \omega_1^{(1)}(t_1) \omega_3^{(1)}(t_4) \omega_1^{(1)}(t_4) \omega_3^{(1)}(t_3) \rangle$$

$$= \left(\lambda_1 \lambda_3 \frac{(kT)^2}{I_1 I_2} + \lambda_1 \lambda_2 \frac{(kT)^2}{I_1 I_3} \right) \int_0^{t_2} dt_3 \int_0^{t_3} dt_4 \times$$

$$\exp\left[-B_1 t_1 - B_1 t_2 + (B_1 - B_2 - B_3) t_3 + (B_1 + B_2 + B_3) t_4\right].$$

On integrating and using (E.1) we find that

$$\epsilon^4 \langle \omega_1^{(1)}(t_1) \omega_1^{(3)}(t_2) \rangle = -\frac{(kT)^2 \lambda_1^2}{I_2 I_3} \frac{\exp\left[-B_1(t_1 - t_2)\right]}{2B_1(B_1 + B_2 + B_3)}. \tag{E.2}$$

On further integration we obtain

$$\epsilon^4 \int_0^{t_1} dt_2 \langle \omega_1^{(1)}(t)\omega_1^{(3)}(t_2)\rangle = -\frac{(kT)^2\lambda_1^2}{I_2I_3}\,\frac{1-\exp[-B_1t_1]}{2B_1^2(B_1+B_2+B_3)}. \tag{E.3}$$

Then we have

$$\epsilon^4 \langle \omega_1^{(3)}(t_1)\omega_1^{(1)}(t_2)\rangle$$

$$= -\frac{(kT)^2\lambda_1^2}{I_2I_3}\int_{-\infty}^{t_1} dv \exp[(B_1-B_2-B_3)v] \times$$

$$\int_{-\infty}^{v} dw \exp[(B_2+B_3)w - B_1|w-t_2|].$$

On dividing the range of integration of v into $(-\infty, t_2)$ and (t_2, t_1) it is found that

$$\epsilon^4 \langle \omega_1^{(3)}(t_1)\omega_1^{(1)}(t_2)\rangle$$

$$= -\frac{(kT)^2\lambda_1^2}{I_2I_3}\left[\frac{(-3B_1+B_2+B_3)\exp[-B_1(t_1-t_2)]}{2B_1(B_1-B_2-B_3)^2}\right.$$

$$\left.+\frac{2B_1\exp[-(B_2+B_3)(t_1-t_2)]}{(B_1+B_2+B_3)(B_1-B_2-B_3)^2}+\frac{(t_2-t_1)\exp[-B_1(t_1-t_2)]}{B_1-B_2-B_3}\right]. \tag{E.4}$$

From this it is deduced that

$$\epsilon^4 \int_0^{t_1} dt_2 \langle \omega_1^{(3)}(t_1)\omega_1^{(1)}(t_2)\rangle$$

$$= -\frac{(kT)^2\lambda_1^2}{I_2I_3}\left\{\frac{t_1\exp[-B_1t_1]}{B_1(B_1-B_2-B_3)} - \frac{(5B_1-3B_2-3B_3)(1-\exp[-B_1t_1])}{2B_1^2(B_1-B_2-B_3)^2}\right.$$

$$\left.+\frac{2B_1(1-\exp[-(B_2+B_3)t_1])}{(B_2+B_3)(B_1+B_2+B_3)(B_1-B_2-B_3)^2}\right\}. \tag{E.5}$$

Next we have to calculate $\langle\omega_1^{(2)}(t_1)\omega_1^{(2)}(t_2)\rangle$. We see from (11,1.12) and (11,1.16) that

$$\epsilon^4\langle\omega_1^{(2)}(t_1)\omega_1^{(2)}(t_2)\rangle = \frac{(kT)^2\lambda_1^2}{I_2 I_3}\exp\left[-B_1(t_1+t_2)\right] S,$$

where

$$S = \int_{-\infty}^{t_1} ds \int_{-\infty}^{t_2} dv \exp\left[B_1(s+v)-(B_2+B_3)|s-v|\right]$$

$$= \int_{-\infty}^{t_2} ds \int_{s}^{t_2} dv \exp\left[(B_1+B_2+B_3)s+(B_1-B_2-B_3)v\right]$$

$$+ \int_{-\infty}^{t_2} ds \int_{-\infty}^{s} dv \exp\left[(B_1-B_2-B_3)s+(B_1+B_2+B_3)v\right]$$

$$+ \int_{t_2}^{t_1} ds \int_{-\infty}^{t_2} dv \exp\left[(B_1-B_2-B_3)s+(B_1+B_2+B_3)v\right].$$

It will follow that

$$\epsilon^4\langle\omega_1^{(2)}(t_1)\omega_1^{(2)}(t_2)\rangle = \frac{(kT)^2\lambda_1^2}{I_2 I_3}\times$$
$$\frac{B_1\exp\left[-(B_2+B_3)(t_1-t_2)\right]-(B_2+B_3)\exp\left[-B_1(t_1-t_2)\right]}{B_1(B_1+B_2+B_3)(B_1-B_2-B_3)} \tag{E.6}$$

and that

$$\epsilon^4\int_0^{t_1} dt_2\langle\omega_1^{(2)}(t_1)\omega_1^{(2)}(t_2)\rangle = \frac{(kT)^2\lambda_1^2}{I_2 I_3}\times$$
$$\frac{B_1^2(1-\exp\left[-(B_2+B_3)t_1\right])-(B_2+B_3)^2(1-\exp\left[-B_1 t_1\right])}{B_1^2(B_2+B_3)(B_1+B_2+B_3)(B_1-B_2-B_3)}. \tag{E.7}$$

We therefore conclude from (11,3.3) combined with (E.3), (E.5) and (E.7) that

$$\epsilon^4\Omega_2^{(4)}(t_1) = (kT)^2\sum_{1,2,3}\frac{\lambda_1^2 J_1^2}{I_2 I_3}\left\{\frac{t_1\exp\left[-B_1 t_1\right]}{B_1(B_1-B_2-B_3)}\right.$$

$$+\frac{2B_1(1-\exp[-(B_2+B_3)t_1])}{(B_2+B_3)(B_1+B_2+B_3)(B_1-B_2-B_3)^2}$$

$$+(1-e^{-B_1t_1})\left(\frac{1}{2B_1^2(B_1+B_2+B_3)}\right.$$

$$\left.-\frac{5B_1-3B_2-3B_3}{2B_1^2(B_1-B_2-B_3)^2}\right)$$

$$-\frac{1-\exp[-(B_2+B_3)t_1]}{(B_2+B_3)(B_1+B_2+B_3)(B_1-B_2-B_3)}$$

$$\left.+\frac{(B_2+B_3)(1-\exp[-B_1t_1])}{B_1^2(B_1+B_2+B_3)(B_1-B_2-B_3)}\right\}, \tag{E.8}$$

where the summation is over the cyclic permutations of 1, 2, 3. For $t=\infty$ equation (E.8) assumes a simple form:

$$\epsilon^4\,\Omega_2^{(4)}(\infty) \;=\; (kT)^2 \sum_{1,2,3} \frac{\lambda_1^2 J_1^2}{I_2 I_3 B_1^2(B_2+B_3)}\,. \tag{E.9}$$

According to (11,3.4) the coefficient of $iJ_1J_2J_3$ in $\epsilon^4\,\Omega_3^{(4)}(t_1)$ is

$$\epsilon^4 \int_0^{t_1} dt_2 \int_0^{t_2} dt_3\ \langle\omega_1^{(2)}(t_1)\omega_2^{(1)}(t_2)\omega_3^{(1)}(t_3)$$

$$+\,\omega_1^{(1)}(t_1)\omega_2^{(2)}(t_2)\omega_3^{(1)}(t_3) + \omega_1^{(1)}(t_1)\omega_2^{(1)}(t_2)\omega_3^{(2)}(t_3)\rangle. \tag{E.10}$$

Now from (11,1.16)

$$\epsilon^4 \int_0^{t_1} dt_2 \int_0^{t_2} dt_3\ \langle\omega_1^{(2)}(t_1)\omega_2^{(1)}(t_2)\omega_3^{(1)}(t_3)\rangle$$

$$=\lambda_1\epsilon^4 \int_0^{t_1} dt_2 \int_0^{t_2} dt_3 \int_0^{t_3} du\ \exp[-B_1(t_1-u)]\ \times$$

$$\langle\omega_2^{(1)}(u)\omega_3^{(1)}(u)\omega_2^{(1)}(t_2)\omega_3^{(1)}(t_3)\rangle$$

$$= \frac{\lambda_1(kT)^2}{I_2 I_3} \int_0^{t_1} dt_2 \int_0^{t_2} dt_3 \int_{-\infty}^{t_3} du \times$$

$$\exp\left[-B_1(t-u) - B_2|u-t_2| - B_3|u-t_3|\right].$$

On performing the integrations with respect to u, t_3 and t_2 we obtain

$$\epsilon^4 \int_0^{t_1} dt_2 \int_0^{t_2} dt_3 \langle \omega_1^{(2)}(t_1)\omega_2^{(1)}(t_2)\omega_3^{(1)}(t_3)\rangle$$

$$= \frac{\lambda_1(kT)^2}{I_2 I_3} \left\{ \frac{1-\exp[-B_1 t_1]}{B_1(B_1+B_2)(B_1+B_2+B_3)} \right.$$

$$+ \frac{\exp[-(B_1+B_2)t_1] - \exp[-B_1 t_1]}{B_2(B_1+B_2)(B_1+B_2+B_3)}$$

$$+ \frac{1-\exp[-B_1 t_1]}{B_1 B_3(B_1+B_2-B_3)} - \frac{\exp[-B_3 t_1] - \exp[-B_1 t_1]}{B_3(B_1-B_3)(B_1+B_2-B_3)}$$

$$- \frac{1-\exp[-B_1 t_1]}{B_1(B_1+B_2)(B_1+B_2-B_3)} + \frac{\exp[-B_1 t_1] - \exp[-(B_1+B_2)t_1]}{B_2(B_1+B_2)(B_1+B_2-B_3)}$$

$$+ \frac{1-\exp[-(B_2+B_3)t_1]}{B_3(B_2+B_3)(B_1-B_2-B_3)} - \frac{\exp[-B_3 t_1] - \exp[-(B_2+B_3)t_1]}{B_2 B_3(B_1-B_1-B_3)}$$

$$\left. + \frac{1-\exp[-B_1 t_1]}{B_1 B_3(B_1-B_2-B_3)} + \frac{\exp[-B_3 t_1] - \exp[-B_1 t_1]}{B_3(B_1-B_3)(B_1-B_2-B_3)} \right\}. \tag{E.11}$$

Similarly it is found that

$$\epsilon^4 \int_0^{t_1} dt_2 \int_0^{t_2} dt_3 \langle \omega_1^{(1)}(t_1)\omega_2^{(2)}(t_2)\omega_3^{(1)}(t_3)\rangle$$

$$= \frac{\lambda_2(kT)^2}{I_3 I_1} \left\{ \frac{1-\exp[-B_1 t_1]}{B_1(B_1+B_2)(B_1+B_2+B_3)} \right.$$

$$+ \frac{\exp[-(B_1+B_2)t_1] - \exp[-B_1 t_1]}{B_2(B_1+B_2)(B_1+B_2+B_3)}$$

$$+ \frac{1-\exp[-B_1 t_1]}{B_1 B_3(B_1+B_2-B_3)} + \frac{\exp[-B_1 t_1] - \exp[-B_3 t_1]}{B_3(B_1-B_3)(B_1+B_2-B_3)}$$

$$+\frac{\exp[-B_1t_1]-1}{B_1(B_1+B_2)(B_1+B_2-B_3)}+\frac{\exp[-B_1t_1]-\exp[-(B_1+B_2)t_1]}{B_2(B_1+B_2)(B_1+B_2-B_3)}\Bigg\}, \tag{E.12}$$

and that

$$\epsilon^4\int_0^{t_1}dt_2\int_0^{t_2}dt_3\,\langle\omega_1^{(1)}(t_1)\omega_2^{(1)}(t_2)\omega_3^{(2)}(t_3)\rangle$$

$$=\frac{\lambda_3(kT)^2}{I_1I_2(B_1+B_2)(B_1+B_2+B_3)}\times$$

$$\left\{\frac{1-\exp[-B_1t_1]}{B_1}+\frac{\exp[-(B_1+B_2)t_1]-\exp[-B_1t_1]}{B_2}\right\}. \tag{E.13}$$

The expression (E.10) is the sum of the right-hand sides of (E.11), (E.12) and (E.13).

On substituting $t_1=\infty$ and collecting terms we find that the limit, as t_1 tends to infinity, of (E.10) is

$$\frac{(kT)^2}{I_1I_2I_3}\left\{\frac{I_2-I_3}{B_1B_3(B_2+B_3)}-\frac{I_1-I_2}{B_1B_3(B_1+B_2)}\right\}.$$

We conclude that

$$\epsilon^4\Omega_3^{(4)}(\infty)=\frac{i(kT)^2}{I_1I_2I_3B_1B_2B_3}\sum\left\{\frac{B_2(I_2-I_3)}{B_2+B_3}-\frac{B_2(I_1-I_2)}{B_1+B_2}\right\}J_1J_2J_3, \tag{E.14}$$

where the summation is now not only over cyclic permutations but over all permutations of 1, 2, 3 as in (11,3.4).

Finally to evaluate $\epsilon^4\Omega_4^{(4)}(t_1)$ from (11,3.5) we note that, since the $\omega_i^{(1)}(t)$'s are Gaussian random variables, eqn (3,4.18) and (11,1.12) yield

$$\langle\omega_r^{(1)}(t_1)\omega_s^{(1)}(t_2)\omega_u^{(1)}(t_3)\omega_v^{(1)}(t_4)\rangle$$

$$=(kT)^2\left\{\delta_{rs}\delta_{uv}\frac{\exp[-B_r(t_1-t_2)-B_u(t_3-t_4)]}{I_rI_u}\right.$$

$$+\delta_{ru}\delta_{sv}\frac{\exp[-B_r(t_1-t_3)-B_s(t_2-t_4)]}{I_rI_s}$$

$$+ \delta_{rv}\delta_{su}\frac{\exp[-B_r(t_1-t_4)-B_s(t_2-t_3)]}{I_r I_s}\Bigg\}.$$

It will follow that

$$\epsilon^4\Omega_4^{(4)}(t) = (kT)^2 \times$$

$$\sum_{i,k=1,2,3}\frac{(J_iJ_kJ_iJ_k - J_i^2J_k^2)I_1(t_1) + (J_iJ_k^2J_i - J_i^2J_k^2)I_2(t_1)}{I_iI_k}, \tag{E.15}$$

where

$$I_1(t_1) = \exp[-B_it_1]\int_0^{t_1}dt_2\int_0^{t_2}dt_3\int_0^{t_3}dt_4\exp[-B_kt_2+B_it_3+B_kt_4]$$

$$= \frac{1}{B_iB_k(B_i+B_k)} - \frac{\exp[-(B_i+B_k)t_1]}{B_iB_k(B_i+B_k)} + \frac{\exp[-B_it_1]-\exp[-B_kt_1]}{B_iB_k(B_i-B_k)},$$

$$I_2(t_1) = \exp[-B_it_1]\int_0^{t_1}dt_2\int_0^{t_2}dt_3\int_0^{t_3}dt_4\exp[-B_kt_2+B_kt_3+B_it_4]$$

$$= \frac{1}{B_i^2(B_i+B_k)} - \frac{\exp[-(B_i+B_k)t_1]}{B_k^2(B_i+B_k)} - \frac{\exp[-B_it_1]}{B_i^2B_k} + \frac{(1-B_kt_1)\exp[-B_it_1]}{B_iB_k^2}.$$

We see that

$$I_1(\infty) = \frac{1}{B_iB_k(B_i+B_k)}, \qquad I_2(\infty) = \frac{1}{B_i^2(B_i+B_k)}. \tag{E.16}$$

The expression for $\epsilon^4\Omega^{(4)}(t)$ may, by (11,3.2) and (11,3.4), be written down from (E.8), (E.10)–(E.13) and (E.15). We deduce its value for $t_1 = \infty$ from (E.9), (E.14)–(E.16):

$$\epsilon^4\Omega^{(4)}(\infty) = \frac{(kT)^2}{I_1I_2I_3}\left\{\frac{I_1\lambda_1^2J_1^2}{B_1^2(B_2+B_3)} + \frac{I_2\lambda_2^2J_2^2}{B_2^2(B_3+B_1)} + \frac{I_3\lambda_3^2J_3^2}{B_3^2(B_1+B_2)}\right\}$$

$$+\frac{i(kT)^2}{I_1I_2I_3B_1B_2B_3}\sum\left(\frac{B_2(I_2-I_3)}{B_2+B_3}-\frac{B_2(I_1-I_2)}{B_1+B_2}\right)J_1J_2J_3$$

$$+\sum_{i,k=1,2,3}\left\{\frac{(kT)^2}{I_iI_k}\left(\frac{J_iJ_kJ_iJ_k-J_i^2J_k^2}{B_iB_k(B_i+B_k)}+\frac{J_iJ_k^2J_i-J_i^2J_k^2}{B_i^2(B_i+B_k)}\right)\right\}. \tag{E.17}$$

For computational purposes it is convenient to express the operators that occur in the second and third terms of the right-hand side of (E.17) as linear combinations of J_1^2, J_2^2, J_3^2 and the symmetric product P defined, as in (11,3.7), by

$$P = J_1J_2J_3+J_1J_3J_2+J_2J_3J_1+J_2J_1J_3+J_3J_1J_2+J_3J_2J_1.$$

To do this we first deduce without difficulty from (7,4.2) that

$$\begin{aligned}
iJ_1J_2J_3 &= \frac{i}{6}P+\tfrac{1}{2}(J_1^2-J_2^2+J_3^2),\\
iJ_3J_2J_1 &= \frac{i}{6}P-\tfrac{1}{2}(J_1^2-J_2^2+J_3^2),\\
J_1J_2^2J_1-J_1^2J_2^2 &= \frac{i}{3}P-J_2^2+J_3^2,\\
J_2J_1^2J_2-J_2^2J_1^2 &= -\frac{i}{3}P--J_1^2+J_3^2,\\
J_1J_2J_1J_2-J_1^2J_2^2 &= \frac{i}{6}P-\tfrac{1}{2}(J_1^2+J_2^2-J_3^2),\\
J_2J_1J_2J_1-J_2^2J_1^2 &= -\frac{i}{6}P-\tfrac{1}{2}(J_1^2+J_2^2-J_3^2).
\end{aligned} \tag{E.18}$$

From these relations and from those obtained by cyclic permutations it may be found that

$$J_2^2J_3^2-J_3^2J_2^2 = J_3^2J_1^2-J_1^2J_3^2 = J_1^2J_2^2-J_2^2J_1^2 = -\frac{2i}{3}P. \tag{E.19}$$

On substituting (E.18) into (E.17) we find after some calculation that $\epsilon^4\Omega^{(4)}(\infty)$ satisfies (11,3.6).

As a consequence of (E.19) we see that

$$\left[\sum_{i=1}^{3} D_i^{(1)} J_i^2, \sum_{l=1}^{3} \frac{D_l^{(1)}}{B_l} J_l^2\right]$$

$$= D_2^{(1)} D_3^{(1)} (B_3^{-1} - B_2^{-1}) [J_2^2, J_3^2] + D_3^{(1)} D_1^{(1)} (B_1^{-1} - B_3^{-1}) [J_3^2, J_1^2]$$
$$+ D_1^{(1)} D_2^{(1)} (B_2^{-1} - B_1^{-1}) [J_1^2, J_2^2]$$
$$= -\frac{2i}{3} P \{D_2^{(1)} D_3^{(1)} (B_3^{-1} - B_2^{-1}) + D_3^{(1)} D_1^{(1)} (B_1^{-1} - B_3^{-1})$$
$$+ D_1^{(1)} D_2^{(1)} (B_2^{-1} - B_1^{-1})\}.$$

On substitution from (11,5.18) we deduce that

$$\left[\sum_{i=1}^{3} D_i^{(1)} J_i^2, \sum_{l=1}^{3} \frac{D_l^{(1)}}{B_l} J_l^2\right]$$
$$= -\frac{2i}{3} P \frac{(kT)^2}{I_1 I_2 I_3} \left\{ I_1 \frac{B_2 - B_3}{B_2^2 B_3^2} + I_2 \frac{B_3 - B_1}{B_3^2 B_1^2} + I_3 \frac{B_1 - B_2}{B_1^2 B_2^2} \right\}. \tag{E.20}$$

References

Abragam, A. (1961), "Nuclear Magnetism". Clarendon Press, Oxford.

Arnold, L. (1974). "Stochastic Differential Equations: Theory and Applications". John Wiley and Sons, New York.

Batchelor, G. K. (1953). "The Theory of Homogeneous Turbulence". Univ. Press, Cambridge.

Bateman, H. (1954). "Tables of Integral Transforms". Bateman Manuscript Project Vol. I. McGraw-Hill Book Co. Inc., New York, Toronto, London.

Bernard, W. and Callen, H. B. (1959). *Rev. mod. Phys.* **31**, 1017–1044.

Berne, B. J. and Harp, G. D. (1970). *Adv. chem. Phys.* **17**, 63–227.

Berne, B. J. and Pecora, R. (1976). "Dynamic Light Scattering". John Wiley and Sons, Inc., New York, London, Sydney, Toronto.

Bloch, F. (1946). *Phys. Rev.* **70**, 460–474.

Bloembergen, N., Purcell, E. M. and Pound, R. V. (1948). *Phys. Rev.* **73**, 679–712.

Bogoliubov, N. N. and Mitropolsky, Y. A. (1961). "Asymptotic Methods in the Theory of Nonlinear Oscillators". Gordon and Breach, New York.

Brink, D. M. and Satchler, G. R. (1975). "Angular Momentum". 2nd ed. Clarendon Press, Oxford.

Budó, A., Fischer, E. and Miyamoto, S. (1939). *Phys. Z.* **40**, 337–345.

Calderwood, J. H. and Coffey, W. T. (1977). *Proc. R. Soc.* A **356**, 269–286.

Callen, H. B. and Welton, T. A. (1951). *Phys. Rev.* **83**, 34–40.

Case, K. M. (1966). *Suppl. Prog. theor. Phys.* Nos. 37 and 38, 1–20.

Case, K. M. (1972). *Transp. Theory and Stat. Phys.* **2**, 129–176.

Chantry, G. W. (1971). "Submillimetre Spectroscopy". Academic Press, London, New York.

Chantry, G. W. (1977). IEEE *Trans. Microwave Theory and Tech.* (U.S.A.), Vol. MTT-25, No. 6, 496–500.

Cramér, H. (1951). "Mathematical Methods of Statistics". Princeton Univ. Press, Princeton.

Cole, K. S. and Cole, R. H. (1941). *J. chem. Phys.* **9**, 341–351.

Darmon, I. Gerschel, A. and Brot, C. (1971). *Chem. Phys. Lett.* **8**, 454–456.

Davies, M., Pardoe, G. W. F., Chamberlain, J. E. and Gebbie, H. A. (1968). *Trans. Faraday Soc.* **64**, 847–860.

Debye, P. (1912). *Phys. Z.* **13**, 97–100.
Debye, P. (1913). *Ber. dt. phys. Ges.* **15**, 777–793; translated in "The Collected Papers of Peter J. W. Debye". Interscience Publishers, New York, London, 1954.
Debye, P. (1929). "Polar Molecules". Dover Publ. Inc., New York.
Dmitriev, V. A. and Gurevich, S. B. (1946). *Zhur. éksp. teor. Fiz.* **16**, 937–940.
Doob, J. L. (1942). *Ann. Math.* **43**, 351–369.
Duff, G. F. D. (1956). "Partial Differential Equations". University of Toronto Press, Toronto.
Edmonds, A. R. (1968). "Angular Momentum in Quantum Mechanics". 2nd ed. Princeton Univ. Press, Princeton.
Einstein, A. (1905). *Annln Phys.* (4) **17**, 549–560.
Einstein, A. (1906). *Annln Phys.* (4) **19**, 371–381.
Evans, G. J. and Evans, M. W. (1978). *J. Chim. phys.* **75**, 522–528.
Evans, M. W. (1977). "Dielectric and Related Molecular Processess", Vol. 3, ed. M. Davies. Chem. Soc., London.
Favro, L. D. (1960). *Phys. Rev.* **119**, 53–62.
Fick, A. (1855). *Annln Phys.* **94**, 59–86.
Fixman, M. and Rider, K. (1969). *J. chem. Phys.* **51**, 2425–2438.
Fokker, A. D. (1914). *Annln Phys.* (4) **43**, 810–820.
Ford, G. W. (1975). "Notes on Stochastic Differential Equations" (unpublished).
Ford, G. W., Kac, M. and Mazur, P. (1965). *J. math. Phys.* **6**, 504–515.
Ford, G. W., Lewis, J. T. and McConnell, J. (1976). *Proc. R. Ir. Acad.* **76A**, 117–143.
Ford, G. W., Lewis, J. T. and McConnell, J. (1977). *Phys. Lett.* A, **63**, 207–208.
Ford, G. W., Lewis, J. T. and McConnell, J. (1978a). *Physica* **92A**, 630–633.
Ford, G. W., Lewis, J. T. and McConnell, J. (1978b). *Proceedings of the 13th IUPAP Conference on Statistical Physics,* Haifa, 24–30 August 1977, Vol. 2, 726–729.
Ford, G. W., Lewis, J. T. and McConnell, J. (1979). *Phys. Rev.* A, **19**, 907–919.
Fröhlich, H. (1948). *Trans. Faraday Soc.* **44**, 238–243.
Fröhlich, H. (1968). "Theory of Dielectrics" 2nd ed. Clarendon Press, Oxford.
Furry, W. H. (1957). *Phys. Rev.* **107**, 7–13.
Gerschel, A., Dimicoli, I., Jaffre, J. and Riou, A. (1976). *Molec. Phys.* **32**, 679–697.
Gikhman, I. I. and Skorokhod, A. V. (1969). "Introduction to the Theory of Random Processes". W. B. Saunders Co., Philadelphia, London, Toronto.
Goldberg, R. R. (1962). "Fourier Transforms". Univ. Press, Cambridge.
Gordon, R. G. (1968). *Adv. Magn. Reson.* **3**, 1–42.
Goulon, J., Rivail, J. L., Fleming, J. W., Chamberlain, J. and Chantry, G. W. (1973). *Chem. Phys. Lett.* **18**, 211–216.
Gross, E. P. (1955a). *Phys. Rev.* **97**, 395–403.
Gross, E. P. (1955b). *J. chem. Phys.* **23**, 1415–1423.
Hasted, J. B. (1973). "Aqueous Dielectrics". Chapman Hall, London.
Herzberg, G. (1945). "Molecular Spectra and Molecular Structure II. Infrared and Raman Spectra of Polyatomic Molecules". Van Nostrand Reinhold Co., New York, Cincinnati, Toronto, London, Melbourne.

Herzberg, G. (1966). "Molecular Spectra and Molecular Structure III. Electronic Spectra and Electronic Structure of Polyatomic Molecules". Van Nostrand Reinhold Co., New York, Cincinnati, Toronto, London, Melbourne.
Herzfeld, K. F. (1964). *J. Am. chem. Soc.* **86**, 3468–3469.
Hill, N. E. (1963). *Proc. phys. Soc.* **82**, 723–727.
Hubbard, P. S. (1961). *Rev. mod. Phys.* **33**, 249–264.
Hubbard, P. S. (1963). *Phys. Rev.* **131**, 1155–1163.
Hubbard, P. S. (1972). *Phys. Rev.* A, **6**, 2421–2433.
Hubbard, P. S. (1973). *Phys. Rev.* A, **8**, 1429–1436.
Hubbard, P. S. (1974). *Phys. Rev.* A, **9**, 481–494.
Hubbard, P. S. (1977). *Phys. Rev.* A, **15**, 329–336.
Hunter, J. J. (1972). *Am. Statistn* **26**, 22–24.
Huntress, W. T. (1968). *J. chem. Phys.* **48**, 3524–3533.
Ivanov, E. N. (1963). *Zh. éksp. teor. Fiz.* **45**, 1509–1517; *Soviet Physics* JETP, **18**, 1041–1045 (1964).
Jeans, J. H. (1933). "The Mathematical Theory of Electricity and Magnetism". 5th ed. Univ. Press, Cambridge.
Jeffreys, H. (1967). "Theory of Probability". 3rd ed. Clarendon Press, Oxford.
Khinchin, A. I. (1934). *Math. Annln* **109**, 604–615.
Khinchin, A. I. (1949). "Mathematical Foundations of Statistical Mechanics". Dover Publ. Inc., New York.
Kirkwood, J. G. (1939). *J. chem. Phys.* **7**, 911–919.
Kirkwood, J. G. (1946). *J. chem. Phys.* **14**, 180–201.
Kluk, E. and Powles, J. G. (1975). *Molec. Phys.* **30**, 1109–1116.
Kolmogorov, A. N. (1931). *Math. Annln* **104**, 415–458.
Kramers, H. A. (1940). *Physica* 7, 284–304.
Krylov, N. M. and Bogoliubov, N. N. (1947). "Introduction to Nonlinear Mechanics". Princeton Univ. Press, Princeton.
Kubo, R. (1957). *J. phys. Soc. Japan* **12**, 570–586.
Kubo, R. (1966). *Rep. Prog. Phys.* **29**, 255–284.
Lamb, H. (1932). "Hydrodynamics" 6th ed. Univ. Press, Cambridge.
Landau, L. D. and Lifschitz, E. M. (1958). "Statistical Physics". Pergamon Press, London, Paris.
Langevin, P. (1908). *C.r.hebd. Séanc. Acad. Sci.* **146**, 530–533.
Lassier, B. and Brot, C. (1968). *Chem. Phys. Lett.* **1**, 581–584.
Lebowitz, J. L. and Résibois, P. (1965). *Phys. Rev.* **A 139**, 1101–1111.
Lebowitz, J. L. and Rubin, E. (1963). *Phys. Rev.* **131**, 2381–2396.
Leroy, Y., Constant, E., Abbar, C. and Desplanques, P. (1967–68). *Adv. Molec. Relaxation Processes* **1**, 273–307.
Leroy, Y., Constant, E. and Desplanques, P. (1967). *J. Chim. phys.* **64**, 1499–1508.
Lévy, P. (1925). "Calcul des Probabilités". Gauthier-Villars et Cie., Paris.
Lewis, J. T., McConnell, J. and Scaife, B. K. P. (1974). *Phys. Lett.* A, **49**, 303–305.
Lewis, J. T., McConnell, J. and Scaife, B. K. P. (1975). *Proceedings of the 12th IUPAP Conference on Statistical Physics,* Budapest, 25–29 August, 1975, Abstracts of Contributed Papers, 33.

Lewis, J. T., McConnell, J. and Scaife, B. K. P. (1976). *Proc. R. Ir. Acad.* **76 A**, 43–69.
Littlewood, D. E. (1950). "A University Algebra". William Heinemann Ltd., Melbourne, London, Toronto.
McBrierty, V. J. (1974). *Polymer* **15**, 503–520.
McConnell, J. (1960). "Quantum Particle Dynamics" 2nd ed. North Holland Publishing Co., Amsterdam.
McConnell, J. (1962). *Annali Mat. pura appl.* (IV) **53**, 203–210.
McConnell, J. (1977). *Proc. R. Ir. Acad.* **77 A**, 13–30.
McConnell, J. (1978). *Proc. R. Ir. Acad.* **78 A**, 87–97.
McConnell, J. (1979). *Report on the 46th Annual Conference on Electrical Insulation and Dielectric Phenomena, Colonie, N.Y.*, 17–20 October, 1977, 3–12.
Ming Chen Wang and Uhlenbeck, G. E. (1945). *Rev. mod. Phys.* **17**, 323–342.
Morita, A. (1978). *J. Phys.* D: *Appl. Phys.* **11**, L1–L4.
Perrin, F. (1934). *J. Phys. Radium* **5**, 497–511.
Perron, O. (1913). "Die Lehre von den Kettenbrüchen". Teubner, Leipzig.
Planck, M. (1917). *Sber. preuss. Akad. Wiss.* **24**, 324–341.
Poley, J. Ph. (1955). *Appl. sci. Res.* B, **4**, 337–387.
Pomeau, Y. and Weber, J. (1976). *J. chem. Phys.* **65**, 3616–3628.
Pople, J. A., Schneider, W. G. and Bernstein, H. J. (1959). "High-resolution Nuclear Magnetic Resonance". McGraw-Hill Book Co., New York, Toronto, London.
Powles, J. G. (1948). *Trans. Faraday Soc.* **44**, 802–806.
Reid, C. J., Yadav, R. A., Evans, G. J., Evans, M. W. and Davies, G. J. (1978), *J. chem. Soc.* Faraday Trans. II **74**, 2143–2158.
Résibois, P. and Davis, H. T. (1964). *Physica* **30**, 1077–1091.
Rocard, M. Y. (1933). *J. Phys. Radium* **4**, 247–250.
Rose, M. E. (1957). "Elementary Theory of Angular Momentum". John Wiley and Sons, New York.
Sack, R. A. (1953). *Kolloidzeitschrift* **134**, 16.
Sack, R. A. (1957a). *Proc. phys. Soc.* **B 70**, 402–413.
Sack, R. A. (1957b). *Proc. phys. Soc.* **B 70**, 414–426.
Saito, N. and Kato, T. (1956). *Busseiron kenk-yu* **92**, 103–113.
Scaife, B. K. P. (1959). British Electrical Research Association Report ERA-L/T 392.
Scaife, B. K. P. (1963). *Prog. Dielect.* **5**, 143–186.
Scaife, B. K. P. (1971). "Complex Permittivity". The English University Press, London.
Scaife, B. K. P. (1973). "Problems in Physical Electronics", ed. R. L. Ferrari and A. K. Jonscher. Pion Ltd., London.
Smith, M. G. (1966). "Laplace Transform Theory". Van Nostrand Co. Ltd., Toronto, New York, Princeton.
von Smoluchowski, M. (1915). *Annln Phys.* (4), **48**, 1103–1112.
Smyth, C. P. (1955). "Dielectric Behaviour and Structure". McGraw-Hill Book Co. Inc., New York, Toronto, London.

Steele, W. A. (1963a). *J. chem. Phys.* **38**, 2404–2410.
Steele, W. A. (1963b). *J. chem. Phys.* **38**, 2411–2418.
Sutton, L. E. (1958). "Tables of Interatomic Distances and Configuration in Molecules and Ions". Special Publication No. 11, Chem. Soc., London.
Synge, J. L and Griffith, B. A. (1959). "Principles of Mechanics" 3rd ed. McGraw-Hill Book Co. Inc., New York, Toronto, London.
Townes, C. H. and Schawlow, A. L. (1975). "Microwave Spectroscopy". Dover Publ. Inc., New York.
Uhlenbeck, G. E. and Ornstein, L. S. (1930). *Phys. Rev.* **36**, 823–841.
Van Kampen, N. G. (1976). *Phys. Rep.* **24 C**, 171–228.
Van Vleck, J.H. (1932). "The Theory of Electric and Magnetic Susceptibilities". Clarendon Press, Oxford.
Van Vleck, J. H. (1951). *Rev. mod. Phys.* **23**, 213–227.
Wall, H. S. (1948). "Analytic Theory of Continued Fractions". Van Nostrand Co., New York.
Wang Chang, C. S. and Uhlenbeck, G. E. (1970). "Studies in Statistical Mechanics", Vol. V, ed. J. de Boer and G. E. Uhlenbeck, North-Holland Publishing Co., Amsterdam.
Wiener, N. (1930). *Acta math.* **55**, 117–258.
Woessner, D. E. (1962). *J. chem. Phys.* **37**, 647–654.
Wyllie, G. (1971). *J. Phys.* C: *Solid St. Phys.* **4**, 564–568.

Author Index

Subject Index